Women of Science

Women of Science

| RIGHTING |
| THE |
| RECORD |

edited by

G. KASS-SIMON
AND PATRICIA FARNES

ASSOCIATE EDITOR,
DEBORAH NASH

Indiana
University
Press

BLOOMINGTON & INDIANAPOLIS

First Midland Book Edition 1993

© 1990 by Indiana University Press
All rights reserved

Manufactured in the United States of America

Library of Congress Cataloging-in-Publication Data

Women of science : righting the record / edited by G. Kass-Simon and
Patricia Farnes ; associate editor Deborah Nash.
p. cm.
Includes index.
ISBN 0-253-33264-8
1. Women in science.   I. Kass-Simon, G. (Gabriele)   II. Farnes,
Patricia, 1931-1985.   III. Nash, Deborah.
Q130.W67   1989
508.8042—dc19   88-45463
                                                                      CIP
ISBN 0-253-20813-0 (pbk.)

3 4 5 6 7 96 95 94 93 92

To those we didn't find

CONTENTS

Preface

In the late 1960s, when the current women's movement was beginning to gain its first real momentum, academic women, long subjected to institutionalized subordination, became suddenly and acutely aware of their own inferior status—not only in the hierarchy of academic employment, but also in the hierarchy of recognized intellectual achievement. One side of this new awareness found immediate expression as political activism; the other gave rise to questions concerning the origin, nature, and justness of the perceived second rank. But here too, for academic women, answering these questions was not just an interesting intellectual exercise, but a vital political necessity; for in the answers lay also the justification for demanding absolute academic equality.

In science, the questions could be put like this: Were so few scientific achievements attributed to women because so few women had participated in science? Or, were there so few because women had produced so little of any real importance? Or, had women simply not been properly acknowledged for their work? And, if any or all of these things were true, what had produced this state of affairs? Was there systematic discrimination against women scientists, or was this simply an accurate assessment of their inferior abilities?

These questions, whose answers now seem obvious, were not at all rhetorical in the early seventies. Although, on the basis of anecdotal evidence, many of us believed that women scientists had achieved far more than had been acknowledged, and that they had been deliberately excluded from greater participation, it still seemed possible that women in science had been evaluated exactly as they deserved.

In 1973, then, when many of us were discussing these problems, it occurred to one of us (GKS) that the best way to address them was to make a concerted inquiry into the achievements of women scientists in the various disciplines. And since the method of evaluating a scientist's work has traditionally been by peer review, it also seemed essential that such an investigation be carried out by trained women scientists. But in 1973, when these ideas began to develop, it was practically impossible to find women scientists willing to carry out the research.

First, it was very difficult in some areas to locate practicing women scientists (women's organizations in a number of fields were just then being formed). Secondly, most women scientists were too busy trying to stay afloat in their own fields to take time out to write about nontechnical subjects. And thirdly, many women shared the general attitude that matters concerning women, even scientific matters, could never be considered worthy of scholarly endeavor. For these reasons, the idea of producing a book about women's scientific achievements was put aside for almost eight years.

During those intervening years, significant changes occurred in our society. In science this was reflected, first, by the numbers of women scientists who seemed suddenly to appear out of nowhere. Many who had been languishing as superannuated research associates were abruptly granted long-overdue promotions into tenurable faculty positions. Their work, heretofore taken for granted, was now recognized and rewarded. Secondly, some scientific fields once considered inherently inaccessible to women were now opened to them—and if women were not actually welcomed, their presence was at least tolerated. Finally, scholarship concerning women became almost respectable—often given official sanction by academic institutions under the title "Women's Studies."

Thus by 1980, although it was still difficult in some areas to find women scientists who could take time away from chronically over-committed schedules, there were others who had already begun to make independent inquiries into the history of women's scientific achievements. Indeed, some of the chapters in this book had their origin as independently initiated research. By 1980, then, it appeared possible to address the questions that had been raised earlier and to collect the results of these inquiries into the present volume.

As the chapters for this book were being compiled, almost all of the authors were surprised and pleased by the many women they had uncovered who had made significant contributions to their science. This same sense of surprise (should one say incredulity?) is still evident among our students and colleagues when we inform them (as we from time to time do) that a particular discovery was made by a woman. The information contained in this volume, therefore, cannot yet be considered commonplace.

Nonetheless, if the increasing attention paid to women scientists by historians and sociologists is an indication, and if the growing number of references to women's research in textbooks and reviews is predictive, then it is but a matter of time before the subject matter in this book becomes uninterestingly familiar and self-evident. When that happens this book will have served its purpose.

G. Kass-Simon
Patricia Farnes

G. KASS-SIMON

Introduction

At the height of the current feminist movement there was a revealing riddle making the rounds which even now seems to puzzle some who hear it. It goes like this: Dr. Smith, a surgeon, and Dr. Smith's son are driving in a car. On the radio, they hear: "The famous surgeon Dr. Smith has just been killed in an automobile accident." At this Dr. Smith's son cries out, "Oh my God, that's my father." And the question is, how can that be?

That the situation depicted in the riddle represented a puzzle to be solved is a telling testimony of our culture's perception of its women. Whether it is also an accurate assessment of women's medical and scientific achievements or whether it is derived from a socially induced myopia is a question that lies at the heart of the essays in this volume.

In one sense history is the recounting of heroes' deeds. This is as true for the history of science as for the histories of nations. In the same breath that we describe the discovery, we acknowledge the discoverer. Indeed, to write the history of an event is to define its heroes.

In a society those deeds which the cultural ethos designates as important are crowned with the laurel "history"; duties which are just as important for survival or progress but which have not been accorded society's encomium are ignored. In our culture those functions which are potentially historical have been reserved for men, those which are not, allotted to women. With few exceptions, therefore, women have functioned outside the realm of history; they have been generally ignored and have only rarely been allowed to become its heroes.

In science, besides those who discover cures, heroes are people whose work has fundamentally changed our perception of the universe; or, they are those whose ideas and inventions have altered the concepts and methods of their science. They are the Einsteins, Galileos, and Darwins of our textbooks. The work of these individuals has literally defined their respective disciplines, and the recounting of the discipline becomes the recounting of their work. One can open any history of science and find the work of hundreds of men who may have helped to create the substance of their discipline. And just as one can find name after name of men in these books, it is almost impossible to find the names of any women. In Asimov's popular *Biographical Encyclopedia of Science and Technology*[1] only 10 women are listed among the 1,195 scientists whose work is described. And in Singer's classical *History of Biology*,[2] no woman scientist is included—although the book is dedicated to Agnes Arber, the well-known British botanist.

From this, one can draw only two alternative conclusions: Either there really are very few women heroes of science, or there are many more but somehow they have been excluded from its history.

Intellectual forebears are as important to us as hereditary ones, and it is often with anger and sadness that women have had to face their own persistent absence from the cultural ethos. A major aim of this volume, therefore, has been to search for women who have taken part in the formation of our current scientific culture and who have been responsible for the formulation of its tenets and precepts. That is, we have tried in part to seek out the lost and buried women heroes of science.

There are a number of women whose contributions to science are so great that no history or encyclopedia can ignore them. The work of Marie Curie, Irène Joliot-Curie, Maria Goeppert-Mayer, Lise Meitner, Gerty Cori, Dorothy Hodgkin, Barbara McClintock, and Rita Levi-Montalcini was so unmistakably important that their achievements were recognized and acclaimed by their peers. All except Meitner were awarded the Nobel Prize. For the most part, the work of these and other extraordinary women scientists is included in the chapters that follow.[3]

More interestingly, because the authors sought to learn whether there were not more women scientists who had done work of note, many were found who had made truly ground-breaking discoveries but who, almost without exception, are left out of the history books. For example, it is hardly known that Ida Hyde produced the first intracellular microelectrode, that Erica Cremer developed gas chromatography, or that Katherine Foot and Ellen Stroebell pioneered biological photomicrography. The techniques and discoveries of these women are now an integral part of their science; yet, as individuals, they had become dissociated from their achievements and their names were all but lost.

Many such women were found—scientists who had made significant and long-lasting contributions to their fields, but who at best are known only to specialists. In medicine there is Winifred Ashby, who developed the standard technique for determining red blood cell survival. In mathematics, Mildred Sanderson's theorem of invariants is considered a classic. In astronomy, Henrietta Leavitt's investigations led the way to determining the size of the galaxy, and in archaeology, Gertrude Cator Thompson's excavations pushed the origin of Egyptian culture back to 5000 B.C.

Because the extent to which women were allowed to participate in the various sciences differed greatly among the disciplines, the extent and signficance of their contributions also varied widely; in fields like biology and crystallography there were many more important women scientists than could be included in this volume, while in engineering and mathematics their number is conspicuously smaller. Nonetheless, the importance of women's contributions to all the disciplines is attested to in every chapter that follows.

If history often becomes the telling of heroes' deeds and if only men's deeds are worthy of history, it is not difficult to draw the mistaken conclusion that all deeds done by men of history are inherently worthy of commemoration. In science this attitude results in the uncritical acceptance of poorly designed experiments or undocumented theories. In the history of science, it fosters the recounting, sometimes in great detail, of wrong, trivial, or even foolish thoughts and theories. In both cases it is the man rather than the work which makes the work noteworthy.

Similarly, the validity or ultimate significance of a man's work is rarely the deciding criterion for his inclusion in the history of science. It is usually sufficient that he and his work are thought historically interesting. None of this holds true for women.[4]

For women in science to be remembered, not only must their work be thought right, but usually it must have such impact upon scientific thought that exclusion is impossible. If women scientists are wrong, or if they narrowly miss the mark, or if they propound ideas that are ultimately superceded, not only are their ideas quickly forgotten, but as often as not, the women are ostracized by their contemporaries or treated with derision.

Therefore, some of the authors have sought to balance this picture by discussing the work of women scientists who were later shown to be wrong or whose work was quickly overtaken. Maureen Julian discusses the work of Dorothy Wrinch, who is now almost forgotten but whose once much-praised theory of cyclols, though inadequate at best, was very influential in her time. Patricia Farnes discusses Margaret Sanger's support of eugenics—a viewpoint that is now frequently criticized. And Foot and Strobell's arguments against the chromosome theory are included in the biology chapter.

Some of the authors sought to describe work of lesser scientific importance for its historical value. Aldrich cites the work of women who illustrated geological reports; Mack focuses on the work of the women at the Harvard Observatory who compiled the endless catalogues of stars, and Green and LaDuke discuss the achievements of a number of minor women mathematicians.

The inclusion of minor work is important in dispelling the mistaken notion that women's scientific work was the extraordinary exception rather than simply the rarer occurrence.[5] By its mere existence such work forces a readjustment of our perception of the scientific milieu and of our conceptualization of scientific history. To the extent to which such work became incorporated or absorbed, or consciously rejected by contemporary scientists, it must be considered an influential force in the development of science. In this respect, the recognition of such work makes a serious inroad into the all-pervasive exclusion of women from the history of scientific thought.

Because sociology determines both science and its history, and because the science one is permitted to do is narrowly prescribed by one's place in scientific society, the telling of women's science also becomes the telling of their sociology. Chapter after chapter attests to the correlation between women's socially defined role and the nature and extent of their scientific achievements. Thus, Berta Scharrer's focus on invertebrate neurosecretion, Ellen Swallow Richards's work in domestic chemistry, and the many important contributions to pediatrics and to social-environmental medicine made by women can be directly attributed to the social niche the scientific community allowed these women. Similarly, women's social role is found again and again to be a major determinant of how the scientific community has judged the work of its women members and accounts in part for the methods by which these women were lost and buried.

Trescott points out that Emily Roebling is seldom given more than gratuitous acknowledgment for having supervised the building of the Brooklyn Bridge after her husband became incapacitated three years after the start of the project. And Lillian Gilbreth's work as an engineer continued to be subsumed under her husband's name long after his death. Jones points to a similar transposition of credit among simply professional associates when Otto Hahn, but not Lise Meitner, was awarded the Nobel Prize for the discovery of fission—although Meitner's contributions are generally recognized as having been fundamental to it.[6] In the same vein, Ralph Gerard, but not Judith Graham Pool, is repeatedly cited for having (re)invented the microelectrode.

Other methods of burial were also uncovered. Farnes points out that some women in the medical profession were lost because they published in relatively obscure women's journals. Still others gradually disappeared from the literature because, like Nettie Stevens, who discovered sex chromosomes, their work was recapitulated by more famous male scientists.

But often, by the very tokens that obscured women's independent achievements and that forced them to assume a subordinate role, opportunities were created which allowed them to enter previously proscribed areas. Mack explains how Edward Charles Pickering's need for large numbers of astronomical calculators gave women their first real chance to practice astronomy on a large scale in a well-equipped observatory. Trescott tells how Julia Hall, having been granted but a diploma for her four years of college work, became her brother's colleague in the invention of the electrolytic production of aluminum. And, Irwin-Williams tells of several husband and wife teams whose field work gave women some of their first opportunities to participate in archaeological excavations.

Some opportunities arose because there were a few men who seemed to have had a genuine interest in furthering the work of at least some women scientists. Julian discusses the many women who are the direct academic descendants of William Henry Bragg and his son, William Lawrence Bragg, and Trescott tells of Robert Richards's constant efforts in support of his wife's own scientific pursuits.

A recurrent theme that runs through every chapter is the importance of the training women received at the women's colleges. As has been pointed out by others,[7] because excellent university education, especially scientific education, was restricted to men, the seven major women's colleges flourished and became the breeding ground for some of this country's best scientific talent. But here too, although women's colleges nurtured women's scientific endeavors, like all ghettos, they also restricted them. Shortages of funds, equipment, and time frequently served to further prevent women faculty members from joining the mainstream of their discipline.

But despite, or, more properly, because of the restrictions placed upon them, women were from time to time given a unique vantage point from which to survey their science. Forced into a scientific side-stream—to use Joan Didion's apt words—rather than cry over curdled milk, they made cottage cheese of it.[8] And because very often the stuff of superb science is literally but the re-viewing of information already at hand, they were given the opportunity to so change their science as to create a new one. Lillian Gilbreth, Alice Hamilton, Ellen Swallow Richards, and Rachel Carson were all responsible for the invention of new fields. And indeed, these women must be considered among the very best heroes of science. Not only did their work alter our perceptions, it in fact served to create them.

The chapters that follow attest to the ground-breaking work of these and other women, and in so doing impeach the fundamental accuracy of many histories of science.

NOTES

1. Isaac Asimov, *Isaac Asimov's Biographical Encyclopedia of Science and Technology*, revised ed. New York: Avon Books, 1976.

2. Charles Singer, *A History of Biology*, revised ed. New York: Henry Schuman, 1950.

3. Not every famous woman scientist is included in this volume. And some, e.g., Emmy Noether, Caroline Herschel, Hypatia, Mary Sommerville, are only briefly mentioned. In some cases this is because the authors wished to avoid redundancy. In others the authors focused on aspects of their discipline which did not include these women.

4. This accounts for the fact that 5 of the 10 women cited by Asimov (*op. cit*) are Nobel Prize winners, but that not 50 percent of the 1,187 men cited have won Nobel Prizes.

5. Margaret Rossiter's book *Women Scientists in America, Struggles and Strategies to 1940* (Baltimore & London: The Johns Hopkins University Press, 1982) makes it clear that there have been many, many women scientists in this country.

6. A comparison of Asimov's account of the work of Hahn, Meitner, and Strassman with that given by Jones in this volume is truly revealing.

7. M. Elizabeth Tidball and Vera Kistiakowsky, "Baccalaureate Origins of American Scientists and Scholars," *Science* 193 (1976):646–652.

8. Joan D. Didion, *A Book of Common Prayer*, New York: Simon and Schuster, Inc., 1977.

Women of Science

CYNTHIA IRWIN-WILLIAMS

Women in the Field

The Role of Women in Archaeology before 1960

This chapter examines the role of women within the discipline of archaeology from inception to present, but focuses primarily on pioneers in the field before the women's movement of the 1960s–1970s. Originally, the goal of the investigation was simply the identification of women who were either initiators of or participants in major paradigm shifts within archaeology. It soon became apparent, however, that this orientation would summarily exclude the vast majority of women in the field. Therefore the inquiry was expanded to include the significant contributions of women to the data base, theoretical structure, and public policy of archaeology, and to the changing role of women in the evolving profession.

The results of this research are organized into four sections: 1) the cultural perspective; 2) the contribution of women to the development of the field; 3) personal profiles of three prominent pioneering women archaeologists; and 4) recent progress and perspectives. This synopsis of women in archaeology does not begin to chronicle all of the contributions of the countless women within the field. Rather, the intent is to examine the role of women in the discipline of archaeology.

THE CULTURAL PERSPECTIVE: THE EXPEDITION MENTALITY

It is useful at the outset to establish a broad social and cultural perspective. Women have come very late into the field of archaeology. Indeed, a search for *recognized* early women innovators and leaders before 1915 came up virtually empty. The causes of this late entry are tightly bound up with the definition and evolution of archaeology as a discipline, and more importantly with the cultural stereotype of "the archaeologist" in relation to the stereotypical role of women in the larger society. At the core of the problem as it is specific to archaeology is the "expedition mentality."[1]

This academic and cultural attitude is well illustrated by the following incident from the life of a leading English woman archaeologist just before the First World War:

> By the time British archaeologist Dr. Margaret Murray had worked up the courage to even attend a conference with her male colleagues, she was 50 years old. As far as she knew no other woman had *ever* ventured into a meeting of the British Association of Anthropology, but when Dr. Murray arrived for the morning session in Birmingham, England in *1913*, she discovered another woman hiding timidly behind a pile of books and sitting as far away from the men as possible. Dr. Murray, however, boldly took a seat among the men. . . . Before the afternoon session of the conference began, a colleague Dr. Murray knew rather well, took the seat beside her and told her sorrowfully that he did not approve of the statements she had made in the morning, or even of her presence at the conference. "There are many things in this world that a woman shouldn't know," he said, "I certainly would not permit you to attend one of my lectures."[2]

Writing two years later, J. P. Droop expressed in print the generally held opinion of that time about the participation of women in field work:

> I may perhaps venture a short word on the question . . . whether in the work of excavation it is a good thing to have cooperation between men and women. Of a mixed dig I have seen something and it is an experiment that I would be reluctant to try again. . . . In the first place there are the proprieties. . . . (In short) I do not believe that such close and unavoidable companionship can ever be other than a source of irritation. . . . [3]

These scenarios reflect accurately the historic attitude toward the participation of women in the field of archaeology. They represent what C. J. Weber has termed the "expedition mentality," which to a large extent has dominated the discipline since its inception. This concept involves a very central conflict between the accepted definition of "the archaeologist" and the definition of male and female identity in Western society.[4]

> The classic idea of archaeology involved going on an 'expedition'—something you did by girding yourself and going off quest-like to face hardships in faraway places. Going on an expedition meant enduring difficulties like finding water, digging latrines, just surviving. In order to meet these grueling challenges of the planet, the assumption was that you had to be 'manly.'[5]

The stereotype dies hard in spite of any evidence to the contrary. The anthropology department of Harvard University recently (1981) defended itself against a motion of censure by the American Anthropological Association for its discriminatory hiring practices by stating that it was unable to fill any of its positions in archaeology with females because "women as a group are not attracted to the discipline" (American Anthropological Association Annual Business Meeting, 1981). This statement was made in the face of the fact that about one-third of all new doctorates in anthropology-archaeology go to women.

These enduring stereotypes and the built-in conflicts they represent provide a perspective for the consideration of the activities and contributions of women to the discipline of archaeology in the nineteenth and the first half of the twentieth century.

WOMEN'S CONTRIBUTION TO THE DEVELOPMENT OF ARCHAEOLOGY

THE HISTORICAL FRAMEWORK: THE DEVELOPMENT OF ARCHAEOLOGY BEFORE 1915

While archaeology as an academic discipline is a comparatively recent phenomenon, the wellspring of curiosity and speculation from which it stems may be traced back at least to the Classical Greeks and Romans. With the death of the Classical civilizations, interest in the origins of humanity faded and was not revived until the Italian Renaissance. During the fifteenth and sixteenth centuries attention on previous cultures focused on the recovery and collection of art treasures of Greece and Rome. The rediscovery of the ethnographies of Tacitus and others also stimulated the development of interest in antiquity. These writings essentially reacquainted Western Europeans with a fascinating part of their past and induced speculation about some of the curious relics found in Western Europe.

As a result of these trends and the evolving intellectual and political orientation of the seventeenth, eighteenth, and early nineteenth centuries, two forms of "humanistic" archaeology or antiquarianism emerged: 1) nationalistic antiquarianism focused particularly on the relics and monuments of the countries of northern Europe, most especially England, and their relation to the origins of the national ethnic groups; and 2) Classical antiquarianism, which initially attracted wealthy adventurers and scholars to the Classical world and the Near East. These two distinct currents within the mainstream of archaeological thought are still visible today.

Nationalistic antiquarianism led in the early nineteenth century in Denmark to the development of the first systematic archaeology. At about the same time another line of development in the nineteenth century evolution of natural science in Europe came into focus, and was to merge with nationalistic antiquarianism to evolve toward the origins of "scientific archaeology."

In the first half of the nineteenth century antiquarians had made many discoveries of artifacts and human bones in association with extinct animals. Indeed, one of these antiquarians was a Mme. de Christol, a pioneer archaeologist, who in the 1820s presented evidence of the association of man-made tools and extinct animals in the caves in southern France. References to her are either terse or nonexistent.

The significance of these early finds to the subsequent evolution of scientific archaeology can only be understood in the context of the contemporary developments in geology. Two major intellectual trends were crucial to the development of "geologic" archaeology in the mid-nineteenth century: the Catastrophist theory, which prevailed among geologists and the general public, and the Fluvialist (uniformitarianist) theory, which sought another explanation for the evidence observed in the geologic record of the earth.

Founded by Baron George Cuvier, Catastrophism held that at times all life on Earth had been destroyed by cataclysmic events (i.e., the biblical Flood), and a new population then took its place. This theory was acceptable to the prevailing Christian theology which followed Archbishop Ussher in believing that the world was created in 4004 B.C. Uniformitarianism, formulated by James Hutton and Charles Lyell, assumed long periods of geologic time and argued that past changes in the earth's crust resulted from the processes that operate in the present.[6] This theory became the foundation of modern geology and was crucial to the development of archaeology as a science. However, wholesale public acceptance of a literal interpretation of the Bible had stifled any creative intellectualization about the nature of the fossil evidence and the antiquity of humanity.

Publication of On the Origin of Species by Charles Darwin[7] had an incredible impact on the development of geology, biology, and archaeology. The establishment or denial of the antiquity of man now became an issue of vital importance to large numbers of people. The significance of the evidence of human artifacts and extinct fauna in geologic deposits reported earlier in the century by Mme. de Christol, and by M. Boucher de Perthes, Father J. MacEnery, and others, suddenly became of first importance in the international confrontation.[8] This confrontation pitted the evolving new scientific disciplines, including archaeology, against the established structure of Christian theology. The effect of the new evolutionary doctrine, as seen by contemporary authors, "seemed to be devastating, to overwhelm the philosophic and religious landmarks of the human race."[9]

The specific effects of the controversy on the embryonic discipline of archaeology were multiple and profound. Within the intellectual community the basic split between nationalistic/prehistoric archaeology and the archaeology of the Classical world broadened and deepened, and resulted in a duality within the field which continues to the present. The former cleaved with the other emerging natural sciences. The latter continued on a more humanistic bent and generally avoided the evolutionary/biblical controversy. The character of the participants was altered as well.

Previously archaeology had been almost entirely the pursuit of a relatively small group of wealthy intellectuals, adventurers, and art collectors. While these individuals continued to dominate the field until well into the twentieth century, the range of social classes as well as the total number of participants grew rapidly in the late nineteenth century.

The last half of the nineteenth century saw the beginnings of archaeology as an academic discipline with a systematic approach to the field as a whole. In Europe, the lines of inquiry continued to be twofold: a continuing interest in the Classical civilizations as well as the archaeology of the "biblical" countries, and the quest for national origins coupled with intense interest in the antiquity of humanity.

In America, the wellspring of interest was quite different. As in Europe, systematic archaeology in America was very limited prior to the 1840s. America lacked a longstanding academic tradition. Europe was, therefore, the source of most, if not all, of the new field techniques, as well as the source of the overall theoretical orientation for the discipline during the nineteenth century and part of the twentieth.

While archaeology in Europe was stimulated by interest in the Classical civilizations and in the evidence for the antiquity of man in its own national territories, inquiry in the Americas focused on local antiquities and on the origin of the American Indians. Willey and Sabloff[10] identify four principal trends within this early "speculative" period of American archaeological development: 1) the early Spanish chronicles on the contemporary cultures; 2) the natural-science descriptive accounts of early explorers; 3) the roots of the later systematic archaeology which appear as the contributions of isolated and precocious individuals like Thomas Jefferson; 4) the pervasive "speculative" mode of intellectual thought about the American Indians, which envisioned them variously as descendants of the Phoenicians, the Scythians, the Lost Tribes of Israel, and the Lost Continent of Atlantis.

The latter part of the nineteenth century and the pre–World War I period in American archaeology saw a shift in focus from speculation and chronicle to the principal concern with accurate description and rudimentary classification of American antiquities. The intellectual trends of Europe were reflected across the Atlantic, and the impact of *On the Origin of Species* was equally deeply felt. The subsequent increasing alliance of prehistoric archaeology with the natural sciences was paralleled. A major distinction during this period, however, derived from the American concern with the native Indian population, so that archaeology in America developed in tandem with ethnography as parallel subdivisions of a broader field of anthropology. One result of this was the development of numerous scholars who were skilled in both fields, including several prominent women. In Europe, this identification with anthropology was less complete. However, in general, American archaeology before 1914 was methodologically and conceptually very dependent on European prototypes.

In this larger context of rapid growth and internal differentiation within the field of antiquarianism/archaeology in nineteenth-century Europe and America, what was the role and contribution of women? To follow the lead of the principal recent chroniclers and historians of archaeology, the apparent answer is "none at all." Neither Daniel[11] in his massive history of archaeology nor Willey and Sabloff[12] in their treatment of New World archaeology mentions a single contribution by a woman archaeologist before World War I.

In fact, however, there were early women pioneers. For the most part they gained entrance to the field and limited public recognition as the wives and daughters of the famous archaeologists of the time. Best known among these was Sophia Schliemann, wife of Heinrich Schliemann. Their story reflects something of the climate of the time. Heinrich Schliemann, German-born but an American citizen, was one of the most colorful of the late-nineteenth-century adventurers who developed a passion for antiquities. In 1869, at age 47, after amassing a fortune from profiteering during the Civil War, Schliemann decided to divorce his "unsuitable" first wife and to marry a Greek girl specifically to help him in his lifelong passion to find and excavate the lost city of Troy. He selected the photograph of 27-year-old Sophia Engastromenos from a collection of photographs presented to him, and proceeded to court and marry the girl in three weeks. He then tutored her rigorously in the arts, history, geography, philosophy, and archaeology, so as to prepare her to be his helpmate for his long-awaited expedition to Hissarlik in Turkey, which he believed was the site of Priam's Troy. Their work together there, and the fabulous success of the expedition with its hoards of golden diadems, plates, rings, etc., caught the imagination of the world.[13] It touched off a spate of expeditions to the Classical world, both serious research and simple plundering projects. It also led to a tremendous growth of interest in archaeology and prehistory. Throughout, although thoroughly overshadowed by her flamboyant husband, Sophia was recognized as the more diligent, careful, and perseverant of the two. What records were made of the excavations were largely the result of her efforts.[14]

A few women managed to enter the field in their own right. Because educational opportunities, field training, and advanced degrees in archaeology were narrowly restricted, Margaret Murray became an archaeologist by the roundabout route of getting a degree in linguistics first. Since linguistics at the turn of the century was centrally involved with philology and the Classical civilizations, Murray was led to the study of Egyptian hieroglyphics and Egyptology. Her professor, Sir Flinders Petrie of the University College of London, was a notable exception for the day because he allowed not only his wife but also Murray and several other women to participate in his famous excavations at Abydos in Egypt in the late 1890s.

He felt that women were especially good at illustrating artifacts and hieroglyphics and at record keeping. Miss Murray spent a great deal of time helping Petrie in his work, and the University College ultimately "took her on the staff at the stipend of a few pounds per annum. This led her on to lecturing up and down the country which brought her many friends."[15] She continued to participate in field projects up until World War II. She was an ardent feminist, and was particularly interested in women in ancient Egypt. She never directed an independent excavation herself but gained considerable recognition and enjoyed a very long and distinguished career as a researcher and popular writer on archaeology and primitive religion. She published her most famous book on Egyptian archaeology on her sixty-eighth birthday (*The Splendor That Was Egypt*, 1931) and her autobiography in 1963 at the age of one hundred.[16] In a greeting to her on her one hundredth birthday Wainright saluted her: "She has always been very great on not just accepting what others had said about something but on looking for yourself to see whether it was so or not. She opened people's eyes to many unsuspected points of view, and shocked many of those comfortably wrapped in their cocoons of unquestioning orthodoxy."[17]

Another significant pioneer, Harriet Ann Boyd Hawes, broke ground throughout her career. Born in Boston in 1871, Harriet Boyd studied at Smith College and acquired a scholarship to the American School of Classical Studies in Athens. There, she was encouraged toward a career as an academic librarian. Boyd implored the faculty to allow her to assist in the school's ongoing archaeological field work, but was not given permission. A tiny young woman of spirit and energy, she took the remainder of her fellowship and went off on her own to look for archaeological remains on the island of Crete. Both the flavor and the significance of the situation are reflected in the following introduction to a biographical account published by the Archaeological Institute of America.[18] (In 1900 Crete was just emerging from a bloody war and a century of oppressive Turkish rule and was virtually *terra incognita* to archaeologists.)

The British, the French, the Italians quickly grasped the opportunity [for archaeological research], but the *first* American participation in this inviting new field was dependent on the energy and courage of Harriet Ann Boyd, a student of the American School of Classical Studies still in her twenties, who decided, quite on her own, to use some of the money from her fellowship for excavating in Crete. Her story is one of the most interesting in the history of archaeological field work—and it is one that has been ignored by the popularizers of archaeological adventure. Here we have all the elements of romance and danger: a young scholar daring to travel—as few women had yet done—through rugged mountains where the inhabitants had but recently found their freedom. Riding on muleback in Victorian attire, accompanied by the faithful Aristides, a native of northern Greece with his mother as chaperon, she was apparently perfectly

unconscious of doing anything unusual or courageous. The photograph taken
in a Cretan studio in 1901 shows in her casual though proper dress and the
potsherds—which replace the conventional parasol or bouquet—her intense
preoccupation with her chosen work.[19]

Harriet Boyd's pluck and persistence were rewarded with great success.
At Gournia she discovered a supremely important Late Minoan site dating
from the Bronze Age, over 3,000 years ago. She returned to excavate it,
directed crews of over a hundred workers and so became the first woman
to direct a major field project—an expedition. Subsequently she excavated
numerous other Bronze and Iron Age settlements in the Aegean and became a recognized authority on the area.[20] However, in spite of her considerable accomplishments neither Daniel[21] nor Willey and Sabloff[22] makes
mention of Harriet Boyd Hawes.

Hetty Goldman followed Harriet Boyd's lead in pioneering work in
Classical archaeology. After receiving training at Bryn Mawr College and
obtaining a doctorate at Radcliffe in 1916, she became a fellow of the American School of Classical Studies. With the reputation of Boyd Hawes before
her, Goldman became the first woman to lead an officially sanctioned expedition under the auspices of the school. She followed this with a series
of brilliant excavations in Asia Minor, Yugoslavia, Greece, and Turkey.[23]
She finally joined the Institute of Advanced Studies at Princeton in 1936,
where she continued until retirement. Hetty Goldman is widely acknowledged as one of the outstanding scholars in her research area and as a
pioneering woman in the field of archaeology. However, like Boyd Hawes,
Murray, and her other female contemporaries, she does not appear in
subsequent chronicles.

What does this brief review of the contribution of women to the field
of archaeology before ca. 1915 suggest as a whole? First, access to the field
was exceedingly limited and difficult. Opportunity for regular training in
field work, an essential ingredient for professional activity, was denied or
was possible only as the result of unusual circumstances (marriage or an
unusually tolerant dig director). One spirited young woman, Boyd Hawes,
simply took matters into her own hands and trained herself. Classroom
instruction was somewhat more open but severely limited. Most of the
women acquired degrees at colleges for women. Attendance at professional
meetings was discouraged. Where there was a possibility for field research
activity at all, it was limited to the more "humanistic" subfield of Classical
archaeology and did not impinge on the more "scientific" or "historic"
areas of national prehistory either in Europe or in North America. British
and American women archaeologists did essentially all of their work outside of their home countries. In spite of these monumental problems, those
few pioneering individuals who persevered paved the way for improvements in the decades that followed.

The Emergence of Archaeology as a Systematic Discipline: 1915–1935

The period between the two world wars saw a virtual explosion within the field of archaeology as it became better established as a serious and systematic academic discipline. The "Early Classificatory–Historical Period" of Willey and Sabloff[24] and the "Period of the Systematists" of Irwin-Williams[25] suggest the orientation of the developing discipline as follows: Its principal objectives were 1) the development of order within the data base; 2) the development of prehistoric cultural sequences by serious and orderly investigation; 3) the initial development of semitheoretical schemes relating the regional/areal sequences being worked out to larger scale structures; and 4) the creation or adoption of new techniques both in the field and in the laboratory to facilitate these objectives.

In Europe, archaeologists focused on producing increasingly detailed sequences covering European prehistory from the Paleolithic through Roman and later historic times. Sequences derived particularly from southern France became the standard for ordering and understanding technologically similar developments elsewhere. In the Near East, particularly in Egypt and Mesopotamia, attention focused on the *origins* of the Classical civilizations and on pushing those origins farther and farther back. Attempts to fix these specific events into larger theoretical structures were dominated by a social/historic evolutionary perspective most persuasively developed by English archaeologist V. Gordon Childe.[26]

Archaeological systematics in North America generally lagged behind those in Europe, although the essential orientation was comparable. This was particularly due to the late development or use of stratigraphic excavation in the New World (available throughout Europe generally since at least the 1870s). Two significant effects of this were the development of other nonchronologic schemes for ordering the rapidly growing data base, and a focus on the classification (typology) of minute materials as a method of achieving a variety of ends. Control of the data base grew in quantity and detail, and the great antiquity of man in the New World became clear.[27]

In this period of rapid growth of both information content and methodological competence, women *in general* are not very well represented. However, there are outstanding exceptions, both among those who achieved recognition in their own right, and among those who acted as part of a successful husband-wife team.

In this context of rapid expansion in Old World archaeology, with its focus on gathering new data from unknown areas, and on the development of systematics, the names of several women stand out. Glyn Daniel in an obituary for one of these (Dorothy Garrod) salutes them as "a splendid regiment of women" who dispelled the gloomy misgivings and forebodings of the earlier male archaeologists.[28]

Most notable among these women was Gertrude Caton-Thompson, who serves as a symbol of professional achievement for the period. Born in 1888 she was educated at Newnham College, Cambridge. Her first occupation during the First World War was with the British Ministry of Shipping, and she attended the Paris Peace Conference. With this background she plunged into life as a full-time archaeologist. She completed her training at the British School of Archaeology in Egypt. Together with another young British woman archaeologist, E. W. Gardner, she worked on a number of the school's projects in Egypt.[29] As its representative she inaugurated the First Archaeological and Geological Survey of the Northern Fayum in 1924–26, and continued work there as the field director for the Royal Anthropological Institution in 1927–28.[30] The significance of the work cannot be overestimated. It pushed the origins of Egyptian culture in Lower Egypt back into the Neolithic, as far as 5000 B.C. Her subsequent project at Hemamiah near Badar together with that of Guy Brunton established the Badarian as the immediate predecessor of the Predynastic period.[31] These significant discoveries and cogent reports were only the first of a long series of field projects which broke new ground in developing culture-historical frameworks for the Middle East and Africa.

From Egypt she proceeded to Rhodesia (1928–29), where she excavated the famous ruins at Zimbabwe, whose rediscovered splendor contributed to the sense of national definition and pride reflected by the adoption of the name Zimbabwe for the whole nation in the postcolonial period.[32] After this she moved on to do definitive work at the early site of Kharga Oasis in Egypt (1930–33) and to southern Arabia to investigate the tombs and temples of Hureidha in the Hadramaut (1937–38).[33] Unlike many of her archaeological contemporaries, she took the time to fully report and publish her work.[34]

Not only were her contributions to the data base enormous, but her grasp of the evolving methodologic and theoretical structure was both formidable and generally ahead of its time. She became one of the first (1946) to recognize and distinguish between the reality of archaeological "culture" and the oversimplified use of arbitrary "techno-typological devices"[35] for definition which characterized much of European archaeological literature well into the 1950s.[36]

Unlike many of her female counterparts she was both widely recognized and active in the larger professional archaeological network in England. She served as president of the British Prehistoric Society, vice-president of the Royal Anthropological Institute, was governor of Bedford College (University of London) and the School of Oriental Studies. She became a Fellow of Newnham College (Cambridge) from 1934 until her retirement in 1957. Through the years she received an array of prestigious medals and an honorary LL.D. from Cambridge. Evidently a woman archaeologist not only could survive the rigors of expeditions to "far-away places," but could become one of the best known and most productive professionals of her time.

Very nearly as prominent and productive within this distinguished early group of English women archaeologists was Dorothy Annie Elizabeth Garrod. In contrast with Caton-Thompson, who confined her activities to the later periods, Garrod was the first major woman researcher to enter and deal directly with the field of Paleolithic archaeology and its controversies concerning the antiquity of man. Born in 1892 the daughter of a prominent physician, she was educated in France and studied Paleolithic archaeology under the great Frenchmen of the time, Breuil, Begouen, and Peyrony.[37] Her field investigations were both outstanding and wide ranging, including work in England, Palestine,[38] Kurdistan,[39] Bulgaria,[40] Gibraltar,[41] and Lebanon.[42] Her earliest activities in the 1920s focused on producing the most current synthesis of the poorly understood British Upper Paleolithic.[43] In the process she acquired an interest in the faunal remains associated with the archaeological materials which lasted throughout her life and led her into some of the first considerations of the ancient climatic and ecologic conditions which these remains represent.[44] Her detailed account of the variation in faunal content of her most famous excavation at Mugharet et Tabun in Palestine was outstanding in this area. The work at Tabun represents a milestone in the development of Paleolithic research in the Near East. It also provided the basis for a virtual revolution in physical anthropology in providing evidence on the development of *Homo neanderthalensis* (Neanderthal Man) in relation to *Homo sapiens*. The indications from these excavations were that the relatively "advanced" form of Neanderthal found here was at least the contemporary of the more extreme "Classic" Neanderthals of Europe, and this thoroughly upset the relatively simplistic evolutionary sequence which had been developed up to that time.[45] Although she is best known for this and other work on the Paleolithic of the Near East, her contributions to the definition and understanding of the Mesolithic period through her work with the Natufian culture and other cultures are also outstanding.[46] Her active mind also produced contributions on prehistoric migrations, prehistoric irrigation methods, Paleolithic art, prehistoric hunt and butchering techniques, and on larger archaeological semi-theoretical frameworks.[47]

Like Gertrude Caton-Thompson, Garrod was able to enter and became both active and well recognized within the British archaeological community. In 1939 she was elected Disney Professor of Archaeology at Cambridge and so became the first woman in any field to become a professor at either of the great British universities, Oxford and Cambridge. Here she is credited with the substantial improvement in the quality of archaeological education which took place in the late 1940s and the 1950s. She served in women's services in both World Wars and, on the basis of her experience in photographic interpretation in the Second, was very instrumental in the development of aerial photography as a crucial methodological tool in archaeology. She received numerous honorary degrees and honors, especially the Gold Medal of the Society of Antiquaries of London, for which she was the first woman recipient. In Garrod's obituary in 1969, Glyn Daniel praises her as "a pioneer woman archaeologist as well as a field archaeologist of the first rank of that exacting and arduous profession."[48]

Another very significant researcher in the prehistory of Europe and the Near East was Elise J. Bäumgartel. German-born, Bäumgartel contributed to the archaeological literature in three languages (German, Italian, and English) in the 1930s–1950s.[49] She came to England in 1934 to work on cataloguing the large Egyptian Neolithic collection that had been amassed but not reported by Sir Flinders Petrie for the University College of London at the turn of the century. Through the support of several of the professors who ultimately arranged for her to be paid a small salary, she worked through the huge collection and prepared to publish a catalogue. The disruption of World War II intervened, but at its end she ultimately produced her ground-breaking publication *The Cultures of Prehistoric Egypt* (1947). Daniel in his comprehensive archaeological chronicle recognizes it as "a work of extraordinary importance," both for its analysis of the significance of this Neolithic material to the later Predynastic and Dynastic cultures, but most especially for its semirevolutionary focus on the relationship between the Egyptian developments and those of Southwest Asia.[50] Of interest here is that this significant contribution and well-deserved recognition came as a result of insightful *analysis in the laboratory* of material already gathered in the field, rather than through another field project per se.

The field of Americanist studies (archaeology in the Americas) proved perhaps even more resistant to the entry of women than did European prehistory. Until the middle 1930s there were very few women working independently in the field in their own right. Several gifted individuals formed part of successful husband-wife teams. One of the best known of these was Ann Axtell Morris, wife of Earl H. Morris, one of the best known Southwestern archaeologists of this century. Ann Axtell Morris was talented as an archaeologist, author, and artist. A graduate of Smith College, she studied with the American School of Prehistoric Research in France. She married Earl Morris in 1923. From then on she collaborated with him in his work for the Carnegie Institution of Washington, D.C., in extensive research on the Mayan culture of Meso-America and in the southwestern United States. At the great site of Chichen Itza she excavated the famous Temple of Xtoloe Centote and recorded the important wall murals in the Temple of the Warriors.[51] In the Southwest she focused on the recording and study of Indian pictographs. Although she collaborated with her husband on virtually all of his work and is reported to have written significant parts of the technical reports, her own best known credited publications are both nontechnical and designed for the general public (*Digging in Yucatan*, 1931; *Digging in the Southwest*, 1933).

Elsewhere in the Southwest, two other husband-wife teams achieved prominence. In this connection, Elizabeth Crozier Campbell and Winifred Gladwin share many features. As part of successful husband-wife teams, both routinely received joint publication credit. Both pairs worked in the western United States at a time (1920s–1940s) when the prehistory of the area was very poorly understood, and were able to make substantial contributions to the development of the early data base and theoretical frameworks. In spite of the relatively open teamwork neither woman received sufficient recognition to warrant obituary or other biographic treatment in the professional journals, so that details of their lives are skimpy at best.

The Gladwins are the best known of the two teams. Harold S. Gladwin was employed by an influential private foundation, the Amerind Foundation, in the 1920s in southern Arizona. In spite of the success of archaeologists further north, in the San Juan Basin, in working out a chronologic sequence for the last 2,000 years of Indian culture (the Pecos Classification), the archaeology of the southern Southwest proved to be extremely intractable to ordering at all. In order to develop the necessary structure Harold and Winifred Gladwin focused their attention on developing a detailed description and classification of regional pottery types, as indicators of culture change and spatial-temporal variation in culture.[52] They used biologic taxonomy as a guide for establishing a binomial system of pottery type designation, so that the color combination or surface treatment of the pottery functioned at the genus level and the specific detail and geographic find-location at the species level. Their voluminous descriptions produced hundreds of named pottery types.[53] Even though the biologic connotations have long since faded away, the binomial system of ceramic classification and most of their named types are still in use. In the same way, the Gladwins organized and structured the classification of Southwestern cultures as a whole, using a biologically derived and explicitly atemporal root-stem-branch framework.[54] Similarly, although the larger structure has been discarded, much of the terminology still survives. Most of the Gladwins' extensive bibliography on ceramics and typology is jointly authored. However, Willey and Sabloff[55] uniformly refer to these significant methodologic contributions as "his" and except for bibliographic citations do not refer to Winifred Gladwin.

Elizabeth Crozier Campbell and William H. Campbell formed a successful team that explored and brought initial order to one of the least-known areas of American archaeology. Working in the harsh deserts of the Great Basin and California in the 1930s and 1940s, they focused national attention on the great antiquity of the lithic surface assemblages and worked with geologists to establish the relation between these assemblages and the rise and fall of ancient desert lakes.[56]

There were exceptions to this limited role for women. One extremely significant figure was (and is) Florence Hawley Ellis, who has been a dominant force in Southwestern archaeology for nearly half a century, and whose detailed profile follows the next section. Another outstanding individual was Frederica Annis de Laguna, whose long career spanned more than four decades. In common with other early women pioneers, and reflecting the character of anthropology in the 1920s and 1930s, de Laguna contributed equally to the fields of archaeology and ethnography. Born in 1906, Frederica de Laguna was educated at Bryn Mawr College and at Columbia University. After several years of frustration in the mid-1930s in finding appropriate professional employment, during which she worked as a museum assistant and an assistant in the Soil Conservation Service, she entered the faculty of her old alma mater, Bryn Mawr, and remained there, except for interruptions during World War II, until retirement. During the war her leadership capabilities raised her to the rank of Lieutenant Commander in the United States Naval Reserve.

Beginning in the early 1930s, when women in America were still very much restricted from doing independent field work, Frederica de Laguna organized and led major expeditions to Alaska to do archaeology and ethnology. These investigations took place in a virtually unknown setting, and included original studies of the archaeology of Cook Inlet, Yakutat Bay, and the northern Yukon.[57] Together they produced a series of crucial pioneering studies of the prehistory and ethnography of the Eskimo and Northern Athabaskans. In addition to these significant contributions, de Laguna occupied leadership roles in the American Anthropological Association.

On the whole the period 1915–1935 saw a very slow but significant entry of individual women into the field of archaeology. In Europe this transition was dominated by a few exceptional individuals whose contributions simply demanded, and generally received, attention. Educational opportunities improved, and with achievement came some access to the professional network. In America, acceptance was much slower and most women with an inclination or talent for archaeology became part of husband-wife teams. The outstanding individual exceptions, like Frederica de Laguna and Florence Hawley Ellis, in spite of their enormous contributions did not visibly enter the professional-social network. Neither is even mentioned in Willey and Sabloff's *A History of American Archaeology* (1974).

The Culture-Historical Period: 1935/40–1960

The central direction of this enormously productive period within the evolution of the discipline of archaeology continued to be toward the development of additional chronologic sequences and the improvement of method and technique. The majority of the literature was still centrally concerned with the production of increasingly fine local and regional se-

quences, filling in "gaps" in yet uninvestigated regions, fitting these regional schemes together by various means to produce larger frameworks.
At the same time, however, there were strong stirrings both in the Old
World and in American archaeology which portended the quite revolutionary developments of the 1960s and 1970s. The relative simplicity of
earlier schemes which equated typology or technology with culture was
repeatedly challenged.[58]

In European archaeology the watershed publication in England of J. G.
D. Clark's study of the ecology of a Mesolithic site[59] set off a trend toward
increasingly environmental and "scientific" archaeology, which has most
recently come to focus on archaeology as human geography.[60] At the same
time, Near Eastern archaeology shifted its emphasis from specific sequence
to larger topics related to the origins of agriculture and sedentary life.[61]

In American archaeology the culture-historical mainstream reached its
peak development in improving the quantity and quality of the data base
and expanding it by the generation of pan-areal syntheses and semi-
explanatory frameworks (syntheses combining evidence from wide areas,
and attempts at explaining cultural change).[62] However, there developed
a small but growing current of dissatisfaction with the simple clear-cut
taxonomic and chronologic activity. This took place within the framework
of participation in and interaction with the larger field of anthropology.
Particularly in the 1950s questions were raised about the objectives of archaeology, about its utility to the "study of man" as a whole, about the
relation of the discipline to "science" and to "history."

Both in Europe and in the United States the sheer number of archaeologists multiplied enormously. In England, where the most accurate figures are available, the number of professional positions rose from ten in
1932 to more than two hundred in 1970.

In Europe the contributions of women to this rapidly growing and evolving field were substantial, but are clearly dominated by one or a few outstanding figures. Foremost among these is Dame Kathleen Mary Kenyon.
Born in 1906 she was educated at Oxford in the 1920s and 1930s. In the
1930s and 1940s she was the secretary of the Institute of Archaeology of
the University of London and finally acting director of the Institute in the
mid 1940s. Although she worked with Dorothy Garrod at Zimbabwe in
the 1930s and made several minor contributions to the archaeology of Palestine and Britain,[63] she is best known for her monumental excavations at
the famous biblical site of Jericho, and as the director of the British School
of Archaeology in Jerusalem.[64] At the great site of Jericho, layer by layer,
she uncovered more than 30 meters of deposits which chronicled the evolution of the biblical city from its origins far back in the Neolithic. On the
basis of her work there she evolved one of the two principal theories of
the 1950s concerning the origins, location, and and causation behind the
development of agriculture. In her view domestication began near natural

permanent oases as increasing desiccation continued to reduce habitats usable to people after the end of the Pleistocene epoch.[65] The contrasting theory espoused by American archaeologist Robert Braidwood, digging at Jarmo in Iraq, placed the origins of domestication along the hilly flanks of mountain ranges where the principal cereals grow in wild form.[66] For several years in the middle 1950s the archaeological community was absorbed by the hot competition of the two theories and their great projects as each brought up older and older remains, dated by the new radiocarbon method.[67] Kenyon has been able to enter and function within the English archaeological community, and has received appropriate recognition, including medals, fellowships, and honorary doctorates, and finally the title Dame of the British Empire.

Mary Nicol Leakey in some ways harks back to an earlier period inasmuch as most of her work was done as the relatively silent partner of a colorful international figure, Louis S. B. Leakey. Mary Nicol Leakey is the great great granddaughter of John Frere, one of the best known of the early English geologist-archaeologists who reported the association of artifacts and extinct animals in the early nineteenth century. She grew up among the famous painted caves of southern France, and was educated in geology and prehistory at the University of London. She was already active in British Paleolithic archaeology in the early 1930s. After her marriage to Louis Leakey they proceeded to form one of the best-known teams in Old World archaeology, in their long-term and enormously important research on the earliest hominids, ancestors of man, at the Olduvai Gorge near the Kenya-Tanzania border.[68] Mary Leakey's persistence, her good fortune in finding important fossils (e.g., *Zinjanthropus* in 1959), and her exacting attention to detail and to accurate recording are all legendary. Louis Leakey's biographer, Sonia Cole, observes that Mary is a more dedicated scientist than Louis ever was; although she could not rival his genius and flair, she had the perfect temperament for the attention to detail that modern archaeology demands. While he was interested only in the spectacular discoveries, Mary was quite content to make exact measurements and compile elaborate tables of tool types, laboriously uncovering the evidence centimeter by centimeter. Mary Leakey has moved into the light of recognition since their divorce (1968) and Louis Leakey's death (1972). She is now best known as an individual scientist for her excavations at Laetoli, where with her meticulously careful excavations she was able to uncover the actual footprints of an early hominid and to document for this predecessor of humankind the existence of upright posture at about 2.0 to 2.5 million years ago. She has received numerous honors and two honorary doctorates.

Several other women made contributions to Old World archaeology in various special areas. Jacquetta Hawkes was originally involved in important field work on the Paleolithic of the Punjab and Kashmir, and on the

British Channel Islands. She is best known, however, for her important syntheses on British archaeology (with her husband Christopher Hawkes),[69] on world prehistory, and on various regions of the Near East.[70] Germaine Henri-Martin, a young French archaeologist, startled the archaeological community with her wholly unexpected discovery of *Homo sapiens* remains below the Mousterian levels at the site of Fontechevade, thus indicating possible contemporaneity or seniority of *Homo sapiens* and *Homo neanderthalensis*.[71] Marija Gimbutas, a native of Lithuania and a naturalized American citizen, provided a series of excellent studies and syntheses on the Bronze Age and Iron Age of Eastern Europe, an area which had been relatively little accessible to Western scholars.[72]

Women in American archaeology made significant progress in the period from about 1935 to 1960. From a position of very marginal participation these pioneers firmly established the presence of women within the archaeological profession. Their contributions to the data base and to archaeological methodology are substantial indeed. Three of the most prominent are still living and graciously consented to grant the interviews which provide the basis for the personal profiles at the end of this section.

Much of the activity in which women in this period were involved concerned areal or regional prehistory, particularly in the Southwest and the easternmost United States. When Dorothy Cross Jensen died after a career of more than four decades, "American archaeology lost one of its senior thoroughly respected hard working and totally committed professionals."[73] Although she is best known for her work in Americanist studies, her two earliest publications on the Near East were significant in themselves. She produced one of the first typologic descriptions of the pottery of Tepe Gawra at a time when ceramic analysis received very little attention.[74] Her principal contributions, however, were to the archaeology of one region in the eastern United States, New Jersey. During the Depression of the 1930s, she was placed in charge of a huge WPA project, a site survey of New Jersey. Although this, like many large surveys of the Depression, was primarily a job-creation program, Cross developed it into the basis for her great two-volume work on the *Archaeology of New Jersey*.[75] In the process she investigated and laid to rest the issue of the "argillite culture," which for over half a century had been the subject of controversy. From her excavations and analysis at Abbott Farm she demonstrated that this crude lithic assemblage, far from being an astonishingly ancient "Paleolithic" in character, was but a local version of the relatively recent New Jersey Archaic phase.[76] Cross was recognized as a thorough professional in research, in teaching (at Hunter College), and in museum work (New Jersey State Museum).

Another outstanding contributor to the regional archaeology of the eastern United States was Marian Emily White. A native of New York State, Marian "Happy" White received a B.A. at Cornell University and, after

Dorothy Cross Jensen. Courtesy
of *American Antiquity*.

military service in World War II, in 1956 became the first woman to receive
a doctorate in anthropology at the University of Michigan, one of the many
top-class American universities which remained resistant to women students. Marian White spent most of her very productive professional career
at the University of Buffalo (New York), and her numerous professional
and public contributions relate to the prehistory of New York State. She
became one of the dominant archaeological influences in the area, not only
because of her numerous excavations of prehistoric sites, but also for her
very early insights into the use of statistics as an analytical tool, her early
interest in settlement pattern change, in archaeological approaches to the
study of cultural stability, in the development of "conservation archaeology," etc.[77] Her interests spanned all periods but her great love was the
prehistory and ethnohistory of the Iroquois Confederacy in New York, and
particularly the Seneca Nation.[78] At a time in the 1970s when relations
between many Indian groups and the archaeological community were at
their nadir, Marian White maintained excellent relations with the Indians,
helping to protect their prehistoric sites and to integrate their young people

Marian Emily White. Courtesy of
American Antiquity.

into archaeological projects. Her courage and dedication in protecting the archaeological resources were legendary; she was undeterred by long odds and was willing to lie down in front of bulldozers or to sue the Army Corps of Engineers to save a site. She served in various leadership roles in regional professional societies and was active in the National Society for American Archaeology Committee on the Status of Women in Archaeology.

During this expansive period, 1935–1960, several man-woman professional partnerships achieved considerable prominence in American archaeology. These were generally much more nearly equal relationships than in previous years. Together, Madeline Dorothy Kneberg and Thomas M. N. Lewis dominated the archaeology of the Tennessee area of the Southeast from the 1940s through the 1970s. Their excavations at Hiwassee Island, the Eva site, and elsewhere enabled them to develop a detailed understanding of the PaleoIndian and Archaic periods of Southeastern prehistory.[79] Virtually all of their numerous and significant publications were done jointly, with senior authorship varying. In spite of their lifelong professional and personal partnership, because of the rigid nepotism rules at the University of Tennessee, where both were employed, they could not be married until they retired.

Betty Jane Meggars and Clifford Evans constituted for over thirty years one of the most successful professional teams in American archaeology.

Beginning in the 1940s Meggars was involved in the prehistory of South America, particularly that of the almost completely unknown Amazon Basin.[80] Together Meggars and Evans contributed enormously to the development of an archaeological sequence there. Their work was so closely linked that Meggars-and-Evans became virtually a single term in the professional literature. They not only dealt with the specifics of South American prehistory itself, but were deeply involved in discussions on transpacific contact, environmental determinism, and prehistoric migration, to which they contributed both leadership and insight.[81] Meggars and Evans were married to each other throughout most of their long careers, but chose to retain separate names.

In the Southwestern United States archaeology was undergoing both rapid growth and high-level professional activity, and was considered by many to represent the "cutting edge" of the discipline. Several women were centrally involved in this growth and activity, including two of the three whose personal profiles follow this section (Florence Hawley Ellis and Marjorie Ferguson Lambert).

A common characteristic of women entering the Southwestern field at this time was a dual interest in archaeology and ethnography. Dorothy Keur began her career with a distinguished monograph on Big Bead Mesa, a Navaho fortress archaeological site, which she analyzed in terms of ethnographic acculturation.[82] Clara Lee Tanner worked principally as an ethnologist but also undertook studies of prehistoric art.[83] Anna O. Shepard, working in design analysis and temper studies, created the enduring manual *Ceramics for the Archaeologist*.[84] Bertha Dutton is equally at home with archaeology and ethnography, and in addition has dealt with primitive art, museum development, and public involvement in archaeology (through the Girl Scout program). Born in 1903, she received her degree as one of the first archaeology students at Columbia University. Much of her work was done for the Museum of New Mexico or the School of American Research, until she ultimately was made director of the Museum of Navajo Indian Art. Her continuous involvement with public participation in archaeology long predates the more general professional recognition of the importance of this interaction.[85] One of her earliest contributions (1938) is also one of her most substantial, the excavation and reporting of Leyit Kin,[86] one of the then little-known "Small Houses" of the famous Chaco culture of New Mexico. Her ethnographic work was equally wide ranging, including accounts of Pueblos and Navahos in the Southwest and the Huichol in Mexico.[87]

In brief review, by 1960 there were increasing numbers of outstanding women in the field of archaeology both in Europe and in the United States. Their contributions include definitive works on the Paleolithic and Neolithic and on the origins of agriculture and settled life in the Near East; the Paleolithic and Bronze Age of Europe; pioneering regional research in the

New World from the Amazon River to Alaska, from New Jersey and New York through New Mexico and Arizona to California; topical and theoretical contributions to larger conceptual areas in anthropology such as typology, acculturation, migration, evolution, and environmental determinism.

While the contributions of women to the field are very considerable, their actual role remained significantly less visible. Their participation is in fact largely an account of the activity of a very few unusual figures. Academic educational opportunities became more open by the 1950s, but opportunities for field training were still very limited. Particularly in the United States, opportunities for professional employment in mainstream academic positions were very few, and participation in the larger professional network was likewise limited. In historical accounts of the development of archaeology, English women are fairly well represented.[88] In American chronicles, however, women are curiously invisible: in their lengthy consideration of American archaeology Willey and Sabloff mention only one woman archaeologist (Marie Wormington) other than as part of a husband-wife team. The substantial and significant contributions of de Laguna, Hawley Ellis, White, Cross, and others do not appear in the text.

THREE GREAT LADIES IN AMERICAN ARCHAEOLOGY

Three of the most outstanding individuals among the women who pioneered the field of American archaeology in the 1930s and 1940s graciously consented to share their recollections and perceptions about their lives in and out of the field.[89]

FLORENCE HAWLEY ELLIS

Florence Hawley Ellis published her first articles on Southwestern archaeology in the 1920s, thus becoming one of the earliest women to enter the Americanist field in her own right. More than half a century later she is still an active and influential force within Southwestern archaeology.

Florence May Hawley was born in 1906 in Cananea, a copper-mining town in Sonora Mexico, in the violent period just preceding the Mexican Revolution of 1910. After several years of intermittent violence and danger, her father, a mining chief chemist, moved his family back into the United States (to Miami, Arizona), and in 1913 Florence was able for the first time to attend a formal school. She was ready to enter college at sixteen and already enjoyed archaeology as a hobby. She had participated in family "digs," guided by archaeologist Byron Cummings.

When she entered the University of Arizona she had no idea of becoming an archaeologist, but thought she might become a teacher like her mother or go into art. Thus she began her college career as an English major. All went well except for an American history course which produced

Florence Hawley Ellis in the field. Courtesy of Florence Hawley Ellis.

too many dates to remember. Family friend Dean Bryon Cummings, then the only member of the anthropology department, suggested she switch to an archaeology course. She did, and so in this improbable fashion the career of one of the best known Southwestern archaeologists had its origins.

In 1928 Florence followed through with a master's thesis on the ceramics from three closely successive stages found in excavated sites near Miami, Arizona. Depending primarily on associations with other wares, she was able not only to separate out three sequential pottery types (Early, Middle, and Late Gila Polychrome), but to suggest their possible Mexican relationships.[90] Both of these already represent very significant contributions to the development of the embryonic field of Southwestern prehistory.

While working on the thesis, Florence collaborated with her father on the first chemical analyses of black pottery pigments (carbon, carbon-mineral, and manganese).[91] These chemical distinctions in pottery paint subsequently became a basic part of ceramic analytic and cultural interpretational methodology in Southwestern archaeology.

Florence began teaching at the small, young Department of Anthropology at the University of Arizona in 1928, at an annual salary of $1,350. The attitude of the poorly paid young scholar was that the emphasis should be on the great opportunity to make some unique contribution to the field, rather than on the salary. At Arizona she was able to take special training in tree-ring dating (dendrochronology) from the pioneer in the field, A. E. Douglass. This, plus her skill at ceramic analysis, gave her the possibility for developing a specialty, which she now recognizes as crucial to her early success. Meanwhile, she taught, worked at the museum, and analyzed tree-ring specimens and ceramic collections.

In 1929 she began field work with the University of New Mexico at a location which was to result in some of her most important work: Chetro Ketl in Chaco Canyon. She excavated the stratified layers in the trash mount and analyzed the ceramics and the tree-ring specimens. Then in 1933, as the Great Depression deepened, she and other young faculty were told they would be laid off for one year. Florence decided to take her savings and the data from Chetro Ketl and go to the University of Chicago for her doctorate. Being warned that statistics was required at Chicago, Florence took a course in the subject before she left. She then conceived the idea of applying this training to establish the significance of the stratified variations in ceramics which she observed in the Chetro Ketl East Dump. This resulted in one of the earliest uses of form statistics (Chi square) in American archaeology. Ultimately, her dissertation and resulting monograph (1934) combined these data with dendrochronologic studies from nine Chacoan sites to explain the history of Chacoan occupation and desertion in terms of ecologic change. It is unquestionably a milestone in the development of Southwestern archaeology. Subsequently she regularly included statistics in her well-known course in Southwestern archaeology.

After a confusion in correspondence about job availability in Arizona, Florence applied to E. L. Hewett in the small new department at the University of New Mexico. She was accepted to the faculty, and arrived in Albuquerque in 1934, degree in hand but so "broke" that she had to pawn her wristwatch and take a lien on her old Ford. Her teaching career at New Mexico lasted from then until her retirement in 1971. From 1937 to 1941, because of her special skills, she was on loan half-time to the University of Chicago to teach dendrochronology. Here she was able to travel and do extensive field work related to developing a Midwestern dendrochronologic sequence.[92]

Once settled in New Mexico, Florence found she had to learn a whole new field of Southwestern anthropology—that related to the history and prehistory of the Indians of New Mexico, on which only a scattering of brief reports then existed. Given her background in ceramics, she absorbed what there was quickly, and almost immediately produced a volume on Southwestern ceramic typology, her classic *Field Manual of Southwestern Pottery Types* (1936). This detailed and accurate reference work covering what was then known stands as a model of its kind and is still, nearly fifty years later, the standard reference for New Mexico prehistoric pottery.

Shortly after her arrival in New Mexico, Florence met and was married to Donovan Senter (1936) and they had one daughter, Andrea. Briefly they cooperated on ethnologic research on Spanish Americans in New Mexico. A combination of differing interest, the separation occasioned by World War II, and the perception of competition even though the two were not in the same field resulted in a breakup of the marriage in 1947. In 1949 she met Bruce Ellis, a historian from the east, who later became Curator of Collections at the Museum of New Mexico. A close friendship formed, which culminated in a marriage of the two which endures today.

Florence Hawley Ellis was an outstanding archaeological force at the University of New Mexico for nearly four decades. She cooperated in or directed numerous sessions of the university field school and was able to carry out research work virtually every summer, including eleven seasons of research at Chaco Canyon; one season on the few still-existent remains of old Pojoaque Pueblo; three on the first permanent church, dwellings, and barracks of San Gabriel del Yungue, the original Spanish settlement in the Southwest; five of sampling the many plazas of Sapawe, a 21-acre Tewa ancestral site; and a final season of sampling the related but somewhat smaller site of Tsama Pueblo. That these projects received their initial write-ups but not formal presentation for publication has sometimes awakened her at night, she says, but data and summaries are stacked for a final onslaught.

During the 1960s and 1970s, she undertook extensive investigations in connection with the Indian Land Claims Commission, assisting in the definition of ancient tribal areas for most of the New Mexico and Arizona Pueblo Indians and for the Navahos.[93] In the 1950s, she became a significant part of the large Wetherill Mesa project, working to establish relationships between prehistoric culture and living peoples.[94]

Although Florence Hawley Ellis is best known for her archaeological research, her work in ethnography/ethnology and her ability to relate the two disciplines are both unusual and outstanding. Her first personal contacts with the Pueblo people begun in 1929 were aided a decade later by the presence of her own small daughter. She shared work, food, and confidences with them and came to be regarded as a trusted friend. Ultimately her contributions in Pueblo and Navaho ethnography were nearly as significant as those in archaeology, and included studies in acculturation, kinship, religion, witchcraft, social organization, and material culture. She sums up her primary concern as the interplay of patterns in the life styles of native peoples, living and dead. Because of her good relations with Puebloan peoples, she was able to carry out research where most archaeologists would have been prohibited. In her work on Pueblo land and water claims she was able to section and study ash heaps (dump mounds) of Zia,

Taos, and Nambe Pueblos, disturbance of which usually is strictly forbidden on religious principles but in this case was permitted under supervision of elders to the end of deriving a date for initial occupation of present village sites.[95] In short, no aspect of Southwestern archaeology or ethnography was closed to her probing interest. Her total contribution to the literature is enormous, numbering over 150 separate publications.[96]

Florence Hawley Ellis's influence as a first-class research archaeologist-ethnologist is rivaled by that as a university teacher. She has been responsible for the development of and has been the personal mentor of generations of Southwestern anthropologists. The esteem in which she is held by former students is reflected in a recent Festschrift in her honor.[97] Over the years she developed and taught nearly two dozen different courses in every area of anthropology. During World War II she found herself giving not only her own courses but those of her male colleagues who had joined the armed forces. After the war, a survey revealed that she still carried the heaviest teaching load in the department and that her salary was significantly lower than that of male colleagues with less seniority. Despite the efforts of various individuals the salary imbalance persisted. In addition, promotion came much more slowly than for her male colleagues. This was also the common experience of other female faculty.

Since her retirement from the university, as an affiliate of Ghost Ranch, Florence has continued to supervise excavations in the Gallina area. In the local museum, which bears her name, she has helped to develop regionally oriented exhibits representing successive cultures in the northern Rio Grande, from PaleoIndian through Archaic, Pueblo, and finally the Spanish and the Anglo.

Florence Hawley Ellis's perceptions on women in the field are invaluable. She considers that having a specialty so that she did not immediately have to compete on a one-to-one basis with the men in the field was very useful, especially to her early career. She also acknowledges the kindness of specific individuals like Byron Cummings, who helped her at the University of Arizona as a student and gave her her first job, A. E. Douglass, who aided and encouraged her work in tree-ring analysis, and E. L. Hewett, who lent her the $200 needed to complete her doctorate and welcomed her into the department at the University of New Mexico. She salutes those men and others who "all along were trying to be gallant and generous and fair." She also recognizes the presence of potential exploiters, like T. T. Waterman, then temporarily on the faculty of the University of Arizona, who tried to get her to publish her ground-breaking master's thesis jointly with him even though he had had no part whatever in it. She believes that she "had a lot of good breaks, none that were tremendously big, but very convenient ones." Reflecting on the experiences of her own lifetime, she feels that women in a "man's field" should try to dress well and behave and act well because "your male co-workers will appreciate it—they expect you to be part of the landscape and you had better be a pleasant part of it." She recognizes that professional competition is difficult and disruptive to the personal life, particularly to marriage, even for individuals in different fields.

Beyond these perceptions on the position of women in the field, Florence Hawley Ellis has some strongly held feelings about the nature of archaeology and science. She insists on a respect for the data, and scorns easily dreamed up theories which do violence to it. On the value of the individual and his/her work she comments, "The producers are the workhorses, not the flash-in-the-pan types," and concludes that, indeed, "By their works ye shall know them."[98]

MARJORIE FERGUSON LAMBERT

Marjorie Ferguson Lambert has a remarkable range and number of achievements in her long career, not only in archaeology, but also in the history and ethnology of the Southwest and its relationships to Meso-America. Her contributions have included not only very significant field investigations, but also major contributions to the development of the Museum of New Mexico, and to public involvement through public exhibits.

Marjorie Elizabeth Ferguson was born in 1910 in Colorado Springs, Colorado. Her father, a Scottish immigrant, imparted to her both a strong value system and an appreciation of her Scottish heritage, including Robert Burns's poetry and bagpipe music. She became interested in archaeology in high school, but never thought of it as a profession. Indeed, she felt impelled toward psychiatric social work. However, after attending lectures by such famous archaeologists as E. L. Hewett and S. W. Morley during her college years at Colorado College, she was to alter her views. To help humanity, she decided, one had to understand the past, including the remote past of archaeology. Her family would have preferred her to have been a librarian or a teacher. Even within the larger field of anthropology "the ladylike thing to do was to go into ethnology." But Marjorie decided to follow her own inclinations and to become an archaeologist. She was supported in her determination by the same E. L. Hewett who had helped Florence Hawley, and by Colorado College Museum Director W. W. Postlethwaite. Edgar Lee Hewett, a giant in Southwestern archaeology, is remembered with esteem and respect by all who knew him, and was particularly known for his fairness to women. When he heard of Marjorie Ferguson's hopes for archaeology, he found a niche for her as a research-teaching fellow at the University of New Mexico, and even lent her funds to buy books.

Marjorie received her master's degree in anthropology there in 1931, and for several years both taught and conducted field research for the University of New Mexico. Here she produced some of her most outstand-

Marjorie Ferguson Lambert. Courtesy of Marjorie Ferguson Lambert.

ing contributions. Particularly important are her excavations at Paa-ko, Puaray, and Kuaua (1937, 1938, 1939, 1954), where her studies of prehistoric and early historic Rio Grande Pueblos remain outstanding classics in the field. She took over the Paa-ko project from two male colleagues who had not been successful on two closely related projects. When she was placed in charge, these colleagues spread the word that she would do no better and that the native laborers would probably refuse to work for a woman. Marjorie took on the project for $125 per month, a significant salary reduction, and, using the latest stratigraphic techniques, brought the project to a successful conclusion. The laborers adored her, shared their lunch (sugar, coffee, and tortillas) with her, and pooled their meager funds to buy her a Christmas present.

When E. L. Hewett retired from the University of New Mexico in 1937, he continued as director of the Museum of New Mexico and the School of American Research (an influential private foundation). He asked Marjorie to join him as Curator of Archaeology, which she did, making her one of the first women to occupy a major curatorial position in the country. In the museum work, Marjorie became known throughout the country for the collections she acquired and for the number and quality of archaeo-

logical exhibits which she created, as well as for her development of a
supportive docents group and for the upgrading of public relations in general. In the laboratory she became a well-known authority on dating techniques, combining ceramic crossdating, tree-ring dating, minerologic
analysis, and later archaeomagnetic and radiocarbon dating in archaeological analysis.

Meanwhile, her field research continued apace. In 1944, combining archaeological and historical investigations, her preliminary work on Onate's
capitol of San Gabriel probably led to the important excavations of the site
of Yunque near San Juan Pueblo (1952, 1953). In the 1960s she conducted
a series of very significant field investigations in caves in southwestern
New Mexico, yielding valuable new data on the early (Archaic) cultures of
the area and on the Chihuahua Mogollon. This important work was cut
short, she relates, with a rare flash of resentment, by the contemporary
Director of Anthropology of the Museum of New Mexico, who blocked an
already approved grant to her by the School of American Research because
of his personal antipathy toward the school's president and managing
board. He later offered her a place on one of his own "pet projects" to
make up for his earlier actions, but Marjorie refused the offer. Instead, she
continued to conduct her own research at historic and prehistoric localities
in the Southwest, including the important Jemez site of Guisewa and the
Twin Hills site near Santa Fe. Beyond this, shorter-term projects have led
her into Mexico and Central America, and have ranged widely beyond
archaeology into prehistoric art and Pueblo ethnography (1942, 1945, 1947,
1954, etc.).

Marjorie's long association with the School of American Research was
reinforced when she became a member of its managing board. Her public
outreach continued to expand, not only through her lectures and exhibits
but also through her concern and support of archaeological and historical
societies, conservation organizations, and Indian groups. Universally respected for her professional contributions, Marjorie is also much valued as
a loyal friend and an inspiration to a generation of younger archaeologists.

In 1950 Marjorie married E. V. "Jack" Lambert, who supported her
strongly in her profession throughout their long marriage. Marjorie's experiences and perceptions on the role of women in the field are enlight-
ening. She has experienced much kindness and support, as from her
mentors, E. L. Hewett and W. W. Postlethwaite, and from many others.
As an attractive younger woman she also experienced the darker side of
male-female relationships when colleagues wanted kisses and other things
for the privilege of working with them. She has been accosted in dark
hallways and at least once had to defend herself by delivering a "resounding biff in the nose." She remains essentially unaffected by these experiences, and finds her male and female friendships of great and equal value.
Reflecting her own experiences and philosophy, Marjorie believes that

women in her field need not be extra tough or defeminized to achieve their goals. Persuasion and cooperation, in her experience, are more successful than aggression.

Marjorie's feeling toward the archaeological discipline in which she has spent her life reflects a combination of matter-off-fact scientific pragmatism and an almost mystical humanism. Her respect for the data is profound and her enjoyment of working with it both in the field and in the laboratory is considerable. She feels sorry for some of the newer archaeologists who are largely theoreticians and have rarely experienced the pleasure of putting a trowel in the ground or washing a potsherd.

Marjorie feels enriched by both the joys and the sorrows attendant on her productive life, because of the challenge that each presented. She considers that her most important asset is her personal integrity, and feels that no science can prosper without this quality. She enjoys those spiritual times when she can be by herself on an ancient site communing with the people who once lived there, receiving from them what the Indian peoples call the "Breath of Life," which endures through time. At the same time, she recalls and has followed her father's Scottish advice to her when she made her decision to become an archaeologist: "Well, lassie, if you're going to cut off your hair, and if you're going to wear pants, and if you're going to go into a man's field, *be woman enough to take it!*"

HANNAH MARIE WORMINGTON

Marie Wormington is both nationally and internationally recognized as a leader in the field of PaleoIndian or Early Man archaeology. Her field investigations in the New World and her comparative studies have taken her over much of the western United States, Canada, and Mexico, and in the Old World to the classic sites of southern France, to Russia, and to China. She is the author of several classic texts and numerous research investigations.

Hannah Marie Wormington was born in 1914, one year after Margaret Murray was rebuked for attending the anthropological conference in England.[99] Marie's heritage included a strong element of French culture from her mother, and she is still bilingual and frequently visits relatives in southern France. Originally it was planned that she would go to school in France, at the Sorbonne, and that she would ultimately go into some kind of literature. However, the crunch of the Depression intervened and she was often hard pressed to get the ten cents for a bus ride to the local college, the University of Denver. She was still interested in literature, but was considering a career in medicine. When faced with dissecting a cat during the prerequisite biology course, she concluded that this was no place for her. However, she took a course in archaeology from E. B. Renard, a dramatic lecturer who brought several famous archaeologists into the field, and decided that archaeology was to be her field.

H. Marie Wormington. Courtesy of H. Marie Wormington and the Colorado State Historical Society.

Marie received her B.A. in 1935 and signed up to go on a field school project with the School of American Research in Europe. The day before they were to leave New York, the project was suddenly canceled. Undeterred by what might have seemed to be a fatal hitch in her plans, this energetic and determined young woman decided that "there was nothing to do but go anyway." Traveling with her mother, Marie launched herself into the French Paleolithic. She recalls that she was too naive to know that "you just didn't call on famous archaeologists without warning and announce you were here." She did just that with Dorothy Garrod, the famous British archaeologist. Garrod proved to be receptive, kind, and helpful and took Marie to lunch at her private club. On the basis of that brief meeting Garrod became a kind of role model for Marie—this was a woman that was *doing something* in archaeology." Marie's work in the French Paleolithic was really launched by another archaeologist, Harper "Pat" Kelley, whom she met at the Musée de l'Homme in Paris. She recounts, "Kelley took me under his wing and began to teach me something about the Paleolithic." Through Kelley's network Marie was able to meet and work with other great French archaeologists, such as Henri-Martin and Dennis Peyrony. Kelley also arranged for her to meet Edgar Howard, an American archaeologist already known for his work in PaleoIndian archaeology.

Through Pat Kelley, it was arranged that some Paleolithic artifacts be sent to the Denver Museum of Natural History. Some of the trustees then contributed small sums so that Marie could be hired as staff archaeologist at $50 per month to photograph PaleoIndian artifacts to be sent back to France in exchange. Thus began Marie Wormington's long association with the Denver Museum, which lasted from 1935 until 1968. It was frequently a stormy relationship. J. D. Figgins, the fair-minded director who had first hired her, was replaced in 1937 by A. M. Bailey, who did not wish to have a woman on the staff. Bailey was overruled by the president of the board of trustees, Charles Harrington, who made Marie Curator of Archaeology and put her on the regular payroll. However, that was the beginning of a 33-year enmity, which culminated in 1968 in the closing of the archaeology department of the museum as a strategy to terminate Marie's position, after granting her a year's leave of absence. The museum division was then reopened as a department of anthropology with all new personnel.

In 1937 Marie realized that further progress was limited without advanced degrees, and she applied for and received a fellowship from Radcliffe College. Most of the relevant courses were given at Harvard University, and Marie became the second woman to be admitted to study in the Harvard Anthropology Department, and the first to specialize in archaeology for a Radcliffe Ph.D., which she received in 1954.

In 1940 Marie married George D. "Pete" Volk. They had been married almost forty years when Pete died in 1980, and Marie recognizes that without his support and devotion she would not have accomplished all she did. During the war years Pete served in the Army Corps of Engineers, and Marie volunteered her services to the Red Cross. She worked as a nurses aid, and also as a Grey Lady in psychiatric wards.

Since she could not make ends meet on the amount she received from the museum, Marie taught part time at various periods at other local institutions, the University of Denver and the University of Colorado. More regular employment at regional universities in or near Denver was unlikely. When the University of Denver decided to honor its famous alumna in 1952 with the Distinguished Service Award, the only archaeologist (male) then on the faculty objected to giving the coveted prize to a woman. The University of Colorado's anthropology department was censured as recently as 1981 by the American Anthropological Association for its discriminatory hiring practices and its complete exclusion of women from the archaeological faculty.

What the Denver Museum did provide was a base for Marie to formulate and write her several highly regarded books and monographs and a platform from which to launch her numerous field investigations. Shortly after her arrival, she began work on the first edition of what ultimately became her best-known work, a synthesis entitled *Ancient Man in North America* (1939). In 1936 she began her field work with the excavation of a small PaleoIndian site in eastern Colorado. In 1937 and 1938 she conducted investigations at rockshelters in the virtually unknown "Western Slope" of the Rocky Mountains, and followed them with excavations at Fremont culture village sites in Utah. This pioneering work culminated in her definitive monograph on the Fremont Culture (1955).

In 1937 Marie had also developed another major synthetic work, *Prehistoric Indians of the Southwest* (1947), which remained the dominant text on the subject for more than two decades. She continued her field work in western Colorado and, with Robert Lister, produced a definitive monograph on the virtually unknown Uncompahgre Complex (1956). Another monograph, reflecting a wide-ranging field survey of PaleoIndian materials of Alberta, Canada, followed (1956). Richard Forbis was the coauthor. The fourth edition, completely rewritten, of *Ancient Man in North America* appeared in 1957, and will probably remain the central text on the subject until Marie's newest version appears.

This enormous productivity brought increasing recognition and Marie began to be sought as a consultant on many of the great PaleoIndian projects, including the excavation of mammoth remains in Santa Isabel Iztapan in Mexico, the search for very early remains at Tule Springs, Nevada, the early site at Onion Portage, Alaska, and the Scottsbluff butchering site near Chadron, Nebraska. Her own later work included excavations in Colorado at the Frazier Agate Basin site and joint excavations with J. B. Wheat at the Jurgens site. Because of her prominence in the field of Early Man studies, and its relevance to Old World origins, Marie has been asked to represent the United States in several international exchanges.

Another result of Marie's productivity and professional recognition was entry into the national professional network, represented by the Society of American Archaeology. The society had proved quite resistant to the penetration of women. Marie served two terms as its vice-president and in 1958 became the first woman to be president of this, the principal national professional archaeological organization. In 1983 she received the Distinguished Service Award of the Society for American Archaeology.

Although she has encountered some intolerance in the field, Marie has not had significant problems with the several male archaeologists with whom she has teamed up in the field—Robert Lister, William Mulloy, Joe Ben Wheat, and Mott Davis. She feels that these were outstanding individuals and that she was fortunate because all were married to unusually intelligent and productive women with whom she could relate. On her own digs, she has always run mixed male-female crews even when this was virtually unknown. "The generalization of my male colleagues was that, since I was there, the dig was already 'contaminated' and there was nothing to be done about it." Marie characterizes a career in archaeology as one which requires a "long and grueling path, but which can produce an exciting and richly rewarding life, with constant intellectual stimulus, and the opportunity to study and experience the different life ways both of contemporary peoples and those of the historic past." In her travels and in her long experience Marie has developed a network of devoted friends, and has served as the mentor and role model for new generations of archaeologists, including the author of this article.

PROGRESS AND PERSPECTIVES, 1960–PRESENT

The period after 1960 is largely beyond the scope of this article. However, a few perspectives and comments are appropriate. The field of archaeology has continued to evolve and to consider itself increasingly a part of "the sciences." It seeks increasingly for explanation of the data as well as their exposition. It is coming into increasing and fruitful contact with other disciplines, such as geology, paleobotany, ecology, geography, economics.

The role of women in the field has also grown and evolved, particularly after the liberalizing movements of the 1970s. Individual women continue to be important contributors to the data base through major research projects. Although it is not feasible to list all of the able people presently active, there are a number of well-established and particularly productive individuals whose work largely falls within this period. These include but are not limited to Elaine Bluhm, Linda Cordell, Hester Davis, Dena Dincauze, Sylvia Gaines, Shirley Gorenstein, Cynthia Irwin-Williams, Alice Kehoe, Jane Holden Kelley, Elizabeth Morris, Ruth Dee Simpson, Dee Ann Story Suhm, Patty Jo Watson, and Natalie Woodbury. A host of younger individuals is expanding the range and quality of contributions enormously.

In spite of a continuing problem with "invisibility," and exclusion from many professional networks, there is growing professional recognition of these research activities. For example, the American Anthropological Association profiled a woman (the author), as the "outstanding research archaeologist" for 1981, in its newsletter. So also, since the 1970s, entry into the professional network represented by the major organizations such as the Society of American Archaeology has opened considerably. The author became the second woman to serve as the society's president (1977–79) and Dena Dincauze the third (1987–89). Women now serve more often on the executive committee (e.g., Ruthann Knudson, Katherine Deagan, Nan Rothschilde) and also occupy other significant positions such as secretary (e.g., Leslie Wildesen and Lynn Goldstein) and editor (Dena Dincauze, Patty Jo Watson). Other women archaeologists, such as Natalie Woodbury and Jane Buikstra, have served similar roles in the American Anthropological Association. Leaders such as Hester Davis, Ruthann Knudson, and Leslie Wildesen have been central to the effort to develop and implement the enormously important federal legislation and other regulatory action designed to preserve and protect the national archaeological resource base. Their efforts have contributed greatly toward the development of the whole new field of cultural resource management which is now responsible for most archaeological research in the United States.

At the same time, however, the position of women in the field is still far from one of equality. Access to graduate training is difficult, and simply getting into a top-level graduate school may be impossible regardless of qualifications.[100] Access to field training is considerably easier than in the past, but is far from even, with an estimated half as many opportunities available to female students as to males. The most daunting difficulties, however, are encountered in finding professional employment, particularly in mainstream academic positions. Data derived from an informal survey of anthropology departments in 1980–1981 are suggestive (information from the American Anthropological Association Guide to Departments of Anthropology 1982). At that time, of the 84 doctorates in anthropology-archaeology, about one-third (28) went to women. Opportunities for academic employment were found to *vary inversely* with the size and prestige of the university and its anthropology department. The largest, most prestigious departments had no or few women archaeologists on staff; medium-sized Ph.D.-granting institutions had about 30 percent of all women archaeologists (about equal to the proportion of graduating Ph.D.s), and the small non-Ph.D. institutions employed all the rest (nearly two-thirds of the total). Taken as a whole, women made up only about one-fifth of the total number of archaeologists in the universities reviewed.

In another recent survey, Leslie Wildesen notes:

Fewer women are awarded their highest degree from prestigious private universities; of those who enter academic employment, few are in tenure-track positions and few ever receive full professorships; salary levels are often significantly lower. In general women archaeologists perceive themselves to be less well off and less likely to be well off than their male counterparts in terms of training, hiring, promotion, tenure, salary, access to research opportunities and professional credibility.[101]

One effect of this restriction on employment in the traditional university framework has been for qualified women archaeologists to lean strongly toward research or to seek innovative jobs in the private and public sectors in the field of cultural resource management. It is to be noted that the author considers that her rapid development toward "prominence" as a research archaeologist was at least partly motivated by the fact that, despite a prestigious degree from Harvard University (the first Harvard degree ever granted to a woman, in 1963), the best job she could find was a half-time position at a minor university paying only $2,400 per year.[102]

Salmon Ruins, New Mexico, one of the largest archaeological excavations ever conducted in the United States, directed by Cynthia Irwin-Williams, 1970–1978. Courtesy of P. B. George.

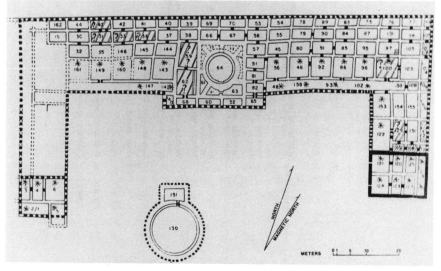

Plan of the Salmon Ruins, New Mexico, an enormous prehistoric Pueblo community excavated by Cynthia Irwin-Williams, 1970–1978. Courtesy of Cynthia Irwin-Williams.

Several of the younger women have already come to occupy senior national positions in the cultural resource management field in public agencies, including individuals such as Janet Friedman and Ernestine Green (U.S. Forest Service), Annetta Cheek (Department of the Interior), Sarah Bridges (Fish and Game Service), and Dianne Gelburd (Soil Conservation Service). There are a host of others in significant positions in cultural resource programs of agencies, universities, and private companies.

On the whole, although there is still a considerable imbalance, the role of women in archaeology, which began in an almost infinitesimal way less than a century ago, has grown to comprise a significant portion of the field. Most important are the contributions of the many women archaeologists to the data base, to the theoretical structure, and to the public policy of the discipline of archaeology. Florence Hawley Ellis sums up the value of these individuals: " 'By their works ye shall know them.' The best that is in us is sweated out onto paper, and lives beyond our lifetimes. It is our sincere contribution to the world, and on this basis the world progresses, bit by bit, in spite of those who see no meaning to life and excuse their laziness by refusing to recognize any ideals."[103]

NOTES

1. C. Weber, "Comments," in Text for Early Man, ed. A. M. Lipinski (Fall 1981).
2. Barbara Williams, Breakthrough: Women in Archaeology (New York: Walker, 1981), p. viii; Margaret A. Murray, "Centenary," Antiquity, vol. 37, no. 146 (1963):92–95.
3. J. P. Droop, Archaeological Excavation (1915), p. 27.
4. Ann M. Lipinski, "Women in Archaeology," in Text for Early Man (Fall 1981).
5. Weber, "Comments."
6. Richard M. Pearl, Geology (New York: Barnes and Noble, Inc., 1975).
7. Charles Darwin, On the Origin of Species by Means of Natural Selection (1859).
8. Glyn Daniel, One Hundred Years of Archaeology (Cambridge: Harvard University Press, 1950); A Hundred and Fifty Years of Archaeology (Cambridge: Harvard University Press, 1976).
9. Daniel, One Hundred Years.
10. Gordon A. Willey and Jeremy A. Sabloff, A History of American Archaeology (San Francisco: Freeman and Company, 1974).
11. Daniel, One Hundred Years; A Hundred and Fifty Years.
12. Willey and Sabloff, History of American Archaeology.
13. Williams, Breakthrough.
14. L. and G. Poole, One Passion, Two Loves: the Schliemanns of Troy (1967).
15. Murray, "Centenary," pp. 92–95; G. A. Wainright, "Dr. Margaret Murray's Hundredth Birthday," Man, vol. 63, nos. 133, 134 (1963):106.
16. Murray, "Centenary," pp. 92–95; Williams, Breakthrough.
17. Wainright, "Hundredth Birthday," p. 106.

18. Introduction to "Memoirs of a Pioneer Excavator in Crete," *Archaeology*, vol. 18, no. 2 (1965):95.

19. Ibid., p. 1.

20. Harriet Boyd Hawes, "Excavations at Kavousi, Crete in 1900," *American Journal of Archaeology*, vol. 5, series 2 (1901):125–157; "Gournia. Report of the American Exploration Society's Excavations at Gournia, Crete, 1901–1903," *Pennsylvania University Free Museum of Science and Art Transactions*, vol. 1 (1904–05):7–44; "Gournia. Report of the American Exploration Society's Excavations at Gournia, Crete, 1904," *Pennsylvania University Free Museum of Science and Art Transactions*, vol. 1 (1904–05):177–188; "Excavations at Gournia, Crete," *Smithsonian Institution Report for 1904* (Washington, D.C., 1905), pp. 559–571; Harriet Boyd Hawes, B. E. Williams, R. B. Seager, and E. H. Hall, "Gournia, Vasiliki and Other Prehistoric Sites on the Isthmus of Hierapetra, Crete: Excavations of the Wells-Huston-Cramp Expeditions, 1901, 1903, 1904," *The American Exploration Society*, vol. 7, no. 60 (1908):44; Charles Henry and Harriet Boyd Hawes, *Crete, the Forerunner of Greece* (New York: Harper and Brothers, 1909).

21. Daniel, *One Hundred Years; Hundred and Fifty Years*.

22. Willey and Sabloff, *History of American Archaeology*.

23. Williams, *Breakthrough*.

24. Willey and Sabloff, *History of American Archaeology*.

25. Cynthia Irwin-Williams, *The Course of American Archaeology: 1800–1975* (ms on file Reno, Nevada, n.d.).

26. V. Gordon Childe, *What Happened in History* (London, 1943).

27. H. Marie Wormington, *Ancient Man in North America*, 4th ed. (Denver Museum of Natural History Popular Series, 1957).

28. Daniel, "Editorial: Dorothy Garrod Obituary," *Antiquity*, vol. 43, no. 169 (1969):2.

29. E. W. Gardner and Gertrude Caton-Thompson, "The Recent Geology and Neolithic Industry of the Northern Fayum Desert," *Journal of the Royal Anthropological Institute of Great Britain and Ireland*, vol. 58 (1926):301–323.

30. Caton-Thompson, "Preliminary Report on Neolithic Pottery and Bone Implements from the Northern Fayum Desert," *Man*, vol. 25 (1925): 153–158; "Recent Excavations in the Fayum," *Man*, vol. 18 (1928): 109–113; "Recent Discoveries in Kharga Oasis, Egypt," *First International Congress for Prehistoric and Protohistoric Sciences Proceedings*, London (1934).

31. Guy Brunton and Caton-Thompson, *The Dadarian Civilization and Predynastic Remains near Badari* (London: Bernard Quaritch, 1928).

32. Caton-Thompson, "Zimbabwe," *Antiquity*, vol. 3 (1929):424–433; "Zimbabwe," *Antiquity*, vol. 4 (1930):491–493; *The Zimbabwe Culture: Ruins and Reactions* (Oxford: Clarendon Press, 1931); "The Zimbabwe Culture: Guesses and Facts," *Man*, vol. 31 (1931):235; "Kharga: Prehistoric," *Man*, vol. 31 (1931):58; "Kharga Oasis," *Antiquity*, vol. 5 (1931):221–226.

33. Caton-Thompson, "Discoveries in Kharga Oasis"; Caton-Thompson and Gardner, "The Prehistoric Geography of Kharga Oasis," *Geographical Journal*, vol. 80 (1932):369–406.

34. Caton-Thompson, "A Temple in the Hadramaut," In *Asia*, vol. 39 (1939):294–299; "The Hadramaut and its Past: Summary," *Man*, vol. 39 (1939):57–58; "The Tombs and Moon of Hureidha (Hadramaut)," *Antiquity*, vol. 19 (1945):187–193; Caton-Thompson and Gardner, "Climate, Irrigation and Early Man in the Hadramaut," *The Geographical Journal*, vol. 93 (1939):18–35.

35. In the 1920s–1950s, archaeological cultures were frequently identified with and confused with the technical descriptions of their characteristic artifacts.

36. Daniel, *Hundred and Fifty Years*, p. 285.

37. Daniel, "Garrod Obituary."

38. Dorothy A. E. Garrod, "Stone Age in Palestine," *El Palacio*, vol. 36 (1929):228–229; "Excavations in the Wady al-Mughara (Palestine), 1931," *American School of Prehistoric Research Bulletin* (1932):6–11, no. 8; "Cave Dwellers in Palestine," *El Palacio*, vol. 33 (1932):236; "Excavations at the Wady al-Mughara (Palestine), 1932–33," *American School of Prehistoric Research Bulletin* (1934):7–11, no. 10; "Four or Five Thousand Years before the Children of Israel Came Seeking Their Promised Land, Palestine Had a Dry Climate and Was Mainly Desert," *Science*, vol. 79 (1934):8, no. 2045; "Excavations at the Cave of Shukbah, Palestine, 1928; with an Appendix on the Fossil Mammals of Shukbah by Dorothy M.A. Bate," *Prehistoric Society Proceedings*, vol. 8 (1943):1–20.

39. Garrod, "The Palaeolithic of Southern Kurdistan: Excavations in the Caves of Zarzi and Hazar," *American School of Prehistoric Research Bulletin*, no. 6 (1930).

40. Garrod, B. Howe, J. H. Gaul, and R. Popov, "Excavations in the Cave of Bacho Kiro, Northeast Bulgaria," *American School of Prehistoric Research Bulletin* (1939):46–126; no. 15.

41. Garrod, Ch. H. D. Burton, G. E. Smith, and D. M. A. Bate, "Excavation of Mousterian Rock-shelter at Devil's Tower, Gibraltar," *Journal of Royal Anthropological Institute of Great Britain and Ireland*, vol. 58 (1928):33–113.

42. Garrod, "The Ancient Shorelines of the Lebanon and the Dating of Mt. Carmel Man," in *Hundret Jahre Neanderthaler; Neanderthal Centenary, 1856–1956* (1958):182–184; "A Pebble Industry of Early Würm Age from the Abri Zumoffen, South Lebanon," in Sen. Dharanidhar, ed., *Studies in Prehistory* (Calcutta, 1966), pp. 41–48; Garrod and Diana Kirkbride, "Excavation of the Abri Zumoffen, A Paleolithic Rock-shelter near Adlun, South Lebanon," *Extrait du Bulletin due Musée de Beyrouth*, vol. 16 (1961).

43. Garrod, *The Upper Palaeolithic Age in Britain* (Oxford: Clarendon Press, 1928).

44. Garrod, "Wady al-Mughara, 1932–33"; "Four or Five Thousand Years"; "Cave of Shukbah."

45. Garrod, "Cave Dwellers"; "Wady al-Mughara, 1932–33"; "Excavations in the Mugharet Et Tabun (Palestine), 1934," *American School of Prehistoric Research Bulletin* (1935):54–58, no. 11; "An Outline of Pleistocene Prehistory in Palestine–Lebanon–Syria," *Quaternia*, vol. 6 (1962):541–546; "The Middle Paleolithic of the Near East and the Problem of Mount Carmel Man," *Journal of the Royal Anthropological Institute of Great Britain and Ireland*, vol. 92 (1962):232–259.

46. Garrod, "Mesolithic Burial from Caves in Palestine," *Man*, vol. 32 (1931):145–148; "Cave Dwellers," p. 236; "Wady al-Mughara, 1931," p. 6–11.

47. Garrod, "A Transitional Industry from the Base of the Upper Paleolithic in Palestine and Syria," *Journal of the Royal Anthropological Institute of Great Britain and Ireland*, vol. 81 (1952):121–130; "Pleistocene Prehistory," pp. 541–546; "The Middle Paleolithic," pp. 232–259; *Primitive Man in Egypt, Western Asia and Europe in Paleolithic Times* (Cambridge: University Press, 1965).

48. Daniel, "Editorial," p. 2.

49. Elise J. Bäumgartel, "Scavo Stratigrafico a Macchia a Marc," *Bullettino di Paletnologia Italiana*, vol. 50–51 (1930–31):119–133; "Funde aus einer gorge Schichtlitchen Station bei Villa Asneros (Rio de Oro)," *Praehistoriache Zeitschrift*, vol. 22 (1931):88–101; "Fragments of Prehistoric Egyptian Pottery," *Man*, vol. 48 (1948):59–60.

50. Daniel, *Hundred and Fifty Years*, p. 198.

51. Earl H. Morris, Jean Charlot, and Ann Axtell Morris, *The Temple of the Warriors at Chichen Itzä, Yucatan*, Carnegie Institution of Washington (Washington, D.C., 1931), pub. no. 406; "Ann Axtell Morris," *American Antiquity*, vol. 11 (1945):117.

52. Winifred Gladwin and Harold S. Gladwin, "The Use of Potsherds in an Archaeological Survey of the Southwest," *Medallion Papers* (Pasadena, 1928); "A

Method for Designation of Ruins in the Southwest," *Medallion Papers* (Pasadena, 1928); "The Ancient Civilization of Southern Arizona," *Medallion Papers* (Globe, Arizona: Gila Pueblo, 1928–35); "A Method for Designation of Cultures and their Variations," *Medallion Papers* (Globe, Arizona: Gila Pueblo, 1934).

53. Gladwin and Gladwin, "The Use of Potsherds"; "Designation of Ruins"; "A Method for the Designation of Southwestern Pottery Types," *Medallion Papers* (Globe, Arizona: Gila Pueblo, 1930).

54. Gladwin and Gladwin, "The Use of Potsherds"; "Designation of Ruins"; Harold S. Gladwin, "Methodology in the Southwest," *American Antiquity*, vol. 1 (1936):256–259.

55. Willey and Sabloff, *History of American Archaeology*.

56. Elizabeth W. Crozier Campbell and William H. Campbell, "The Pinto Basin Site: An Ancient Aboriginal Camping Ground in the California Desert," *Southwest Museum Papers* (1935), no. 9; "The Lake Mohave Site," *The Archaeology of Pleistocene Lake Mohave: A Symposium* (1937):9–44; no. 11; Elizabeth W. Crozier Campbell, "A Folsom Complex in the Great Basin," *Masterkey*, vol. 14 (1940):7–11; "A Museum in the Desert," *Masterkey*, vol. 3 (1929):5–10; "Two Ancient Archaeological Sites in the Great Basin," *Science*, vol. 109 (1949):340.

57. Frederica de Laguna, *The Archaeology of Cook Inlet, Alaska* (Philadelphia: University of Pennsylvania Press, 1934); "The Prehistory of Northern North America as Seen from the Yukon," *Memos of the Society of American Archaeology* (1947).

58. Willey and Sabloff, *History of American Archaeology*.

59. J. G. D. Clark, *Excavations at Star Carr* (Cambridge, 1954).

60. I. Hodder and H. Orton, *Spatial Analysis in Archaeology* (1976).

61. Daniel, *One Hundred Years*.

62. Willey and Phillip Phillips, *Method and Theory in American Archaeology* (Chicago: University of Chicago Press, 1958).

63. Kathleen M. Kenyon, "Excavations at Jericho, 1952," *Palestine Exploration Quarterly* (January/April 1952): 62–82; "A Survey of the Evidence Concerning the Chronology and Origins of Iron Age A in Southern and Midland Britain," *London University Institute of Archaeology Annual Report*, 8 (1952):29–78; "Excavations in Jerusalem 1961–1963," *Biblical Archaeologist*, vol. 27, no. 2 (1964):34–52; *Palestine in the Middle Bronze Age* (Cambridge: University Press, 1966).

64. Kenyon, "Excavations at Jericho"; "Iron Age A"; *Digging Up Jericho* (London: Ernest Benn, Ltd., 1957); "Jericho," *Archaeology*, vol. 20, no. 4 (1967):268–275.

65. Kenyon, "Jericho and Its Setting in Near Eastern History," *Antiquity*, vol. 30 (1956):184–195.

66. R. J. Braidwood and Willey, *Courses toward Urban Life* (Chicago: Viking Fund Publications in Anthropology, 1962), no. 32.

67. Kenyon, "Jericho."

68. Louis S. B. Leakey and Mary D. Leakey, "Archaeological Excavations at Olduvai Gorge, Tanzania," *National Geographic Society Research Reports* (1963), 179–182; "Recent Discoveries of Fossil Hominids in Tanganyika: at Olduvai and near Lake Natron," *Current Anthropology*, vol. 6, no. 4 (1965):422–424; "Excavation at Olduvai Gorge, Tanzania, 1968," *National Geographic Society Research Reports: Abstracts and Reviews of Research during the Year 1968* (1976):201–203.

69. Jacquetta Hawkes and Christopher Hawkes, *Prehistoric Britain* (Middleton: Penguin Books, 1958).

70. Jacquetta Hawkes, "The Achievements of Paleolithic Man," in Gabel Creighton, ed., *Man Before History* (Englewood Cliffs, New Jersey, 1965), pp. 21–35; "The City of Croesus," *Harvard Today* (1965):14–148; *Prehistory* (The New American Library, 1965).

71. Daniel, *Hundred and Fifty Years*, p. 396.

72. Marija A. Gimbutas, *Bronze Age Cultures in Central and Eastern Europe* (The

Hague: Mouton & Co., 1956); "European Prehistory: Neolithic to the Iron Age," *Biennial Review of Anthropology* (1963):69–106; "The Bronze Age Culture of Central Europe. An Interpretation of the Unétice, Tumulus and Urn Field Culture," *International Congress of Prehistoric and Protohistoric Sciences*, vol. 2, 6 (1965):388–395.

73. Robert Ehrich, "Dorothy Cross Jensen, 1906–1972," *American Antiquity*, vol. 38, no. 4 (1972):407–411.

74. Dorothy J. Cross, "The Pottery of Tepe Gawra," in E. A. Spriser, ed., *Excavations at Tepe Gawra I, Levels I–VIII* (University of Pennsylvania Press, 1935), pp. 38–61.

75. Cross, *Archaeology of New Jersey*, vol. 1 (Trenton: Archaeological Society of New Jersey and New Jersey State Museum, 1941); *Archaeology of New Jersey, Vol. 2, The Abbott Farm* (Trenton: Archaeological Society of New Jersey and New Jersey State Museum, 1956).

76. Cross, *The Abbott Farm*.

77. Marian E. White, "Niagara Frontier Iroquois Village Movements," *Eastern United States Archaeological Federation Bulletin*, no. 19 (1960), p. 16; "Settlement Pattern Change and the Development of Horticulture in the New York–Ontario Area," *Pennsylvania Archaeologist*, vol. 33, no. 1–2 (1963):1–12; Elisabeth Tooker and Marian E. White, "Archaeological Evidence for Seventeenth Century Iroquoian Dream Fulfillment Rituals," *Pennsylvania Archaeologist*, vol. 34, no. 3–4 (1968):1–5.

78. White, "Ethnic Identification and Iroquois Groups in Western New York and Ontario," *Ethnohistory*, vol. 18, no. 1 (1972):19–38.

79. T. M. N. Lewis and Madeline Kneberg, "The Archaic Horizon in Western Tennessee," *Tennessee Anthropology Papers*, no. 2 (1947); "An Archaic Autobiography of an Eva Warrior," *Tennessee Archaeologist*, vol. 7 (1951):1–5; "The Autobiography of a Bone House Indian," *Tennessee Archaeologist*, vol. 8 (1952):37–41; "The Camp Creek Site," *Tennessee Archaeologist*, vol. 13 (1957):1–48; "The Archaic Culture in the Middle South," *American Antiquity*, vol. 25 (1959):161–183.

80. Betty J. Meggars and Clifford Evans, "Archaeological Investigations at the Mouth of the Amazon," *Bureau of American Ethnology Bulletin 167* (Washington, D.C., 1957).

81. Meggars, "Environment and Culture in the Amazon Basin: An Appraisal on the Theory of Environmental Determinism," in Angel Palerm, ed., *Studies in Human Ecology* (1957), pp. 71–89; Meggars and Evans, "Archaeological Evidence of a Prehistoric Migration from the Rio Napo to the Mouth of the Amazon," in Raymond H. Thompson, ed., *Migrations in New World Culture History* (Tucson, 1958), pp. 9–19; Emilio Estrada, Meggars, and Clifford Evans, "Possible Transpacific Contact on the Coast of Ecuador," *Science*, vol. 135, no. 3501 (1962):371–372.

82. Dorothy L. Keur, "Big Bend Mesa; an Archaeological Study of Navaho Acculturation, 1745–1812," *Society for American Archaeology*, no. 1 (Menasha, Wisconsin, 1941).

83. Clara L. Tanner, "Life Forms in Prehistoric Pottery of the Southwest," *The Kiva*, vol. 8 (1943):25–32.

84. Anna O. Shepard, *Ceramics for the Archaeologist*, pub. no. 609 (Washington, D.C.: Carnegie Institution of Washington, 1954).

85. Bertha P. Dutton, "Archaeological Mobile Camps, Senior Girl Scouts, 1950," *El Palacio*, vol. 57 (1950):366–371.

86. Dutton, "Leyit Kin, a Small House Ruin, Chaco Canyon, New Mexico: Excavation Report," *Monograph of the University of New Mexico and the School of American Research* (Albuquerque, 1938).

87. Dutton, *Happy People, the Huichol Indians* (Santa Fe: The Museum of New Mexico Press, 1962).

88. Daniel, *Hundred and Fifty Years*.

89. Quotes in the following profiles are directly from the interviews, given in 1981 and 1982.

90. Florence M. Hawley, "Prehistoric Pottery and Culture Relations in the Middle Gila," *American Anthropologist*, vol. 32 (1930):522–536.

91. Hawley, "Prehistoric Pottery Pigments in the Southwest," *American Anthropologist*, vol. 31 (1929):731–754; "Chemical Examination of Prehistoric Smudged Wares," *American Anthropologist*, vol. 32 (1930):500–502; "Chemistry in Prehistoric American Arts," *Journal of Chemical Education*, vol. 8 (1931):35–42.

92. Florence H. Senter (Ellis), "Dendrochronology: Can We Fix Prehistoric Dates in the Middle West by Tree Rings?" in Proceedings of the Nineteenth Annual Indiana History Conference, *Indiana History Bulletin*, vol. 15, no. 2 (1938):118–128; "Dendrochronology in Two Mississippi Drainage Tree-Ring Areas," *Tree Ring Bulletin*, vol. 5 (1938):3–6; Hawley, "Tree-Ring Analysis and Dating in the Mississippi Drainage," vol. 2, *University of Chicago Publications in Anthropology, Occasional Paper* (1941); Hawley and E. J. Workman, "Protocol: A New Dendrochronograph," in "Tree-Ring Analysis and Dating in the Mississippi Drainage," vol. 2, *University of Chicago Publications in Anthropology, Occasional Paper* (1941):101–103.

93. Florence H. Ellis, "Anthropological Data Pertaining to the Taos Land Claim," in *American Indian Ethnohistory, Pueblo Indians*, vol. 1 (New York: Garland Publishing, Inc., 1974), pp. 29–105; "Anthropology of Laguna Pueblo Land Claims," in *American Indian Ethnohistory, Pueblo Indians*, vol. 1 (New York: Garland Publishing, Inc., 1974), pp. 29–105; "Anthropological Study of the Navajo Indians," in *American Indian Ethnohistory, Pueblo Indians*, vol. 1 (New York: Garland Publishing, Inc., 1974).

94. Ellis, Review of "The Archaeological Survey of Wetherill Mesa: Mesa Verde National Park—Colorado" by A. C. Hayes, *American Anthropologist*, vol. 69, no. 1 (1967):101.

95. Ellis, *A Reconstruction of the Basic Jemez Pattern of Social Organization, with Comparison to Other Tanoan Social Structures* (University of New Mexico Publications in Anthropology, 1964), no. 11; "Archaeological History of Nambe Pueblo, 14th Century to Present," *American Antiquity*, vol. 30 (1964):34–42; "The Immediate History of Zia Pueblo as Derived from Excavations in Refuse Deposits," *American Antiquity*, vol. 31 (1966):806–811.

96. T. R. Frisbie, *Collected Papers in Honor of Florence Hawley Ellis* (Albuquerque: Papers of the Archaeology Society of New Mexico, 1975), no. 2.

97. Ibid.

98. Ibid.

99. Williams, *Breakthrough*.

100. Ruthann Knudson, "Comments," in A. M. Lipinski, ed., *Text for Early Man Magazine* (Fall 1981).

101. Leslie Wildesen, "Comments," in A. M. Lipinski, ed., *Text for Early Man Magazine* (Fall 1981).

102. Irwin-Williams, "Comments for Anthropology Newsletter" (May, 1981).

103. Frisbie, *Collected Papers*.

I wish to thank the women who helped make this chapter possible by their personal interest and conversation: H. Marie Wormington, Florence Hawley Ellis, and Marjorie Ferguson Lambert. This chapter was written at the Desert Research Institute, University of Nevada System, in Reno, Nevada. The text was originally typed by Jo Janowski, and the final manuscript was produced by Shirley Garcia.

MICHELE L. ALDRICH

Women in Geology

The *History of a Science*, or in other
words, of the progress of the human
mind in relation to that science, is
always interesting, and furnishes hints
for further improvement.
Almira Hart Lincoln Phelps, 1830.[1]

Geology is a field science, a laboratory and museum science, and a theoretical science. It is also highly practical, since geological investigations
commonly have direct uses in mining and petroleum exploration. Finally,
geology is usually practiced collectively by several researchers rather than
by a lone scientist working on a natural landscape or laboratory problem.
The history of women in geology reflects these general features of the field.

Civilizations had creation stories to account for the origin of the earth
from the early days of humanity, but these were not scientific in the sense
of being based in systematic investigations of nature. In Europe during the
last several centuries, the collection of evidence and the refinement of creation accounts gradually became natural history, which dealt with plants,
animals, and rock formations (including fossils). Geology as a separate
science is largely a product of the nineteenth and twentieth centuries; it is
a young science dealing with old rocks. Consequently, it has had fewer
theoreticians than the older sciences. Given the relative newness of the
science and the low numbers of women who have participated in it until
recently, little significance should be attached to the fact that the major
theoreticians in geology have been men. Historians of science emphasize
the development of grand theories and sweeping syntheses in the scientific
disciplines, but this bias, when applied to geology, overlooks the contributions of women to the field. Female geologists—and male geologists as
well—devoted themselves more to the empirical than to the theoretical
aspects of the science. To ignore their work is to distort the record of geology
for the past two centuries.

The historical analysis below emphasizes women in geology in the
United States since 1830. Thanks to the bibliographic work of Robert and
Margaret Hazen,[2] one can build a list of women who published books or

articles in geology before 1850. For this early period, state geological surveys were a major site for geological researches. Accordingly, the survey reports have been scanned for what they reveal about women's contributions to American geology in the years before the establishment of the United States Geological Survey (USGS) in 1879. By the late nineteenth century, women geologists were at work in the Geological Survey, in women's colleges, and at state surveys. In the early twentieth century, they begin to contribute also in jobs with oil companies, museums, and universities. Published biographies of these pioneer professional women geologists summarize their careers and evaluate their contributions to the science. Women geologists who are alive and, for the most part, still working in science usually have not yet been assessed by their peers and hence are not included in this chapter. Thus, contemporary women geologists, such as Ursula Marvin, who has contributed to the very recent work on a theory of plate tectonics, do not appear in this discussion of women's roles in the history of American geology.

NINETEENTH-CENTURY AMERICAN WOMEN'S CONTRIBUTIONS TO GEOLOGY AS SCIENTIFIC ILLUSTRATORS

In the United States, women first appeared in geology in significant numbers as scientific artists. Starting in the 1830s, they did many of the drawings which accompanied state geological survey reports and other geological publications. To understand why this work was valuable to early science requires a word about the history of printing. While photography was introduced in the same decade that these pioneering women started their drawings, illustrations in books were not commonly printed from photographs until the 1870s. Instead, drawings were used as the basis for woodcuts, lithographs, and copper or steel engravings. Indeed, for some scientific illustrations, such as those of microfossils, drawings are still used today to avoid the distortions of the camera when dealing with three-dimensional objects. Furthermore, early geological reports conveyed much information in drawings which are now handled in other media. For example, before contour maps were widely used, scenic drawings of the landscape provided topographic evidence not easily rendered in prose. Finally, early geologists regarded expression of the beauty of the earth as part of their business of instructing the public in the wonders of nature. Their reports, especially those of the state surveys, carried numbers of scenic plates—sometimes an entire volume of them.

There was some variation in the organization and conduct of early geological surveys, but most employed a chief geologist and several assistants. The surveys were paid for by government: in the United States before the

Civil War, by the states, but after 1865, more and more by federal agencies. Results appeared in annual reports (narrative accounts commonly devoted to findings of economic importance) and final reports (technical scientific monographs) published at public expense and distributed free or at low cost. In the first year of a survey, the head geologist would review the published work on the geology of the terrain to be studied and do a rapid reconnaissance of the entire area to gain a general idea of its stratigraphy and geological structure. Thereafter, the chief geologist sent assistants into the field to map and describe in more detail the geology of certain locations of economic or geologic interest, and supervised the work of assistants in the laboratory on specific problems which came up during field studies. Women contributed to geological surveys during this second phase of intense study of specific terrains and problems, by doing the drawings needed to illustrate locales and specimens under discussion in the reports.

Women were qualified as scientific illustrators thanks to the curriculum of the early secondary schools for girls. Sketching was taught at such institutions as the Troy Female Seminary, and drawing was considered a genteel skill which young ladies could attain without threatening their feminity. The schools also taught science.[3] For example, Amos Eaton, the geologist at Rensselaer Polytechnic Institute, taught the students of Emma Willard's Troy seminary.[4] Mount Holyoke taught geology from its earliest days as a secondary school and on into the years when it became a college.[5]

Orra White Hitchcock was among the first women whose drawings illustrated geological publications. Her husband, Edward Hitchcock, was the first state geologist of Massachusetts (1830s and 1840s), and, incidentally, an advisor to Mary Lyon, founder of Mount Holyoke. Orra Hitchcock did both scenic and technical drawings for his books. The atlas of plates which accompanied his reports[6] was in good part a portfolio of her work. Her scenic plates often contained human figures or houses; this was not mere artistic convention, but rather a traditional and subtle way of conveying scale in a scene. Her technical plates, such as drawings of fossils, were not so successful in showing scale, but neither were those of her contemporaries—men and women—in scientific illustration. Her technical drawings did not provide as much geological detail as did those by later women and men. Nevertheless, she deserves recognition as an early woman contributor to American geology.

Edward Hitchcock also hired a Miss Doolittle, orphan daughter of New Haven's famous Revolutionary War general, to color the maps for the Massachusetts survey. Before the advent of color lithography, the watercolor tints used to differentiate geological formations on a map had to be brushed in by hand. According to letters between Hitchcock, Benjamin Silliman, and Doolittle (found in Hitchcock's papers at Amherst College), Miss Doolittle colored four hundred maps for Hitchcock at a little over three dollars per hundred; she made her living in art work and had taught drawing in

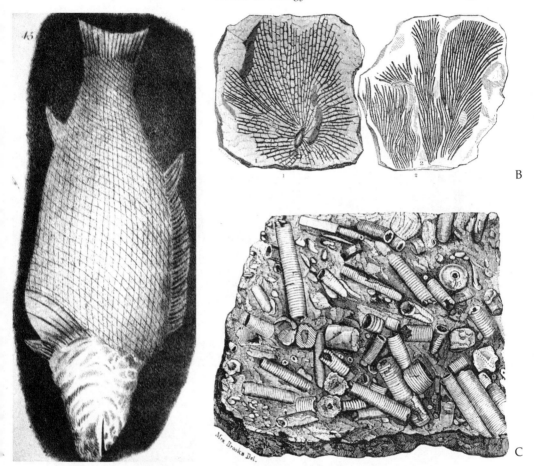

Illustrations of fossils, drawn by women. (A) Orra White Hitchcock, for
Edward Hitchcock's report on the geology of Massachusetts, 1841. (B)
Sarah Hall, for James Hall's report on the geology of the western part
of New York State, 1843. (C) Mrs. Brooks, for James Hall's report.

South Carolina. She was not unique among American women in cartographic employment. Sarah A. Campbell listed herself in the Philadelphia
street directory as a "map colorist" from 1846 to 1860 and even placed an
ad about her work in the 1856 city business guide.

Two talented women geological artists drew for the state survey of New
York, which published its reports starting in the 1840s. Little is known of
Mrs. Brooks, an Albany woman who drew for James Hall, a survey district
geologist who later became state paleontolgist. Hall also commissioned his
wife, Sarah, to draw for him; she did scenic as well as technical plates for
his geology report[7] and later for his paleontology volumes.[8] Their plates

A

B

C

Topographical drawings for state geological surveys. (A) Orra White
Hitchcock. (B) Sarah Hall. (C) Eben Horsford, for James Hall's report.

Advertisement in Philadelphia City Directory for 1856.

succeeded in portraying more data about specimens than had Orra Hitchcock's. James Hall also hired men to draw for his report, and some of their work was inferior to that of the women. The scenic sketches by Eben Horsford, who—mirabile dictu—taught drawing at Rensselaer Polytechnic Institute, showed serious problems with perspective. Fortunately for the history of scientific illustration, Horsford went on to a distinguished career in agricultural chemistry and published no more geological drawings.

Women's drawings appeared in later American state surveys. Mrs. Myers (presumably the wife of the printer of the Illinois reports with the same last name) collaborated with Fielding Bradford Meek, a noted specialist in invertebrate fossils, on a plate for an Illinois report in 1866.[9] Kate Andrews drew for the Ohio survey volumes published in 1875[10] and Harriet Huntsman's work appeared in a Kansas survey report[11] in 1898.[12]

After the Civil War, the federal government became deeply involved in geological surveys, financing four of them between 1867 and 1879, when the United States Geological Survey was started. The four surveys concentrated on western lands organized into territories rather than states. At least two women drew for these surveys. Miss or Mrs. H. Martin of Albany did several paleontological plates for the survey led by Clarence King; her work appeared in a volume published in 1877.[13] Edward Drinker Cope, a paleontologist on Ferdinand Hayden's survey, hired Cecelia Beaux of Philadelphia to draw for his volume on the Cretaceous vertebrates of the American West.[14] Beaux, who was twenty at the time, went on to an important professional career as an artist, specializing in portraiture.[15] Cope himself had a low opinion of women's intelligence, from which he apparently exempted Beaux.[16]

Women, then, first appeared in American geology in the role of scientific artists. Their function reflected women's schooling at the time and seemed consistent with femininity, which made it easier to get a toehold in the science and accustomed men to women's involvement in geology. Most importantly, women contributed significantly to the science itself by conveying evidence visually about specimens and landscape through proficient and attractive drawings.

NINETEENTH-CENTURY WOMEN'S CONTRIBUTIONS TO GEOLOGY AS COLLECTORS AND AUTHORS

Hard as they worked, there was no way early geologists could cover an entire landscape in great detail and find all the minerals and fossils necessary to describe the terrain accurately and fully. Rather, they supplemented their work and that of assistants with materials sent in by collectors, usually amateurs who resided in the region under study, but sometimes professionals who gathered rare specimens and sold them for a living. Many new species of fossils were first discovered by collectors, and the geological record was much fuller because of their activities. The phenomenon of collectors was not unique to geology; they could also be found in botany and zoology, where their contribution to science was much the same—supplying additional data that could be used by one side or the other in the more lofty theoretical scientific controversies of the day.

Fossils are used to assign relative ages to the geological formations in which they occur. Minerals are most often valuable economically, as the raw material for manufactures, but also are sometimes used in field studies to correlate strata, especially where fossils are absent. Furthermore, fossil collections are important evidence in the arguments over the evolution of organisms in geological time and over variation in climate during past eras. The activities of mineral and fossil collectors who were women merit attention for these reasons.

Some British women helped advance the field of geology through their collecting activities. Mary Ann Mantell is reputed to have discovered Iguanadon fossils, which her husband, Gideon, described in his highly regarded, widely read books on the older rocks of England.[17] Lady Gordon Cumming (nee Eliza Maria Campbell), a horticulturist and gardener who propagated plant hybrids in her efforts to improve her husband's estate in Scotland, in 1839–1842 collected ancient fish fossils systematically and extensively on their lands. She and her daughter prepared careful drawings of many of them, but science did not fully benefit from their labors, due to the sudden death of Lady Gordon Cumming and due to the dilatory nature of geological publications at the time. Her work did provide some assistance to Louis Agassiz, who eventually published the definitive monographs on the fish species with which she worked.[18] While Lady Gordon Cumming and Mrs. Mantell were amateur collectors, Mary Anning, daughter of a Lyme Regis man who supported his family in part by selling fossils, provided for herself and the family by continuing her father's trade after his death. In about 1812, she and her brother Joseph discovered and sold an important, nearly intact ichthyosaur skeleton, the study of which helped British paleontologists immensely in understanding this marine creature.[19] She continued her sales of fossils for several decades thereafter.[20] Her portrait today hangs near an ichthyosaur skeleton in the British Museum of Natural History.

American women also collected geological material which was written up by male scientists, but those whose efforts have been documented so far collected in later decades than did their British sisters. Mrs. Okeley donated specimens to the Mississippi state geological survey in the 1850s.[21] Miss Errington's collections proved valuable enough to the California survey in the 1860s to inspire paleontologist William Gabb to name a new species after her.[22] Collections by Mary P. Haines helped the Indiana geologists in their work in the 1870s.[23]

In evaluating the early written work by women in geology, some perspective on their productivity is required. First, in the geological literature published before 1850, their output is a tiny fraction. Of over 11,000 citations in Margaret and Robert Hazen's comprehensive bibliography[24] of early geological books and articles published in America, only about a dozen are by women. Second, women published far more in other sciences, notably botany, than they did in geology. Third, the early women who did publish in America on geological topics issued only one or a few items; their productivity in geology was not sustained, but episodic. They wrote school textbooks or articles on isolated natural phenomena not necessarily connected with theoretical issues of the day. With these caveats in mind, it is instructive to examine what women did write on geology before 1850.

The earliest items published in America by a woman on geology, according to the Hazens, were extracts on an earthquake from a book of travels by Hester Lynch Piozzi.[25] Between 1789 and 1793, at least four American magazines reprinted her description. She was reporting the experiences of an Italian woman who lived through the disaster but lost a son to it, and the narrative stressed the human toll of the episode rather than the natural phenomenon. Hester Piozzi was a friend of Samuel Johnson and was thus known in English literary circles of her time.[26]

Several later women reported on geological phenomena in their neighborhood or travels. Mary Austin Holley's *Texas. Observations: Historical, Geographical, and Descriptive* (1833 and 1836)[27] remarked on the soils, water resources, minerals, and mountains of the region as well as on settlements, Indians, and politics. Mary Holley had worked in botany while at Transylvania University in Kentucky, and learned geology from the eccentric genius Constantine Rafinesque. The map which accompanied her volume showed topographic features (rivers and mountains) and the location of mines. In 1843, Lydia Maria Child, an important American literary figure and activist in abolition, published an account of her visit to Mammoth Cave in Kentucky, giving considerable detail about the geology of the site as well as some interesting reflections on the mulatto slave Stephen who guided the tourists; in making a point in favor of abolition, she also documented a contribution by a black person to geology, a member of another group whose work for the science is just starting to be told.[28] Finally, in *Rural Hours* (1850),[29] a book which sold reasonably well and thus reached a more sizable audience than did the work of the earlier women, Susan Fenimore Cooper (daughter of the novelist) used the annual cycle of life in the New York countryside to present information on weather, soil, mineral springs, rivers, and other features of the geologic landscape, although she emphasized birds and plants over rocks.

Two women published poems on geology, not a unique medium for discussing science at the time.[30] Felicia Dorothea Hemans's "Epitaph for a Mineralogist"[31] appeared in 1836 in a magazine edited by Nathaniel and Elizabeth Hawthorne. Hannah Flagg Gould published "The Mastodon" in a popular journal in 1847.[32] She concentrated on the public display of the skeleton, but also noted the occurrence of stone hunting implements with it at the excavation site.

Some women worked on textbooks in geology and cognate sciences. Emma Hart Willard published *Ancient Geography*[33] in 1824 as a sixty-five-page supplement to William Woodbridge's *System of Universal Geography*.[34] Their combined book went through more than ten "editions" (really reprintings) in the next twenty years. The *American Journal of Science* praised their book as a scientific arrangment of material which had previously been presented as a jumble of facts. While her section of the text stressed human geography in older civilizations, Emma Willard also had sections on rivers, mountains, volcanoes, and other geological phenomena which determined cultures in the ancient world. Her sister, Almira Hart Phelps, edited two texts by men. She revised and enlarged *The Child's Geology*[35] by the prolific Samuel Griswold Goodrich (better known as "Peter Parley" and the author of scores of best-selling history, science, and language texts). She translated and revised Louis Vauquelin's reference work, *Dictionary of Chemistry*, which included many entries germane to mineralogy. She added a history of chemistry and an elementary introduction to her version, abridged some of the long sections on manufacturing, and spliced in the latest findings in chemistry from French, English, and American writers.[36]

Two women published texts in geology as single authors. Miss D. W. Godding produced *First Lessons in Geology*[37] in 1847, a simple, short book aimed at children. Jane Kilby Welsh targeted a slightly older audience in her two-volume work, *Familiar Lessons in Mineralogy and Geology* (1832 and

1833).[38] She cast the text as dialog among the members of a family; while the father did most of the instructing, the mother and girls were active participants in the discussions and experiments. The first volume centered on mineralogy, and took the reader through the minerals in chemical groups rather than on crystallographic feature, although Welsh included crystallographic data and techniques in the descriptions. The second volume dealt with geology, starting with the oldest granite rocks and working toward the youngest sediments, or "alluvial" formations. She used the standard English and French authorities and reported the latest finds of American state geological surveys. Edward Hitchcock and other New England scientists reviewed her manuscript before she went to press.

In 1835, the *American Journal of Science* published an article on the dispute between Maria Graham Callcott, an Englishwoman, and George Greenough, the president of the Geological Society of London, over the elevation of land in Chile after a severe earthquake in 1822. Maria Callcott, not a trained scientist, took extensive notes at the time of the earthquake and wrote them up later for publication in the GSL *Transactions*. She had recorded the time, frequency, severity, and effects of the shocks. Greenough challenged her results; in particular he disputed her statements about the land rise in connection with the quake (at the time earthquakes were not associated with geologic faults; that became a common correlation only after the 1906 San Andreas episode). Callcott rebutted his protests one by one, using details from her general journal of travel and corroborating evidence from other observers. Greenough also challenged the worthiness of her account because she was not a geologist. This she turned against him: "As to ignorance of the science of Geology, Mrs. Callcott confesses it; and, perhaps, that circumstance, and her consequent indifference to all theories connected with it, render her unbiassed testimony of the more value."[39] Of the handful of women who published in geology in the early nineteenth century, Callcott stood out as an original observer who contributed directly to one of the great disputes of the time, over whether land elevations occurred in modern times due to any phenomenon other than volcanoes.

In 1859, Hepsa Ely Silliman of Brooklyn, New York published her theory of the origin of meteorites.[40] Mrs. Silliman was uneasy with theories of extraterrestrial sources and hypothesized instead that the stony bodies aggregated from dust in the atmosphere. Her pamphlet showed considerable reading in the scientific literature of the time; she cited Faraday's work on chemical affinity and electromagnetism to account for the forces which drew the dust together in her theory. For the benefit of other researchers, she included a list of meteorites which had been recorded since the Greeks. While Silliman's theory does not agree with modern notions of the origins of meteorites, it was plausible until later in the century when petrographic microscopes revealed more about their structure.

Though the number of women who published in geology was small, they made contributions worth summarizing here. They popularized the discipline though textbooks, articles in magazines, and even poetry. Several of them worked on the borders between geology and other disciplines, such as chemistry and geography, or wrote about geology as part of a general view of the natural history of a landscape. They excelled at descriptive work, written up in the lush language of the Victorians, or simplified for children. At least one of them, Mrs. Callcott, found her work challenged by one of the authorities in the field, and vigorously defended it.

These women who were geologic artists, collectors, and writers were not professional geologists who did research and published it under their own name for wages. Individually, their contributions to the field of geology may seem minor, especially to a modern scientist used to large-scale research projects. As Sally Gregory Kohlstedt remarked about women in nineteenth-century American science generally, these women were at the periphery of the science, not in its center.[41] Nonetheless, their collective presence was a significant precedent for the more intense, professional involvement of women in geology in the 1890s and into the twentieth century.

THE FIRST TWO GENERATIONS OF WOMEN GEOLOGISTS IN AMERICA: FLORENCE BASCOM AND HER STUDENTS

For women to join the ranks of professional geologists, they had to be trained in the discipline, which by 1860 had become highly technical. The women whose work was described in the earlier sections of this chapter benefited from the schooling they received in secondary schools open to females. For the professionals, access to higher education was required, and this came about through the admission of women to universities and the establishment of women's colleges in the second half of the nineteenth century. An examination of the education of women in geology is outside the scope of this book, but Margaret Rossiter has published on the general issue of women's training in science in a recent book[42] and on geology in the women's colleges in particular in an article in the *Journal of Geological Education*.[43]

Florence Bascom (1862–1945) was the daughter of Emma Curtiss Bascom, an artist and feminist, and John Bascom, a professor and college administrator. In 1874, he became president of the University of Wisconsin, where Florence enrolled and in 1884 received a bachelor of science degree. After training in particular with Roland Irving of the geology department, she worked on her master's degree under the supervision of Charles Van

Hise from 1884 through 1887. She taught at a college in Illinois for two years and from 1889 until 1893 enrolled at Johns Hopkins, where she received the first Ph.D. awarded to a woman from that school. Her doctoral thesis centered on the rocks of South Mountain in Maryland, which she demonstrated by use of the petrographic microscope were metamorphosed volcanics rather than sediments. In 1893–1895, she taught at Ohio State University at the behest of Edward Orton.[44,45]

The United States Geological Survey hired Florence Bascom as an assistant geologist (a standard entry position) in 1896. With that job came all the advantages of survey employment—laboratory facilities, reference collections of rocks and fossils, library access, cartographic and other technical assistance, civil service status, and inspiring colleagues. Bascom's intellectual contributions to geology were made in connection with her USGS assignments. At the time she joined the staff, the Survey was involved in mapping selected areas of the United States which were believed to hold great promise for economic development and scientific understanding of the physical history of the continent. Bascom was part of this effort; she mapped some very difficult areas of the Mid-Atlantic piedmont, using stratigraphic principles, petrographic analyses done in the laboratory, and endless hours in the field to understand the structure of the terrain. Most of the area required dealing with metamorphic rocks, arguably the most obscure of the three classes of formations. In 1906, on a sabbatical from her Survey and college duties, she studied optical crystallography with Victor Goldschmidt in Heidelberg, bringing back the latest techniques to the United States and teaching them to others.[46,47] Petrology was then a relatively new science, and Bascom was important for transmitting the latest advances in it to co-workers and for demonstrating their usefulness in her own field studies. Her mapping, although subsequently challenged by some of her own students and modified somewhat by later workers, was crucial because it dealt with the rocks underlying the Appalachians and their foothills. Bascom sought to untangle the initial stratigraphic sequence and later deformation of these rocks; in so doing, she contributed to understanding the mountain-building processes of the entire range.

Bascom also contributed to geology by building up a department at Bryn Mawr College. When she started there, geological facilities at Bryn Mawr could not have been more different from the situation at the U.S. Geological Survey. She had to create the library, collections, and laboratories, and had to battle for a solid place in the curriculum for geology. As for colleagues, she produced them herself, in her students. Four of them went on to receive doctoral degrees and three worked for the Geological Survey as she had done. The geologic problems they studied sometimes overlapped hers, and they did not hesitate to disagree with her interpretations on some points, but that fate she shared with other mentors in all disciplines and it reflects no ill on her or them.

Florence Bascom and others in Goldschmidt's laboratory, 1906. Courtesy of Sophia Smith Collection, Smith College.

The earliest Bascom student of note was Ida Ogilvie (1874–1963), who
received a Bryn Mawr degree in 1900 and did graduate work at the University of Chicago before going to Columbia University to complete her
Ph.D. in 1903. Like Bascom, she enjoyed working in petrology, but wishing
to teach graduate students at Columbia, she offered instead to lead the
glacial geology course, which no one else had taken on there. Ogilvie taught
mainly at Barnard and gained a reputation for her superb lectures. Also
like Bascom at Bryn Mawr, she worked hard to build up the collections
and curriculum of the Barnard geology department, a kind of contribution
to geology not often recognized by professional honors. Ogilvie published
only a few papers in geology before the demands of patriotism led to a
deflection in her career. During World War I, she worked with the Women's
Land Army, and after the war she returned to teaching geology but devoted
most of her administrative and research efforts to agriculture, especially
dairying.[48,49] Ogilvie's work in geology was marked by efforts on behalf
of students and institutions rather than by the extensive research and publications which distinguished Bascom's later students.

Julia Gardner (1882–1960) graduated from Bryn Mawr with a bachelor's
degree in 1905 and a master's in 1907. At Bascom's urging, she did her
doctoral work in paleontology at John Hopkins under William Clark and
Edward Berry. These were major figures in American geology at the time.

After Gardner finished her degree in 1911, Clark and Berry hired her to teach at Hopkins and as a researcher on invertebrate paleontology at the Maryland geological survey. Gardner joined the U.S. Geological Survey in 1920, working on the stratigraphy and fossils of Gulf and Coastal Plain deposits from Texas to the Northeast. While these sediments did not present the structural difficulties of the rocks on which Bascom worked, their areal extent was vastly greater and their economic relevance more immediate, since these were oil-bearing formations. In analyzing them, Gardner contributed not only to regional stratigraphy but also to the paleontology of mollusks, the fossils on which she concentrated in her correlations of the Coastal Plain sediments. During World War II, Gardner worked on military geology problems at the Survey. Her interest in Pacific Island fossils stemmed from the war and postwar years and kept her active even after her retirement from the Geological Survey in 1952.[50,51]

The geologic interests of Bascom's third distinguished student, Eleanora Bliss Knopf (1883–1974), more closely paralleled Bascom's own than did Gardner's. Knopf was the daughter of a general important in the military history of World War I and the peace negotiations thereafter. She received her Bryn Mawr B.A. in 1904 and worked for the college for five years. In 1910–11, she attended Berkeley and then finished her doctoral thesis at Bryn Mawr jointly with Anna Jonas Stose, the fourth important Bascom student of the early twentieth century. Knopf and Stose received their doctorates in 1912 for their study of metamorphic rocks near the college. In 1912, Knopf joined the USGS as a geologic aide and studied the chrome ores, metamorphism, and erosional history of the region of her dissertation, publishing singly or with Stose. In 1920, she married Adolph Knopf, another survey geologist. They and his three children by a previous marriage moved to New Haven, where he taught at Yale and she continued on WAE ("When Actually Employed," a common contract research arrangment) with the USGS. From 1925 until her retirement in 1955, she concentrated on the Stissing Mountain area near the New York–Connecticut border, a site of thrust faults, overturned rocks, and metamorphism which had puzzled geologists since the 1840s. Having exhausted the methods of analysis then available to American geologists for dealing with such terrains, she mastered petrofabrics, a technique developed by Bruno Sandler of Innsbruck, in which geologists use texture, grain orientation, and optical properties of minerals to explain the origin and geologic history of the rocks in which the minerals are embedded. Her book on the subject, *Structural Petrography* (1938), brought this system of analysis to America and won her considerable distinction in geology.[52] Her application of petrofabrics to Stissing Mountain rocks helped establish the applicability of this difficult process.[53]

Like Knopf and Bascom, Anna Jonas Stose (1881–1974) worked on metamorphic rocks. Jonas received a Bryn Mawr B.A. in 1904, her M.A. in 1905, and a Ph.D. in 1912 for her joint dissertation with Knopf. She worked for the American Museum of Natural History in 1916–1917, the family glass business in 1918, the Maryland and Pennsylvania state geological surveys in 1926–1945, and the U.S. Geological Survey in 1930–1954. Her forte was mapping the Appalachian mountains and their piedmont, initially with Eleanora Knopf and then with George Stose, a USGS geologist whom she married in 1930. Stose and Stose worked quickly across large areas, and their interpretations were sometimes highly schematic and controversial. Their mapping was revised by later workers in Appalachian geology, but the Stoses' structural analyses, even the ones which sparked disagreement at the time, often proved sound. R. V. Dietrich, the judicious biographer of Mrs. Stose, details the vindication of her hotly debated ideas about the Martic thrust fault and Wissahickon schist (worked on with Eleanora Knopf), the Brevard Zone rocks (largely Anna Stose's own work), and the Carterville and Fried thrust faults, Reading klippen, and Lynchburg gneiss (with George Stose). Dietrich declared that "she gave all subsequent Appalachian geologists a basic picture to modify and perfect—a picture that was largely blank when she began her work."[54] Thrust faults had been recognized in these mountains since the 1840s, but Knopf and Anna Stose documented the fact that they were much more extensive and important to the history of the range than had previously been supposed. Modern theories of plate tectonics assign a crucial role to thrust faults, which are the mechanism by which the rocks of one plate are pushed over that of another. Plate tectonics in the Appalachian region also goes a long way toward explaining the widespread metamorphism which these women detailed; such metamorphism occurs because when plates ram into each other, not all the pressure is relieved by faulting.

Bascom, Ogilvie, Gardner, Knopf, and Jonas formed the first critical mass of professional women geologists in America. Primary credit for their many accomplishments must be given to their intelligence and training, but the four who worked for the United States Geological Survey also had the many resources of the Survey upon which to draw for a significant part of their research careers. Fortunately for later women in geology, these five were sufficiently diverse in personality, research interests, and career paths that they did not create a "woman geologist" stereotype. Gardner specialized in sedimentary rocks and paleontology, Ogilvie in glaciology, and the others worked on metamorphics and petrography; all were skilled in the classical geologic techniques of stratigraphy and cartography. All five successfully undertook arduous field work. Ogilvie and Bascom had more contact with students, while Gardner attained more administrative experience than the others. Stose was known for her salty and blunt language while the others were described as ladies by those who knew them. Knopf married and acquired three young stepchildren; Stose married a childless widower; the others remained single. To women of their own and later generations, they offered a high standard of accomplishment and a healthy pattern of diversity.

CONTRIBUTIONS OF OTHER WOMEN GEOLOGISTS IN FEDERAL AND STATE GEOLOGICAL SURVEYS

The women profiled in the preceeding section worked mainly at the U.S. Geological Survey and also at various state surveys and colleges and universities. During the twentieth century, American women geologists could also be found in museums and oil companies. Like the men in the field, women geologists worked for a variety of employers over the years, but their experience in certain work settings warrants separate analysis. As with Bascom and her students, these women geologists in different kinds of jobs offer students of history a convenient grouping for assessing their contributions to the science of geology.

Women paleontologists in addition to Gardner worked at the USGS for a substantial part of their career. Jean Hough (1903–1961) received a bachelor's degree from the University of Chicago in 1929 but interrupted her graduate work for family concerns and received her Ph.D. in 1946. After a few years with the American Museum of Natural History, she joined the USGS from 1949 to 1960, where she specialized in fossil vertebrates, doing extensive field work in the Northern Plains and Rockies states.[55] Helen Duncan (1910–1971) achieved a distinguished reputation for her work on fossil corals and bryozoa at the USGS, where she worked from 1942 until 1971. Her bachelor's (1934) and master's (1937) degrees were from the University of Montana; she taught at the University of Cincinnati before joining the Survey to work on military geology projects. Although Duncan published important papers under her own name, much of her work was done to help colleagues who needed fossils identified in her area of expertise and was acknowledged in notes in their publications, according to her biographers.[56] As noted, the study of fossils is of practical and theoretical value to geologists: paleontologists assemble the evidence for writing about plant and animal evolution, and they provide evidence for assigning relative ages to rocks, an exercise crucial in economic geology.

Some women mineralogists also worked for the USGS for long periods. Margaret Foster (1895–1970) was trained in chemistry (B.A., Illinois College) and joined the Survey in 1918 as an analyst of water resources. She developed methods for measuring trace minerals in water and wrote several papers on ground water in the Coastal Plain formations of the Gulf states. During World War II, Foster was on the Manhattan Project, where she devised new methods for uranium and thorium analysis. When she returned to the Survey, she switched to the geochemistry of clay minerals and micas.[57] Jewell Glass (1888–1966) joined the USGS in 1930, after work-

ing for the War Department and the Department of Agriculture. Her M.A.
and Ph.D. degrees were received for graduate work done while with the
USGS. Like Duncan, she was generous in helping colleagues, who acknowledged her mineralogical contributions in notes in their publications.
Glass's own papers centered on pegmatite minerals and rare earths, many
of which were of economic importance, such as beryllium.[58] Marjorie
Hooker (1908–1976) had a bachelor's degree from Hunter and an M.A.
degree from Syracuse. She worked at Columbia while studying for a doctorate, but left in 1943 to work for the military and then the State Department, and transferred to the USGS in 1946, where she was employed for
the next thirty years. Much of her survey work involved abstracting, translating, and compiling geologic lierature, but she also assembled chemical
data useful in correlating metamorphic and igneous rocks. She was unusually energetic in geological organizations as treasurer, secretary, com-
mitteewoman, and delegate to international meetings.[59]

As noted in the previous section, some of Bascom's students worked
for state geological surveys during part of their careers. Many other women
geologists also found employment with the states, and at least three of
them worked at state surveys for a decade or more of their professional
lives. State surveys had much smaller staffs than the USGS and fewer
laboratory and other resources, but a geologist at a state survey could make
a significant contribution in a specialized area, especially one related to the
practical concerns of economic geology peculiar to the state.

Winnie McGlamery (1887–1977) had a bachelor's degree in English but
switched to geology for graduate work at Johns Hopkins. After brief jobs
with Humble Oil, the American Museum of Natural History, and the museum of the University of Rochester, she joined the Alabama state geological survey as paleontologist in 1931 and remained with them for thirty
years. She and Walter B. Jones developed the collections of core samples
from oil and water drilling in Alabama and described the geological sequences of the subsurface formations from these reference materials. Much
of her work remained unpublished but readily accessible to geologists who
needed it. In the field, she devoted herself to gathering fossils to build up
the paleontologic collection at the survey. The materials gathered, arranged, cataloged, and described by McGlamery and Jones still constitute
major baseline data in the Gulf area for oil geologists and for scientists
wishing to reconstruct the geologic and evolutionary history of the region.[60]

Virginia Kline (1910–1959) worked for the North Dakota survey and the
Mississippi survey briefly, and for the Illinois survey from 1944 to 1959.
She received her Ph.D. from the University of Michigan in 1935 and worked
for several oil companies as a stratigrapher and paleontologist in the 1930s
and 1940s. At the Illinois survey, she wrote the monthly report on gas and
oil drilling and coauthored the annual compilation on Illinois petroleum
and gas production. These seemingly routine documents included assess-

ments of any geological discoveries made during commercial exploration that year. According to her biographer, Kline mapped the oil-bearing zones of Illinois geological formations during her work for the survey.[61]

Like Kline, Louise Jordan (1908–1966) held a variety of jobs before joining the Oklahoma state survey from 1955 to 1966. She earned a B.A. in geology from Wellesley in 1929 and for ten years worked for colleges, the Turkish government, and an oil company while studying toward her doctorate, which she received in 1939 from MIT. She worked for Sun Oil from 1940 to 1949 and for the Florida survey in 1950–51. Her years with the Oklahoma survey were marked by extensive publications on the petroleum productivity of strata, mainly on topics of economic relevance to the gas and oil industries, and by serving as mentor to University of Oklahoma students and younger professionals in economic geology.[62]

The women geologists at the state and federal surveys made noteworthy contributions mainly on topics related to economically important formations, those which were known or hypothesized to contain ores, water, or petroleum. In this feature of their work they were similiar to their male colleagues. Also like many of the men, they shared their work through building reference collections of fossils, chemical data, geological bibliographies, and well cores, and by sharing their knowledge of specialized aspects of geology with colleagues who wrote papers in which the women were acknowledged but did not appear as coauthors. The women brought to their survey work substantial training in science and in many cases, especially the women on the state surveys, considerable experience in oil companies and teaching.

CONTRIBUTIONS OF WOMEN IN MUSEUMS, COLLEGES, AND UNIVERSITIES

Several women paleontologists did most of their professional work at museums. After her training at Berkeley, Billie Untermann (1906–1973) was associated with the Utah Field House of Natural History, eventually becoming its director. As a museum technician, she helped assemble the dinosaur skeletons which were the hallmark of the museum. She and her husband coauthored popular works on the geology of the Uintah Basin.[63] Angelina Messina (1910–1968) received a doctorate for work at New York University and joined the university staff in 1936 to work on a massive catalog of foraminifera, microfossils which are especially useful in paleontologic correlation of strata from oil-drilling samples. She went with the project when it transferred to the American Museum of Natural History. The catalog brought together and analyzed taxonomic literature from virtually all countries and attested to Messina's foreign language and administrative skills. She successfully undertook an analogous work on ostracods.

In 1955 she cofounded the journal *Micropaleontology*, of which she was editor until her death.[64]

Winifred Goldring (1888–1971) worked at the New York State Museum in Albany starting in 1914. She had a B.A. and an M.A. from Wellesley and did graduate work at Johns Hopkins and Columbia. Although she specialized in Devonian fossils, Goldring's assignments were centered on popularizing geology through innovative museum exhibits and publication of handbooks for the general public. Her contributions to the technical literature included mapping two quadrangles in the state and doing research on crinoids and Devonian plants. From 1939 until retiring in 1954, she served as State Paleontologist of New York.[65,66] Tilly Edinger (1897–1967) came to the United States in 1940 and worked at the Museum of Comparative Zoology at Harvard except for one year when she taught at Wellesley. She had been trained at Heidelberg and Munich and received a doctorate from Frankfurt in 1921. Until the Nazis forced her out, Edinger had worked in the Senckenberg Museum, where she published an important monograph on the brain in fossil vertebrates, a topic on which she continued to work at the MCZ.[67] According to Stephen Gould, "she virtually established the field of paleoneurology. . . . [and] proved that the brain's evolution could and should be studied directly from fossils, not from a misleading hierarchy of modern species. . . . "[68] He noted that her work promoted the idea of evolution as branching, rather than progressive and linear, as most evolutionists of her time and earlier had assumed. Edinger was arguably the woman geologist in America who made the most substantial theoretical contributions to geology.

Mignon Talbot (1869–1950) was, like the women geologists in museums, a paleontologist. She received a Ph.D. from Yale in 1904 and then joined the faculty at Mount Holyoke College, where she taught geology until her retirement in 1935. She published few technical papers but instead poured her energy into building a department at Mount Holyoke. A fire in the science hall in 1916 wiped out the collections she had gathered, and she had to start gathering minerals, books, and fossils again from scratch.[69] In contrast, Grace Anne Stewart (1893–1970) found a stable, well established department at Ohio State University during her years there (1923–1954). A Canadian by birth and training through the master's degree, she received her Ph.D. from the University of Chicago. She had worked briefly for the Canadian national survey, despite prejudice there against women geologists.[70] Her many publications in geology during her Ohio years centered on Devonian and Silurian microfossils and invertebrates of the state. Stewart was a conscientious teacher and administrator who worked to improve the university museum as well as her department, and in that regard she resembled Talbot, but the university's resources and the presence of colleagues enthusiastic about research enabled her to publish more than Talbot did.[71] While Stewart's work on invertebrates was in a long tradition harking back to James Hall, her studies of Paleozoic microfossils were unusual among geologists of her time. Microfossils had mainly been worked on in younger formations, where their relevance for petroleum-bearing beds had given geologists a special incentive to concentrate on them.

Two women geologists made important contributions at Columbia University in the role of research associate. Margaret Rossiter[72] has outlined the difficulties of this assignment generally for women scientists: the pay was low and credit was not always given for work done. The two women geologists seemed to have escaped the worst of these features. Mary Garretson (1896–1971) was thanked by Amadeus Grabau for her contributions to his publications. She also did popular writing in science and worked as a consultant in geology and engineering in the 1950s.[73] Her younger counterpart in mineralogy, Peggy Kay Hamilton (1922–1959), divided her time between advising graduate students and doing her own research. She specialized in clay minerals, which led her to several Atomic Energy Commission projects on the association of clays with uranium minerals. As with Grabau, Hamilton's mentor, Paul Kerr, was scrupulous about acknowledging her role in their work, and they copublished several papers.[74]

Of course other women geologists worked for colleges and universities, and many of them are profiled elsewhere in this chapter. Nonetheless, it is striking that, historically, most women in geology (like men in the field) made their contributions in nonacademic settings. In this feature, geology differs from other sciences discussed in this book.

CONTRIBUTIONS OF WOMEN GEOLOGISTS IN PETROLEUM COMPANIES AND OTHER INDUSTRIAL WORK

Most women geologists in the private sector worked for oil companies, but at least two had productive careers elsewhere. Helen Bartlett (1901–1969) was a ceramic scientist with AC Spark Plug from 1931, when she received her Ph.D. from Ohio State, until her retirement in 1966. She applied her training in mineralogy and petrology to the study of alumina ceramics, especially in regard to their uses as insulation for spark plugs. Bartlett was unusual among women geologists for having some of her contributions to the firm marked in the form of patents. Her biographer also noted that she was the first woman to attain high technical status in General Motors, of which AC was a division.[75] In the next generation of women geoscientists, Marjorie Korringa (1943–1975), who received a Ph.D. from Stanford in 1972, started a promising career with Woodward-Clyde that was cut short by her death in an airplane crash. She had already gained a reputation for

Mrs. H. H. Adams at a plane table in the Batson oil field, ca. 1903. Courtesy of Southwest Collection, Texas Tech University.

careful studies of active fault systems in relation to decisions about where to run oil pipelines and where to build nuclear reactors. Like Bartlett, she seemed headed for considerable managerial responsibility.[76]

Contributions of women geologists in oil companies took two forms— practical, economic results which increased productivity or found new petroleum resources, and scientific, theoretical contributions which grew from laboratory and field studies that benefited academic as well as applied geology. Women have been involved in several dramatic money-making discoveries in American oil fields since 1920. For example, Fanny Carter Edson (1887–1952) was in charge of analyzing heavy minerals in the cores which came out of the first well drilled in the Marshall pool of Oklahoma. The well hit modest production in what the field men contended was the Wilcox sand, but she insisted the Wilcox was deeper, and after many more feet of drilling done on her advice, the well came in with vastly more productive flows than it would have produced, had operations stopped at the shallower level.[77] This discovery was crucial for the development of the Marshall field. Edson worked for Roxana Oil (later part of Shell) from 1924 until 1938, setting up and administering the Tulsa laboratory for the company. She had two tasks before her—keeping up with the daily arrivals of core material to be analyzed, and detecting the regional stratigraphic and tectonic patterns that could guide the company's future operations in the Mid-Continent. The second assignment led her to publish several papers which benefited geologists inside and outside the petroleum industry, as it enabled them to understand the subsurface geological structure of the region.[78] Margaret Fuller Boos (1892–1978), who received a Ph.D. from the University of Chicago in 1924, also applied her knowledge of tectonics and mineralogy to problems of petroleum geology. She worked for several oil companies from 1932 to 1970, and she and her husband became noted for their research on the structure and stratigraphy of the Front Range in Colorado. Their publications assisted oil companies operating in the region and, like Edson, helped geologists studying the general, scientific problems associated with mountain-building and geological change.[79]

Most of the other women geologists who worked for oil companies specialized in paleontology, especially micropaleontology. The drilling methods of the 1920s mangled rocks to the point where macrofossils were difficult to identify. When women paleontologists started oil company laboratories in the 1920s, most paleontologists believed the chronologic range of microfossils was too wide to make them useful in age determinations. Two women petroleum geologists, Esther Applin (1895–1972) and Alva Ellisor (1892–1964), were among those who proved that foraminifera and other microfossils could indeed be used with some precision to correlate rocks from drilling cores. Their work helped advance the scientific study of micropaleontology and provided badly needed index fossils for oil drilling operations. Ellisor, trained at the University of Texas (B.A. in geology, 1915), worked for Humble Oil as early as the summer of 1918. In 1920, after a job with the Kansas state geological survey, she came back to Humble (now EXXON) to set up a paleontologic laboratory and worked there until retiring in 1947.[80] Applin received an M.A. degree from the University of California at Berkeley in 1919 and worked for Rio Bravo Oil during 1920–1927 and as a consulting geologist for various petroleum firms until 1944, when she moved to Florida and took a job with the U.S. Geological Survey.[81]

In 1921, Applin presented a paper at a Geological Society meeting by her supervisor at Rio Bravo suggesting that microfossils could be used to date subsurface Gulf Coast formations which marked oil-bearing beds. Professor J. J. Galloway offered the classic putdown: "Gentlemen, here is this chit of a girl, right out of college, telling us that we can use Foraminifera to determine the age of a formation. Gentlemen, you know that it can't be done."[82] Four years later Applin coauthored a paper with Ellisor and Hedwig Kniker demonstrating conclusively that it could and had been done; they detailed the sequence and oil-bearing zones in the Gulf Coast using microfossils.[83] Microfossil correlation remained indispensable to oil drilling until the advent of electric logs[84] and was a specialty in which women geologists gained considerable stature. Like many other new fields of science, it was more open to women than established branches of the discipline, which remained largely male enclaves.

Applin had at least two women colleagues at Rio Bravo who also made significant contributions to geology. Laura Weinzierl received a geology B.A. from the University of Texas in 1923. She worked for Rio Bravo in the early 1920s and then set up a paleontologic laboratory for an oil geologist who later went with her to Marland Oil Company, where she worked until her untimely death in 1928.[85] Similarly, Dorothy Palmer (1897–1947) joined Rio Bravo in 1924, two years after receiving an M.A. degree from Berkeley. From 1930 to 1933 and 1940 to 1947, she worked for Atlantic Refining. She and her husband specialized in Caribbean geology, and their studies of the fossils and formations of the region, Cuba in particular, assisted in the exploration for petroleum there, as well as advancing understanding of the geology of the area, then barely studied.[86]

Carlotta Maury (1874–1938) obtained a Ph.D. in geology from Cornell in 1902. After jobs at Columbia, the Louisiana state geological survey, and a college in South Africa, she worked for Royal Dutch Shell from 1910 to 1937. Maury prepared sixteen confidential reports for the firm on Venezuelan invertebrate fossils and stratigraphy, material which proved useful to the firm in exploring the oil resources in that country. She specialized in mollusks and was known in the more academic circles of geology for her work on Brazilian fossil fauna, done in connection with geological surveys undertaken for the Brazilian government.[87]

PROBLEMS FACED BY WOMEN WHO WERE PROFESSIONAL GEOLOGISTS AND HOW THEY DEALT WITH THEM

The collective record of achievement by the women geologists profiled in this chapter is indeed impressive. However, it is important to remember that women constituted a tiny fraction of the population of professional geologists during the first several decades of this century. Statistics are few, but we do know that for 1920–1970, women obtained less than 4 percent of all doctorates awarded in the earth sciences. Only physics, engineering, and agriculture posted worse records for women entering the field on this level.[88] Thus, women such as Winifred Goldring felt isolated at times, and she once wrote that young women contemplating a career in science should consider picking a specialty such as botany where there were enough women to make them more comfortable than she felt in geology.[89]

Doing field work was something of a vexing issue early in the era of women's entry into professional geology. The problems ranged from the trivial (what to wear) to the serious (safety). Winifred Goldring was urged

Geology field trip, Mount Holyoke College, ca. 1900. Courtesy of Mount Holyoke College Library/Archives.

by Ray Bassler of the U.S. National Museum not to imitate Marjorie O'Connell's example of arming herself with a revolver and heading into the field.[90] The disapproval was not voiced by men only, however. Mignon Talbot of Mount Holyoke College, who included field trips in training her students, nonetheless believed that laboratories, offices, or museums were more proper spheres for women than most field locales.[91] Many women geologists simply ignored this taboo and went into the field to do research, despite bad health, the inconvenience of dress, and the head-shaking of colleagues. If women who worked for oil companies had to do on-site identification of fossils and minerals, they went out on the drilling rigs.[92]

The women sometimes faced discrimination in getting jobs or, once they had them, in obtaining equitable salaries. Winifred Goldring, on asking about a job with the U.S. Geological Survey, was told they wanted a "he-man" paleontologist.[93] On another occasion, Goldring found out her pay ($2,300) was less than that of clerks and stenographers at the New York State Museum, where she worked. She also stated that Dolly Radler was making about half as much money as men doing comparable work in the oil industry in the 1920s.[94]

The career of Marjorie O'Connell demonstrated the twin difficulties of hiring and salaries for women in geology. She was trained by Amadeus Grabau at Columbia, who imparted to her his zeal for research. She declined teaching at women's colleges because time required in the classroom was too great to permit research as well. Instead, she took temporary secretarial and sales jobs while waiting for a research opening, and pawned her clothes when the money ran out. Finally hired for a specific paleontologic task at the American Museum of Natural History, she found herself earning far less money than men doing similar assignments. O'Connell guessed that museum administrators had assumed she was living at home and thus needed less, when in fact she was estranged from her family because of her career and had to support herself.[95] Finally, she abandoned geology for a public relations job with a bank which paid her $5,000 a year, over twice what Goldring received for full-time paleontological work.[96] She later married William Shearon and shifted her career to social work, geriatrics, and public health, holding jobs in state and federal government through the Depression.[97] Since detailed biographical records of those who leave a field are hard to find, one cannot evaluate how typical O'Connell's experiences in geology were, but the low numbers of women in geology suggest that there may have been many who chose another life rather than battle the problems of women in the discipline.

On the positive side are two institutional arrangements peculiar to geology which actually assisted women who tried to maintain a professional career and fulfill traditional roles as wives and mothers, although this benefit was an accidental byproduct of the procedure. One was WAE ("When Actually Employed") at the U.S. Geological Survey. The Survey hired many geologists for specific tasks and locales rather than keeping them as permanent employees stationed in Washington, D.C., or at another Survey headquarters. When Eleanor Knopf moved to New Haven to accompany her husband on his appointment at Yale, the Survey continued to hire her on WAE to work on the Stissing Mountain project. Knopf was able to raise three young stepchildren, fulfill the demands of being a faculty wife, and continue as a geologist.[98] In the petroleum industry, the status of consulting geologist provided the same flexibility for Helen Plummer, who worked as a paleontologist on a contract basis with a number of firms from 1925 until 1948, while also maintaining a traditional home for her geologist husband.[99] Of course, given the low numbers of women in geology, most of the beneficiaries of WAE and consulting work were men, particularly academic geologists who took on government contract work or petroleum exploration as adjunct to their teaching tasks.

Also helpful to women geologists were many men supportive of women's professional aspirations. Among them were E. T. Dumble in the oil industry, Edward Orton of the Ohio state survey, Charles Schuchert of Yale, and Amadeus Grabau at Columbia. They and other open-minded men welcomed women as students and employees; some lobbied hard to find them jobs and see to their professional advancement. Often the husbands of women geologists accompanied them in the field and served as coauthors on publications. The women were elected occasionally to office in predominantly male geological societies; large numbers of men must have supported their candidacy for them to win. As far as the women were concerned, those men who were prejudiced—such as the one who told Dolly Radler he "wouldn't talk to any woman about the oil business"[100]— hurt themselves more than they hurt the women, because they shut themselves off from the valuable economic and technical expertise gathered by women scientists in the field and in the laboratory.

The careers of early women geologists offer lessons for women's history as well as for the history of geology. These women worked through the Depression and into the 1940s and 1950s, times which historians have portrayed as especially discouraging for women, and in a field in which men dominated. There seemed to be no specifically female corner of geology; even in micropaleontology, where the contributions of women were especially numerous, men still accounted for most of the practitioners. Cumulatively and individually, the biographies of these women make inspiring reading. Despite low wages, restricted job openings, taboos against field work, and an occasional sense of isolation, women geologists advanced both practical and academic aspects of the science.

NOTES

1. Phelps, A. H. L. 1830. Ed. of L. N. Vauquelin, *Dictionary of Chemistry*. New York, Carvill, 531 p.

2. Hazen, R. M. 1982. *The Poetry of Geology*. London, George Allen and Unwin, 98 p.

3. Warner, D. J. 1978. Science Education for Women in Antebellum America. *Isis*, v. 69, pp. 58–67.

4. McAllister, E. M. 1941. *Amos Eaton Scientist and Educator, 1776–1842*. Philadelphia, University of Pennsylvania Press, 587 p.

5. Talbot, Mignon. 1922. The Department of Geology. *Mt. Holyoke College Alumnae Quarterly*, v. 6, pp. 128–132.

6. Hitchcock, Edward. 1841. Final Report of the Geology of Massachusetts. Amherst, J. S. and C. Adams, 831 p.

7. Hall, James. 1842. Geology of New York, Part IV, Comprising the Report of the Fourth Geological District. Albany, Carroll and Cook, 683 p.

8. Hall, James. 1847. *Paleontology of New York*, v. 1. Albany, Van Benthuysen, 339 p.

9. Meek, F. B. 1866. Illinois State Geological Survey (Worthen Survey), *Paleontology*. Springfield, Steam Journal Press, 470 p.

10. Newberry, J. S. 1875. Report of the Geological Survey of Ohio, v. 2, part 2. Columbus, Nevins and Myers, 435 p.

11. Williston, Samuel. 1898. Kansas Geological Survey—Paleontology. Topeka, J. S. Parks, 594 p.

12. Aldrich, M. L. 1982. Women in Paleontology in the United States 1840–1960. *Earth Sciences History*, v. 1, pp. 14–22.

13. United States Geological Survey of the Fortieth Parallel (King Survey), 1877, Report, v. 4, *Paleontology* (by James Hall and R. P. Whitfield). Washington, D.C., Government Printing Office, 667 p.

14. Cope, E. D. 1875. Vertebrata of the Cretaceous Formations of the West (U.S. Geological Survey of the Territories [Hayden Survey]): Washington, D.C., Government Printing Office, 302 p.

15. Grafly, Dorothy. 1971. Cecelia Beaux, *in* E. T. James and J. W. James, eds., *Notable American Women*, v. 1. Cambridge, Harvard University.

16. Gould, S. J. 1981. *The Mismeasure of Man*. New York, W. W. Norton, p. 155.

17. Simon, Cheryl. 1983. Dinosaur Story—Who Found the Tooth? *Science News*, v. 124, p. 312 [reporting research by Dennis Dean on M.A. Mantell].

18. Andrews, S. M. Lady Eliza Maria Gordon Cumming, *in* Andrews, *Discovery of Fossil Fishes in Scotland up to 1845*. Edinburgh, Royal Scottish Museum, pp. 72–73.

19. Rowe, S. R., T. Sharpe, and H. S. Torrens. 1981. *Ichthyosaurs—A History of Fossil "Sea Dragons."* Cardiff, National Museum of Wales, 31 p.

20. Woodward, B. B. 1901. Mary Anning, *in* Leslie Stephen, ed., *Dictionary of National Biography*, v. 22. Oxford, Oxford University Press, p. 401.

21. Harper, Lewis. 1857. Preliminary Report of the Geology and Agriculture of Mississippi. Jackson, E. Barksdale, 350 p.

22. Whitney, J. D. 1865. Geological Survey of California. Philadelphia, Sherman, 498 p.

23. Cox, E. T., 1879. Eighth, Ninth, and Tenth Annual Reports of the Geology of Indiana. Indianapolis, Indianapolis Journal Co., 541 p.

24. Hazen, R. M. and Hazen, M. H. 1980. *American Geological Literature, 1669–1850*. Stroudsburg, Pa., Hutchinson Ross.

25. Piozzi, H. L. 1793. Affecting Picture of an Earthquake Scene. *Lady's Magazine and Respository of Entertaining Knowledge*, v. 2, pp. 187–188. For other versions, see Hazen and Hazen 1980, p. 303.

26. Stephen, Leslie. 1896. Hester Lynch Piozzi, *in* Leslie Stephen, ed., *Dictionary of National Biography*, v. 15. Oxford, Oxford University Press, p. 196.

27. Holley, M.A. 1833. *Texas. Observations, Historical, Geographical, and Descriptive*. Baltimore, Armstrong and Plaskitt, 167 p.

28. Child, L. M. F. 1842. Mammoth Cave. *Anglo-American*, v. 2, pp. 7–9.

29. Cooper, S. F. 1850. *Rural Hours*. New York, Putnam, 521 p.

30. Hazen 1982, *op. cit.* (n. 2).

31. Hemans, F. D. 1836. Epitaph on a Mineralogist. *American Magazine of Useful and Entertaining Knowledge*, v. 3, p. 72 (repr. in Hazen 1982, pp. 49–50).

32. Gould, H. F. 1847. The Mastodon. *Sartain's Union Magazine*, v. 1, p. 8.

33. Willard, E. H. 1824–1844. *Ancient Geography*. Hartford, var. pub., 67 p. This item went through several reprintings for at least twenty years.

34. Woodbridge, W. C. 1824. *A System of Universal Geography*. Hartford.

35. Aldrich, M. L. 1980. Eleanora Knopf, *in* Barbara Sicherman and C. H. Green, eds., *Notable American Women*, v. 4. Cambridge, Harvard University Press, pp. 401–403.

36. Phelps 1830, *op. cit.* (n. 1) p. 13.

37. Godding, D. W. 1847. *First Lessons in Geology. Comprising Its Most Important and Interesting Facts, Simplified to the Understanding of Children*, Hartford, H. S. Parson, 142 p.

38. Welsh, J. K. 1832–1833. *Familiar Lessons in Geology and Mineralogy Designed for the Use of Young Persons and Lyceums.* Boston, Clapp and Hull, 2 vols.

39. Callcott, M. G. 1835. On the Reality of the Rise of the Coast of Chile in 1822. *American Journal of Science,* v. 28, pp. 236–247.

40. Silliman, H. E. 1859. *On the Origins of Aerolites.* New York, W. C. Bryant, 31 p.

41. Kohlstedt, S. G. 1978. In from the Periphery—American Women in Science: 1830–1880. *Signs,* v. 4, pp. 81–96.

42. Rossiter, M. W. 1982. *Women Scientists in America—Struggles and Strategies to 1940.* Baltimore, Johns Hopkins University Press, 439 p.

43. Rossiter, M. W. 1981. Geology in Nineteenth Century Women's Education in the United States. *Journal of Geological Education,* v. 29, pp. 228–232.

44. Rosenberg, C. S. 1971. Florence Bascom, *in* E. T. James, J. W. James, and P. S. Boyer, eds., *Notable American Women,* v. 1. Cambridge, Harvard University Press, pp. 108–110.

45. Smith, I. F. 1981. The Stone Lady, A Memoir of Florence Bascom. Bryn Mawr, Bryn Mawr College, 49 p.

46. Ibid.

47. Bascom, Florence. Papers. Sophia Smith Collection, Smith College (Boxes 7 and 8).

48. Arnold, L. B. 1978. Ida Ogilvie, Geologist. *Barnard Alumnae Magazine,* v. 67, pp. 12–13.

49. Wood, E. A. 1964. Memorial to Ida Helen Ogilvie (1874–1963). *Geological Society of American Bulletin,* v. 75, pp. 35–39.

50. Nelson, C. M., and M. E. Williams. 1980. Julia Anna Gardner, *in* Barbara Sicherman and C. H. Green., eds., *Notable American Women,* v. 4. Cambridge, Harvard University Press, pp. 260–262.

51. Ladd, H. S. 1962. Memorial to Julia Anna Gardner (1882–1960). *Geological Society of American Proceedings for 1960,* pp. 87–92.

52. Rodgers, John. 1977. Memorial to Eleanora Bliss Knopf 1883–1974. *Geological Society of America Memorials,* v. 6, 4 p.

53. Aldrich 1980, *op. cit.* (n. 35).

54. Dietrich, R. V. 1977. Memorial to Anna I. Jonas Stose 1881–1974. *Geological Society of America Memorials,* v. 6, 6 p.

55. Patterson, Bryan. 1961. Margaret Jean Ringier Hough 1903–1961. *Society of Vertebrate Paleontology News Bulletin,* v. 62, p. 36.

56. Flower, R. H., and J. M. Berden. 1977. Memorial to Helen M. Duncan 1910–1971. *Geological Society of America Memorials,* v. 5, 3 p.

57. Fahey, J. J. 1971. Memorial of Margaret D. Foster [1895–1970]. *American Mineralogist,* v. 56, pp. 686–690.

58. Jesperson, Anna. 1968. Memorial to Jewell Jeannette Glass (1888–1966). *Geological Society of American Proceedings for 1966,* pp. 225–227.

59. Jesperson, Anna. 1978. Memorial to Marjorie Hooker 1908–1976. *Geological Society of America Memorials,* v. 8, 4 p.

60. Copeland, C. W. 1979. Memorial to Josie Winifred McGlamery, 1887–1977. *Geological Society of America Memorials,* v. 9, 3 p.

61. Oros, M. O. 1962. Memorial to Virginia Harriett Kline (1910–1959). *Geological Society of America Proceedings for 1960,* pp. 115–117.

62. Nicholson, Alexander. 1968. Louise Jordan (1908–1966). *American Association of Petroleum Geologists Bulletin,* v. 52, pp. 2058–2060.

63. White, Alice. 1973. Billie R. Untermann 1906–1973. *Society of Vertebrate Paleontology News Bulletin,* no. 99, pp. 65–66.

64. Ellis, B. F. 1971. Memorial to Angelina Rose Messina (1910–1968). *Geological Society of America Proceedings for 1968*, pp. 212–214.

65. Fisher, D. W. 1974. Memorial to Winifred Goldring 1888–1971. *Geological Society of America Memorials*, v. 3, pp. 96–102.

66. Kohlstedt, S. G. 1980. Winifred Goldring, *in* Barbara Sicherman and C. H. Green, eds., *Notable American Women*, v. 4. Cambridge, Harvard University Press, pp. 282–283.

67. Romer, A. S. 1967. Tilly Edinger, 1897–1967. *Society of Vertebrate Paleontology News Bulletin*, no. 81, pp. 51–53.

68. Gould, S. J. 1980. Tilly Edinger, *in* Barbara Sicherman and C. H. Green, eds., *Notable American Women*, v. 4. Cambridge, Harvard University Press, pp. 218–219.

69. Haff, J. C. Memorial to Mignon Talbot. *Geological Society of America Annual Report for 1951*, pp. 157–158.

70. Rossiter 1982, *op. cit.* (n. 42) pp. 236–237.

71. Spieker, E. M. 1973. Memorial to Grace Anne Stewart 1893–1970. *Geological Society of America Memorials*, v. 2, pp. 110–114.

72. Rossiter 1982, *op. cit.* (n. 42).

73. Behre, C. H. Jr. 1975. Memorial to Mary Welleck Garretson 1896–1971. *Geological Society of America Memorials*, v.4, pp. 72–73.

74. Kerr, P. F. 1960. Memorial to Peggy-Kay Hamilton (1922–1959). *Geological Society of America Proceedings for 1959*, pp. 133–136.

75. Schwartzwalder, Karl. 1971. Memorial of Helen Blair Bartlett [1901–1969]. *American Mineralogist*, v. 56, pp. 668–670.

76. Packer, D. R., W. R. Dickinson, and K. M. Nichols. 1977. Memorial to Marjorie K. Korringa 1943–1974. *Geological Society of America Memorials*, vol. 6, 3 p.

77. Sheldon, Ruth. 1941. The Ladies Find Oil. *Scribner's Commentator*, v. 10, pp. 110–115, p. 78.

78. Leiser, J. B. 1953. Fanny Edson Carter (1887–1952) *American Association of Petroleum Geologists Bulletin*, v. 37, pp. 1182–1186.

79. Byers, Virginia, and Doris Osterwald. 1981. Memorial to Margaret Fuller Boos 1892–1978. *Geological Society of American Memorials*, v. 11, 3 p.

80. Teas, L. P. 1965. Alva Christine Ellisor (1892–1964). *American Association of Petroleum Geologists Bulletin*, v. 49, pp. 467–471.

81. Berdan, J. M. 1975. Memorial to Esther Richards Applin 1895–1972. *Geological Society of America Memorials*, v. 4, pp. 14–18.

82. Maher, J. C. 1973. Esther Richards Applin (1895–1972). *American Association of Petroleum Geologists Bulletin*, v. 57, pp. 596–597.

83. Applin, E. R., A. E. Ellisor, and H. T. Kniker. 1925. Subsurface Stratigraphy of the Coastal Plain of Texas and Louisiana. *American Association of Petroleum Geologists Bulletin*, v. 9, pp. 79–122.

84. Owen, E. W. 1975. Trek of the Oil Finders. *American Association of Petroleum Geologists Memoir*, v. 6, 1008, p. 751.

85. Applin, E. R. 1928. Memorial to Laura Lane Weinzierl. *Journal of Paleontology*, v. 2, p. 383.

86. Palmer, K. V. 1948. Dorothy K. Palmer, 1897–1947. *Journal of Paleontology*, v. 22, pp. 518–519.

87. Reeds, C. A. 1939. Memorial to Carlotta Joaquine Maury. *Geological Society of America Proceedings for 1938*, pp. 157–168.

88. Vetter, B. M., E. L. Babco, and Susan Jensen-Fisher. 1983. *Professional Women and Minorities*, 4th ed.. Washington, D.C., Scientific Manpower Commission, 280 p.

89. Goldring, Winifred. 1929. Letter to Walter Bucher of February 29. *Goldring Papers*, New York State Archives, Albany.

90. Bassler, Ray. 1922. Letter to Winifred Goldring of April 22. *Goldring Papers*, New York State Archives, Albany.

91. Talbot 1922, *op. cit.* (n. 5) p. 128.

92. Berdan 1975, *op. cit.* (n. 81).

93. Schuchert, Charles. 1928. Letter to Winifred Goldring of May 1. *Goldring Papers*, New York State Archives, Albany.

94. Goldring 1929, *op. cit.* (n. 89).

95. O'Connell, Marjorie. 1920–1921. Letter to Christine Ladd Franklin. *Ladd Franklin Papers*, Butler Library, Columbia University, New York.

96. Goldring 1929, *op. cit.* (n. 89).

97. *American Men of Science*, 1921–1944, 3rd–7th eds. Var. places, Science Press, entries on M. O'Connell Shearon.

98. Aldrich 1980, *op. cit.* (n. 35).

99. Adkins, M. G. M. 1954. Helen Jeanne Plummer. *American Association of Petroleum Geologists Bulletin*, v. 38, pp. 1854–1857.

100. Sheldon 1941, *op. cit.* (n. 77) p. 78.

Sally Gregory Kohlstedt (Syracuse) and Margaret Rossiter (Harvard) encouraged me to undertake this study and shared their files and ideas about women in American science. Alan Leviton (California Academy of Sciences) critiqued the manuscript, helped with the research, and provided the plates.

Curators at several archives made collections relating to women in geology available: New York State Archives (Winifred Goldring Papers), New York State Library (James Hall Papers), Amherst College (Edward Hitchcock Papers), Sophia Smith Collection at Smith College (Florence Bascom Papers), Library of Congress (Eleanora Knopf letters in Tasker Howard Bliss Papers), Cushman Foundation at the Smithsonian Institution (Esther Applin Papers), and Columbia University (Marjorie O'Connell letters in Christine Ladd Franklin Papers). Jack Marquardt and Harry Heiss of the Smithsonian Institution, as well as Peter Rodda and Ray Brian of the California Academy of Sciences, were especially helpful in the search for sources.

PAMELA E. MACK

Straying from Their Orbits

Women in Astronomy in America

Women have made major contributions to American astronomy from the late nineteenth century to the present. In the late nineteenth century women astronomers in the United States gained acceptance in the astronomical community. This acceptance was made possible by changes in the type of science done by astronomers in the late nineteenth and early twentieth centuries. Despite this acceptance, the contributions of women were limited by the community, which deemed as appropriate for women only certain fields involving the large-scale processing of data. Within those fields some women made important discoveries and others provided support essential to the development of the field. Up until the mid-1920s, however, women who wanted to make contributions to astronomy in areas outside of what had evolved as women's work found their paths blocked. This chapter will discuss only American women astronomers and will concentrate on the period before 1920.

THE STATE OF ASTRONOMY

Women were able to gain a role in astronomy because of changes in both the content and the organization of the field in the late nineteenth century. By the late nineteenth century astronomers had analyzed the orbits of the planets and other objects in the solar system using Newton's theory of gravitation and were turning their attention to stars. Scientists then began to try to understand the differences between stars. The first step toward analyzing the variety and arrangement of stars was to classify and catalog them. Bigger telescopes and new techniques for using them, together with an increased interest in discovering unusual stars, led to projects for larger and more complex catalogues.

New developments in physics led to new ways of observing and classifying stars. The development of spectroscopy made it possible to learn something about the composition and temperature of stars and to classify them in a way that had physical meaning. Many astronomers believed that each star progressed through the whole range of spectral types during its lifetime, so a number of different classification systems were developed representing various theories of the life cycles of stars. Classification of the spectra of large numbers of stars was necessary to refine theories of stellar evolution. Astronomers also realized that this research would improve understanding of the structure of the universe, because different spectral types of stars were not uniformly distributed in the depths of space.[1]

The late nineteenth century saw changes not only in the scientific theories and priorities of astronomy but also in the institutions where astronomy was done and in the techniques used. The first permanent observatory in the United States was established in 1838 by Albert Hopkins at Williams College. The first observatories capable of competing with those in Europe were Harvard College Observatory, founded in 1839, the Naval Observatory, founded in about 1842, and the Cincinnati Observatory, founded in 1843. Small observatories grew up quickly at various colleges, but when women first entered astronomy in the 1870s, astronomical work in the United States had not yet caught up with that being done in Europe.

The first development to change the balance of power in astronomy in the late nineteenth century was photography. The first photograph of a star was a daguerreotype taken at Harvard College Observatory in 1850.[2] After the development of the dry plate photographic process in the 1870s, photography became truly useful to astronomy (the earlier wet plate process could not be kept wet during the long exposures necessary for astronomical photography). Soon photographic plates were being taken in large numbers, particularly at Harvard, and they were available as a permanent record which could be observed and compared at leisure. This influenced the kind of astronomical work being done, because the new techniques were most helpful for the study of large numbers of faint objects, such as unusual stars and nebulae.

The second major trend in institutions and instruments in this period was the development of large observatories supported by private philanthropy. Larger and larger new telescopes were built at a rapid rate with the support of rich benefactors who each wanted to fund the largest telescope in the world. This resulted in the 36-inch refractor at Lick Observatory in 1889, the 40-inch refractor at Yerkes Observatory in 1897, and the 60-inch and 100-inch reflectors at Mount Wilson in 1908 and 1917.[3] These telescopes brought the United States to leadership in observational astronomy. They stimulated an interest in the size and structure of the universe because they could see fainter and more distant objects than ever before. Work on such topics resulted in the better understanding of the size of our galaxy and the nature of the spiral nebulae, allied problems central to astronomy in the 1920s.

All these observatories were of semi-independent status. Cincinnati Observatory was funded by public subscription, the Naval Observatory mostly by Congress (giving it less independence). The staff of Harvard College Observatory had no teaching responsibility, despite its connection to the university, and graduate students were not accepted until the 1920s. Yerkes was similarly attached, to the University of Chicago. The Carnegie Foundation supported Mount Wilson, so it was independent of any university. Because observatories were not yet closely connected with universities, the academic status of astronomy was uncertain. Many astronomers were primarily self-educated, and even the young, rising astronomers in the late nineteenth century learned physics and mathematics in colleges but learned astronomy by working as assistants in observatories.

The scientific state of astronomy reflected this recent transition to professional status, which had happened only a few decades earlier in Europe. Amateur astronomers had concentrated on individual interesting objects: professional astronomers had the resources to undertake the massive compilations of data that were increasingly necessary. One of the major research projects of most observatories in the third quarter of the nineteenth century was determining star positions. However, in the last quarter of the nineteenth century a new sort of astronomy grew up. Using the spectroscope as a tool, astronomers studied the spectra of stars. This work led to the development of a new approach to astronomy, astrophysics, in the early twentieth century.

These new institutions and the large projects they sponsored on new scientific questions provided an opportunity for women to move into astronomy without trying to take over work previously done by men.[4] Certain methodologies and areas of study within the new type of research became defined as women's work. Astronomy is not the only science in which this occurred in the late nineteenth century; Margaret Rossiter argues in her study of women in science:

> Advocates of such "women's work" had no trouble developing a rationale for separate kinds of jobs for women. They had merely to urge women to capitalize on the relatively warm welcome that they were already receiving in the marketplace for two kinds of jobs: those that were so low paying or low ranking that competent men would not take them (and which often required great docility or painstaking attention to detail) and those that involved social service, such as working in the home or with women or children (and which were often poorly paid as well).[5]

In astronomy, women moved into low-paying jobs that required painstaking attention to detail. But in astronomy, perhaps more than in other fields, women's work was not defined only by low pay and low prestige. Some women were rewarded for outstanding work with substantial prestige as scientists. A whole research area in astronomy came to be defined as women's work. Most of this work required repetitive, painstaking attention to detail, but it resulted in important contributions to astronomy.

WOMEN'S COLLEGES

Starting in the late nineteenth century, the newly created women's colleges revolutionized opportunities for women in science by providing women students with education in science and providing women hired as professors with opportunities for research. By 1890 women had made a place for themselves in astronomy; women wrote 4 percent of the papers (excluding minor notes and regular columns) in the three major astronomical journals in the United States between 1890 and 1920.[6] The importance of women's colleges was shown by the fact that of the 238 papers by women, 48 percent were written by women from these institutions. This figure may exaggerate the contribution of women's colleges, however, because much of the work done by women at major observatories was published under the name of a supervisor or recorded in the publications of those observatories rather than in journals. Still, the research work of women at women's colleges is of special significance because they could select their own research topics and were free to develop their results on their own. The fact that even with these freedoms they did scientific work of limited significance reflects not necessarily on their talent but on the limitations placed on them by the job they were hired to do and the advice they were given. Even at the independent women's colleges women astronomers tended to stay within the limitations of what had become accepted as women's work.

Vassar

The most significant early astronomy program at a women's college developed at Vassar College. Maria Mitchell (1818–1889), the first director of the Vassar College Observatory, created the earliest opportunities for women to study astronomy on a level which prepared them for professional careers. From the founding of Vassar in 1847, she set up a program of rigorous astronomical training and commitment to research that served as a model for the other women's colleges. The development of the program at Vassar was typical of the women's colleges. Initially, the major limitation was equipment. Growing numbers of students convinced the administration to improve the equipment but resulted in a teaching load that left little time for research. Finally, the 1890s brought the hiring of additional faculty and a new commitment to original research. Vassar also provides an example of the growth of ties between the women's colleges and the large observatories that allowed the women's colleges to place their graduates in astronomical jobs.

In 1847, Maria Mitchell, the 29-year-old librarian of the Nantucket Atheneum, discovered a comet. She had learned astronomy from her father, a noted amateur, had studied extensively on her own from books, and was an able and dedicated observer.[7] Her father sent word of her discovery to his friend William Bond, the director of Harvard College Observatory, who confirmed it and sent a report to the president of Harvard, Edward Everett. It happened that the King of Denmark had offered a gold medal for the first discovery of a new comet that was invisible to the naked eye at the time of the discovery. The committee decided initially to give the award to Father Francesco de Vico of Rome, who had discovered the same comet two days later (transatlantic mail was so slow that the decision was made before the news of Mitchell's discovery arrived). But, "President Everett, who believed that George Bond had been unjustly deprived of the medal in 1846, determined that an American astronomer should not be outdone a second time."[8] After some diplomatic wrangling, Mitchell received the prize that made her famous.

The combination of this honor and her father's personal connections enabled Mitchell to find professional employment. Her position as a dutiful unmarried daughter caring for her aging father made it necessary for her to continue to work. A friend of hers, Commander Charles H. Davis, hired her as a computer (one who does computations) for the Coast Survey in 1849, at a salary of $300 a year. She worked at home computing tables of the positions of planets (because she was a woman she was assigned Venus) for the *American Ephemeris and Nautical Almanac*, an annual compilation of astronomical tables for mariners.[9] This was not the first time a group of people, usually men, had been hired to do the routine calculations necessary to produce astronomical tables, but Mitchell's employment set a particularly important precedent because the *Nautical Almanac* was a large-scale enterprise and operated on a continuing basis.

In 1865, Matthew Vassar founded Vassar College as the first women's college that sought to offer women the same course of studies that men received at the best men's colleges. Matthew Vassar sought to hire the foremost women scholars as professors, but the trustees successfully opposed him for most of the faculty positions. In the end the only important woman hired was Maria Mitchell, who was invited to be professor of astronomy and director of the observatory.[10] Vassar promised Mitchell an observatory equipped for research and a house for herself and her father. She accepted the position, and she and her father lived and worked there until his death in 1869 and her retirement in 1888.

Vassar supplied Maria Mitchell with a telescope that could have been sufficient for worthwhile scientific research. When minor problems arose, however, the college was reluctant to spend the money to solve them. With a 12-inch objective lens, the equatorial telescope was one of the largest in the United States in 1865, but the objective was of poor quality. Mitchell wrote:

Maria Mitchell at her telescope in Nantucket, from a painting by Hermione Dassel (1851). Courtesy of Vassar College Archives.

After nearly three years I have come to the conclusion, that the telescope's illuminating power is good; its defining power bad. It picks up minute points of light very well: it does not give them a sharp outline. . . . It is exceedingly trying to an observer, to have in charge a telescope ranking as fourth or fifth in the country, and to be unable to compete with those of much less aperture.[11]

Alvin Clark and Sons (the most famous telescope maker in the country and possibly the best in the world at that time) refigured the objective in 1868, and in her next annual report Mitchell proudly claimed that the telescope ranked third in the country, exceeded only by those in Chicago and Cambridge.[12] This problem solved, however, difficulties arose with the mounting of the telescope. An inadequate mounting made comet and asteroid position work impossible at this time, so Mitchell concentrated on observations of the surface features of Jupiter and Saturn. These were published in the *American Journal of Science*. She finally persuaded the college to provide the money to have the mounting rebuilt in 1887.[13] This allowed comet

The Hexagon, senior astronomy class at Vassar College in 1868. Mary
Whitney is seated at the table. Courtesy of Vassar College Archives.

and asteroid position measurements to become the major research interest
at Vassar as at most other women's colleges.

Despite these problems, Maria Mitchell gave her students training and
inspiration to become professional astronomers. A gifted teacher, she encouraged her students to question what they were taught and to learn for
themselves.[14] Brought up a Quaker and deeply involved in the women's
movement (she was president of the fourth Women's Congress in 1876 in
Philadelphia), she wrote: "I consider it one of my duties to the young
women who come into my department to encourage a respect for remu-
nerative occupation. Why should girls be brought up with an idea that
paid labor is ignoble?"[15] A letter from one of her students to another shows
that Mitchell communicated these beliefs to students and made them want
to seek astronomical careers even though they were discouraged by attitudes toward women in the outside world. Gertrude Mead, a student
studying for her M.A. at Vassar, wrote in 1870 to Mary Whitney, who later
became Mitchell's successor:

You think I am brave. Well, if I am, what are you? To be sure I am back here alone, but then I have Miss Mitchell and all these grand instruments, and nobody makes fun of it all here. But when I get home, no one there will take any interest in Astronomy. I shall have no telescope at first, and there will be no one there to help me on. Do you think I shall be brave enough then to hold on tight to what I have begun? When I think of it so, I get discouraged. Presently I think of it this way. There you have been away from the college two years. You have your telescope, and have commenced work, and now you have even gone out to Stafford. [Professor T. H. Stafford at the Dearborn Observatory of the University of Chicago] If you have held out, why shouldn't I? Then I feel better. . . . I believe in you thoroughly, and that is what I don't in many. You said last commencement day, "It is worth working for." I believed you, and I shall always cling to that sentence. . . . Will you help me to hold on, and take an interest in what I do in it? I never expect to accomplish much. I have thought it all over, and don't think I care at all to get fame. I want to work at it because I love it, and because it is so much better to have a purpose to live for than to have nothing.[16]

Two of Maria Mitchell's students, Mary Whitney and Antonia Maury, went on to become notable astronomers, and a number became professionals in other fields.

Maria Mitchell trained Mary Whitney (1847–1921) as her successor. Whitney graduated from Vassar in 1868 and received her M.A. there in 1872. She studied at Harvard in 1869–1870 and in 1872. Maria Mitchell persuaded the famous Harvard mathematician Professor Benjamin Peirce to invite Whitney to attend first his lectures on Quaternions (vectors) and then his senior course on celestial mechanics.[17] After two years of teaching high school in Massachusetts, Whitney accompanied her sister to Zurich in 1874. They remained in Zurich for two years, while Mary Whitney studied mathematics and her sister studied medicine. In 1881 Whitney became Maria Mitchell's assistant, and in 1888, when Mitchell retired, Whitney became professor of astronomy and director of the Vassar College Observatory.

Whitney inherited a heavy teaching load. By 1906–1907 there were eight different astronomy courses given, with a total of 160 students.[18] With her own funds, Whitney hired an assistant, Caroline Furness, who helped with the teaching and research from 1894 to 1910, and later succeeded Whitney as professor of astronomy. In an annual report, Whitney described their schedule: a graduate student "observed every evening when the telescopes are not in use by undergraduates. Frequently, when clear evenings have been few, she uses the glass up to 10 o'clock: then Miss Furness and I begin our special work, and carry it on to 1 or 2 a.m."[19]

Maria Mitchell had given most of her energy to obtaining the necessary equipment and to teaching, so it was left to Mary Whitney to establish scientific work at Vassar. Maria Mitchell had made some observations of surface features of Jupiter and Saturn and had determined the longitude of the Vassar College Observatory by exchange of clock signals with Harvard in 1877. Whitney started out at Vassar eager to begin original research, but shortly after her arrival, her sister's health broke down. Care of her sister, who became a permanent invalid and lived with her at Vassar, took up what free time Whitney had for research. Within a few years Whitney's mother and sister both died, and by 1894 Whitney had time to devote to science and to start an ambitious research program.[20] Whitney was an excellent mathematician, and therefore chose to concentrate on calculating orbits and making observations of minor planets. Benjamin A. Gould, editor of the *Astronomical Journal*, encouraged this work by publishing the results, and writing to Whitney: "I wish to congratulate you on the useful service your observatory is doing in these observations of asteroids which you are so persistently following."[21] Vassar could make a greater contribution by observing regularly a minor planet or variable star that was not regularly followed elsewhere than by observing comets, which attracted the attention of astronomers all over the world.

The strong leadership in the astronomy department of Vassar resulted in the largest amount of scientific work of any of the women's colleges. A bibliography of the publications of the staff of the Vassar College Observatory while Whitney was director lists 102 articles and other publications.[22] Fifty-nine of these papers were published in the *Astronomical Journal*, 18 in *Astronomische Nachrichten*, and 16 in *Popular Astronomy*, an impressive body of scientific work for any small college. Thirty of the notes and papers listed are observations of comets, 27 are observations of minor planets, and 22 concern variable stars. These subjects were taken up approximately consecutively: the most important work at Vassar was comet observations until the mid-1890s, then minor planets were emphasized, then in 1901 observations of variable stars were started. In addition to visual observing, the equipment was upgraded sometime in the 1890s by the purchase of a Repsold machine for making accurate measurements of stars on photographic plates. This allowed research on plates borrowed from other observatories (the equatorial telescope at Vassar was not suited for photographic work).[23] The observations necessary to calculate the orbits of comets and minor planets and the pattern of variation of variable stars were pieces of the puzzle astronomers were working on, although not an area of quick results or high prestige.

Since the women's colleges hired only a handful of their own graduates, the professors needed to develop not only their own research programs but also connections with the large observatories that could provide jobs for students. The large observatories started to employ women in the 1880s and 1890s. The women did the routine computation and measurement of photographic plates that became increasingly necessary as astronomy became more mathematical and more dependent on photography. Alumnae records show that Vassar was particularly successful in placing its graduates in this sort of work; at least nine found jobs at major observatories.

The basis of these institutional connections was personal friendships. Cooperation between Vassar and Columbia started in 1896 when Professor Harold Jacoby suggested that Caroline Furness make a detailed study of some plates of the region around the North Pole.[24] In 1900, Furness got her Ph.D. from Columbia for this work. The Yerkes Observatory accepted women for positions as volunteer summer assistants, and Furness worked there under Hale in 1900. She made many useful friendships; in 1903 Frank Schlesinger wrote her that she should encourage one of her students to apply to Professor Hale for a job at Yerkes as a computer on parallax work supported by the Carnegie Foundation. He added that conditions for boarding had improved since Furness's stay.[25] A connection with the Mount Wilson Observatory was established when Hale left Yerkes to found Mount Wilson, where he became the first director.

Letters from the large observatories show that they looked to Vassar when they wanted to hire women. George E. Hale, then at Yerkes, wrote to Mary Whitney in 1901 to ask: "Can you recommend any young women for work here in measuring and computing?"[26] Simon Newcomb of the Naval Observatory wrote in 1905: "I now have work for one or two good computers; if you have any that you can recommend for intelligence, accuracy and speed I should be very much pleased."[27] An engineer from the American Telephone and Telegraph Company inquired if Vassar graduates were trained for, and interested in, computing work in industry. He wrote: "Last winter, in talking with Mr. Walter S. Adams about the work at the Mount Wilson Solar Observatory, he said that a large part of the computing was done by Vassar graduates who had received their training in your department of Astronomy which had especially fitted them for such work."[28]

OTHER WOMEN'S COLLEGES

Institutional patterns at the other women's colleges similar to those at Vassar led to a number of strong teaching and research programs. Teaching of astronomy at Mount Holyoke actually began even earlier than at Vassar, but no original scientific work was done there until about 1900. Mount Holyoke was founded in 1837 as a female seminary, and only in 1888 became a college, although it strove from its founding to offer an education comparable to that offered at a men's college. The teacher of astronomy, first at the seminary and later at the college, was Elizabeth M. Bardwell. Mount Holyoke had been given a six-inch telescope in 1853, and in 1881 a trustee gave money for an eight-inch equatorial telescope (the last one made by the elder Alvin Clark), a three-inch meridian circle, and a building to house them, called the John Payne Williston Observatory.[29] Bardwell studied at Dartmouth in 1873–1874 and Charles A. Young, a notable astronomer whom she had known at Dartmouth and who later worked at Princeton, visited Mount Holyoke every year from 1869 to 1908 to give a course of lectures.[30] Young wrote of Bardwell:

> While perhaps she was not particularly rapid in her mental processes she was clear in her understanding and had an excellent ability to communicate her knowledge to others. She was therefore an admirable teacher, inspiring, diligent, conscientious, and "faithful unto death."[31]

After the death of Elizabeth Bardwell in 1899, Anne Sewell Young became professor of astronomy. The niece of Charles Young, Anne Young had graduated from Carleton College in 1892, taught school for three years, and received an M.S. degree from Carleton in 1897. Well trained in research, she had in fact published in astronomical journals before she was hired by Mount Holyoke.[32] In 1900, Young started a program of daily sunspot observations at Mount Holyoke, and after 1907 these observations were sent regularly to Zurich as part of a worldwide cooperative project. Other scientific research at Mount Holyoke included asteroid positions, comet orbits, and extensive work on variable stars. This was all that Young could do with the equipment available at first, except during some summers which she spent at Yerkes, and work she undertook at Columbia for which she received a Ph.D. in 1906. In 1912, Mount Holyoke purchased a Gaertner measuring engine for making measurements on photographic plates, which were borrowed from Yerkes.[33] Young was able to hire an assistant to help with the teaching and research starting in 1902.

Astronomers at Smith College pursued similar work. The Smith College Observatory was founded in 1887, and Mary Emma Byrd became the first director.[34] A graduate of the University of Michigan, Byrd had been the principal of a high school in Indiana, had studied for a year under Edward C. Pickering, the director of the Harvard Observatory, and had worked for five years as an assistant at the Goodsell Observatory of Carleton College. The Smith Observatory had an 11-inch equatorial telescope and a small meridian circle, but no regular income for maintenance and for purchase of books until an endowment was given by Elizabeth Haven in 1898. The equatorial lacked electric lighting for the micrometer (an instrument attached to the telescope eyepiece for measuring the angular distance between two objects such as a comet and a star) until students raised money for it by voluntary contributions in 1891–92.[35]

Byrd was a dedicated teacher who published two textbooks and many articles on the teaching of astronomy, but she also undertook scientific research. The worst problem, as at most small colleges, was teaching load; Byrd wrote in an annual report in 1892–93:

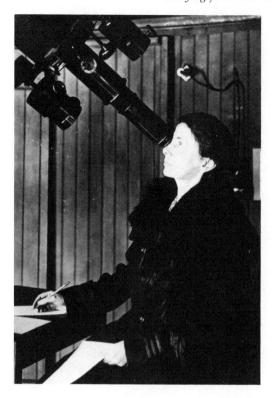

Anne S. Young. Courtesy of Mount Holyoke College Library/ Archives.

Except in bitter midwinter weather this year, there have not been more than three or four nights a term when I could handle a telescope for any purpose save to adjust for students until after ten o'clock at night; and my work begins in the morning at eight or half after eight. May I beg you to consider how short, under such circumstances, must inevitably be the time during which a teacher can keep fresh springs of inspiration for her students. I have put off the evil day by working late at night, on very cold nights, and by spending since last Commencement, seven weeks of vacation time at the observatory.[36]

An assistant was hired in 1893, but the number of students continued to grow. The main subject of research was comet positions, four sets of which were published in the *Astronomical Journal* from 1894 to 1904. Many observations of comet positions were necessary to calculate orbits, but comets generated so much excitement that astronomers all over the world made more than enough observations. When the *Astronomical Journal* did not publish observations that she sent them in 1905, Byrd wrote Pickering for help; "The truth is that we work so hard here to carry on a little independent work, living, I might almost say, for the sake of comets that come within

Astronomy class at Smith. Mary Emma Byrd is second from the right.
Courtesy of Sophia Smith Collection, Smith College.

the reach of our glass until observations and reductions are completed, that this attitude on the part of the *Journal* is no small disappointment."[37] With the help of a letter of introduction from Pickering the comet observations were finally published in the *Astronomische Nachrichten.*[38]

Byrd retired in 1906 and was succeeded by Harriet Williams Bigelow. Byrd's retirement came at the early age of 57 as a protest against the acceptance by Smith of what she considered to be tainted money, gifts from Rockefeller and Carnegie.[39] Bigelow had worked at Smith as an assistant since 1898 and had earned a Ph.D. from the University of Michigan in 1904. She kept a research tradition going at Smith and published seven papers, mostly on comets, in the *Astronomical Journal.*[40] Smith produced five women who pursued astronomical careers at women's colleges, but only two who are known to have worked for more than five years at a large observatory.

Two more eastern women's colleges, Radcliffe and Wellesley, offered instruction in astronomy and produced some important women astronomers, but these colleges produced little scientific research. At Radcliffe, astronomy was taught by Arthur Searle and John Edmands of Harvard College Observatory, and some students who showed scientific talent were hired by the observatory.[41] At Wellesley, Sarah F. Whiting developed one of the first undergraduate teaching laboratories in physics in the country, with the help of E. C. Pickering. She wrote of her experiences: "For many years I was almost alone in college work in this line meeting the somewhat nerve wearing experience of constantly being in places where a woman was not expected to be, and doing what women had not at that time conventionally done."[42] Whiting taught a course in astronomy, but Wellesley did not have an observatory until 1900, by which time Whiting was more dedicated to her teaching than to astronomical research.

The professors at Vassar, Mount Holyoke, and Smith all pursued some scientific research. That this research did not produce scientific work of enough value to bring them fame was due to other demands on their time and energy and to their choice of research topics. Observations of variable stars were valuable, but like comet and asteroid positions and spectral classification, large amounts of work could be put into determining light curves that contributed only a small amount of data toward the advancement of astronomical theories. An astronomer of the next generation, Cecilia Payne-Gaposchkin, said that in the first quarter of the twentieth century variable stars were considered a second-class problem.[43] The women at the women's colleges contributed small pieces to large research projects rather than pursuing new ideas.

The professors at the women's colleges chose research problems in the areas stereotyped as women's work because those topics could be pursued with the equipment they had available and because the male astronomers they consulted about their research programs suggested those topics. The advice women received is typified in a pamphlet written by Pickering to encourage women astronomers:

> The criticism is often made by opponents of the higher education of women that, while they are capable of following others as far as men can, they originate almost nothing, so that human knowledge is not advanced by their work. This reproach could be well answered could we point to a long series of observations as are detailed below, made by women observers.[44]

Pickering wished to show that women could contribute to human knowledge, but the work he advised them to do was still not very original. Pickering's advice to women reflected his belief in the importance of data-gathering and his distrust of theory, but it was also typical of the expectations that limited women astronomers, even at women's colleges.

The astronomy programs at the women's colleges succeeded at least in that women had the opportunity to acquire scientific training and a few women professors had opportunities for scientific research. However, the majority of graduates of women's colleges who went on to work in astronomy found jobs at observatories. The question, then, is whether those jobs allowed them to make good use of their training and to do original scientific work.

THE OBSERVATORIES

Women astronomers found two types of careers open: as professors in women's colleges, or as assistants in large observatories. The large observatories—Harvard, the Naval Observatory, Yerkes, Columbia, Mount Wilson—all hired women assistants. Harvard hired women first and employed the largest number of women, so it probably served as an example for the others. Harvard employed women as computers to do routine calculations and make measurements on photographic plates. Most of the women had limited scientific training, but those with more ability found some opportunities for creative work open to them. The most significant scientific contributions by women astronomers came from Harvard.

HARVARD

The example Maria Mitchell set by computing navigation tables at the Coast Survey may have encouraged Harvard to hire women. Mitchell was still working for the Coast Survey in 1857, when the director who had hired her, Commander Davis, was ordered to sea and replaced by Joseph Winlock. Winlock later became the third director of Harvard College Observatory, and Harvard hired the first three women in 1875, the year of his death. Although it appears that the women did not start work at Harvard until after his death, Winlock may have been influential in their hiring. More generally, the example of Maria Mitchell probably made the idea of hiring women to do computing at Harvard seem reasonable.

At least two of the first three women employed by Harvard had long connections with the observatory and were probably hired because of those connections. Anna Winlock grew up in the observatory residence while her father was director. Two months after her graduation from high school in 1875 she started work at the observatory, to help support her family after her father's sudden death that June.[45] Mrs. R. T. Rogers was also hired sometime in 1875. She was a relative of William A. Rogers, who supervised the women's work.[46] A special arrangement between William Rogers and Harvard's president, Charles W. Eliot, resulted in the hiring of the third woman, R. G. Saunders, in November of 1875. On November 23, Rogers wrote President Eliot: "In accordance with your request, I wrote to Miss Saunders, and this morning I have her reply."[47] On November 27, Eliot wrote to Arthur Searle, acting director of the observatory: "Will you have the kindness to engage Miss Saunders as a computer at the Observatory for one year at a salary of $600 a year."[48]

These three women were hired to reduce observations taken with a meridian circle that had been installed in 1870. The observatory used this instrument to measure star positions in the zone of the sky assigned to Harvard as part of the international project to revise the basic *Durchmusterung* star catalogue. Such catalogues had long been an important part of astronomical research, but they were growing larger and larger, requiring huge amounts of repetitive work. The women performed the calculations necessary to reduce the observations to standard coordinates. This involved repetition of a particular formula for each star, using tables of logarithms for multiplication. Women assistants were limited to this kind of work until photography became important. Men did similar computations as part of other work, but it appears that after 1880 more women than men were hired for jobs restricted to routine calculations.

Credit for Harvard's leadership in opening opportunities to women goes in large measure to Edward Charles Pickering, director of Harvard College Observatory from 1877 until his death in 1919. Pickering came from a wealthy family and had studied physics at the Lawrence Scientific School of Harvard University. He held a professorship of physics at the Massachusetts Institute of Technology when President Eliot invited him to become the director of Harvard College Observatory. Pickering had a Baconian concept of science; he believed that the Harvard Observatory could make the largest contribution to science by collecting large amounts of data. He was also a skillful entrepreneur who successfully raised money for the large projects he conceived. The need for many people to reduce data for these projects encouraged the hiring of women.[49]

Also, Pickering appears to have been sincerely interested in increasing opportunities for women. Margaret Harwood, who worked at the observatory during the last five years of his directorship, remembered Pickering's concern and courtesy with the women. When asked why he acted this way, she said: "Only because he could get them less expensively. Well, he was like the old-fashioned country gentleman, you might say."[50] A skillful administrator, he no doubt hired women because they worked for lower salaries and he believed they were more patient than men with the work he wanted done. But he took pride in the women at Harvard, and he encouraged other women astronomers by advising the professors at the women's colleges, by encouraging amateurs, and by founding a fellowship in 1916 for women to study at the Harvard Observatory.

In the next ten years the number of women at the observatory grew slowly. Pickering next hired Selina Cranch Bond, daughter of the first director of the observatory and sister of the second. She needed work because the money left to her by her father had been mismanaged, and she sent Pickering many letters asking for employment before he hired her in 1879. Pickering hired only two more women before 1885, Nettie A. Farrar and Williamina P. Fleming, hired in 1881. All of these women remained at

Harvard at least until 1885, but the opportunities for women to make important scientific contributions came later.

The year 1886 brought the founding of the Henry Draper Memorial, which resulted in a major expansion of opportunities for women at Harvard. Henry Draper, a wealthy New York doctor who spent most of his time doing astronomical research, in 1872 had taken the first photograph of the spectrum of a star (spectra had been observed visually for many years but never before photographed). When he died in 1882, his widow, who had always served as his assistant, attempted to carry on his astronomical investigations, with the cooperation of Pickering. However, the work had grown to a scale that required a major observatory, so in 1886 she agreed to fund research at Harvard which would continue her husband's work.[51]

Pickering set up the Draper Memorial as a large-scale project to classify the spectra of stars.[52] The data were recorded on photographic plates, and then the spectra of each of hundreds of stars on each plate had to be classified. This work was started by Nettie A. Farrar, continued by Williamina P. Fleming when Farrar left to get married at the end of 1886, and finally completed by Annie Jump Cannon. A contemporary account describes the other tasks involved: "In addition to this they record their observations, reduce the coordinates of objects examined, identify the objects photographed with the stars in various catalogues, and finally check the results by a direct comparison of the chart with the photograph."[53]

Because of this new line of women's work, amply supported by Anna Draper, and because of Pickering's success in raising money for more assistants to do computing, the number of women working at Harvard increased rapidly between 1885 and 1900. Twenty-one women were hired between these dates, leaving, when a few retirements are counted in, a staff of nineteen women in 1900. The previous training of only eight of the twenty-six women hired before 1900 is known; but of these only two, Antonia Maury and Annie Cannon (both of whom became famous), had graduated from college. Mabel Stevens took the Radcliffe entrance examinations, but financial circumstances forced her to find employment at the observatory instead.[54] Annie Masters had previously been a teacher and a bookkeeper,[55] and Florence Cushman had graduated from Charleston High School and worked for a business firm.[56] Desertion by her husband forced Imogen Willis Eddy, who was described as a lady of good family, to seek work.[57] When she applied to the observatory she had previous experience in astronomical work, as she had worked for two years for Benjamin A. Gould at the Coast Survey measuring stellar photographs, but she had never studied astronomy.[58]

The backgrounds of these women and the pay they received show that the jobs carried a fairly low status, but they were desirable compared to the alternatives open to women at the time.[59] Pickering hired a few women for their special skills, but most got their jobs by connections or on the basis of their character. Searle wrote one applicant about the skills required: "A knowledge of ordinary arithmetic and a legible handwriting are all the necessary qualifications of a computer, although, of course, the more that is known of languages and mathematics the better."[60] The observatory paid most women 25 cents an hour, and the women worked seven hours a day, six days a week, with one month paid vacation a year for full-time employees.[61] R. S. Saunders, and probably a few other full-time employees as well, received a salary of $600 a year, very slightly higher. The majority of women, however, were paid by the hour. No regular schedule of raises for seniority existed, although Pickering gave raises to a few women who showed particular skill.[62] The base rate remained 25 cents an hour at least through 1906.[63] The observatory usually paid men the same rate for computing as women, but men considered this an unacceptable wage, and few applied for jobs. Male assistants who did the observing at night and other mechanical work received a higher salary, $800 a year in 1874.[64]

Between 1900 and 1920 the staff of women at Harvard remained large. The telescopes at the Harvard Observatory could not compete with the larger telescopes being built in the Midwest and particularly on mountains on the West Coast, such as Yerkes and Mount Wilson. Therefore, to maintain its reputation, Harvard moved into the new field of astrophysics and put even more stress on photographic research. Some women computed for the male staff researchers, but most worked in the photographic department doing computing and analysis of objects on photographic plates.

Even after 1900, a graduate of one of the women's colleges who wished to pursue a career in astronomy would not find work at a large observatory an ideal next step, but at least it provided some possibility of a career.[65] Following the Harvard example, similar jobs for women became available at other observatories. Even without complete staff lists for some of the major observatories, I found 164 women who worked at observatories between 1875 and 1920. Most of them performed routine tasks, but a few published scientific papers and gained entrance to scientific societies.

LIMITATIONS OF OBSERVATORY WORK

Despite these opportunities, not all women were successful in pursuing astronomical careers. Many women started out with promise but did not end up in astronomical jobs. Patterns are hard to establish, except that women's college alumni records suggest that some women left to get married. However, the example of one unsuccessful woman, whose story happens to be well documented, reveals limitations of the opportunities open to women in this period.

Mary Wagner had studied at Vassar for two years when in 1893 she wrote to Pickering at Harvard College Observatory asking for a job. She wrote:

Before entering college, I taught five years and received a fair salary. I have no doubt that I can get nine or ten hundred dollars a year in some high school, but I would much prefer a six or seven hundred dollar position in an observatory, and if there is an opportunity to study or to advance myself in any way, I might be willing to take less.[66]

Pickering wrote back that there was a position, but it might not be suitable because the work required was partly clerical and partly routine computing, at a salary of $500 a year.[67] Wagner replied that she was willing to compute or do whatever else was required of her, but "I am more fond of practical work and have a good eye for seeing; I have used the telescope, microscope, and spectroscope a great deal."[68] Arthur Searle wrote back to her (because she was to be employed in his department), "I fear that there will be little opportunity for you here in the way of telescopic work," and suggested that she visit the observatory to find out more about the job.[69] After some deliberation, Wagner accepted the job, writing, "It hurts my pride to work for five hundred dollars, but I have a great desire to see your observatory and to work there for a year must surely be instructive to me."[70]

Mary Wagner worked at Harvard from September to the first week of December of 1893. The correspondence resumes in December, when she suddenly returned to her home in Minneapolis to take care of her seriously ill mother. From her home she wrote Searle to ask about her pay. She thanked him for his kindness and wrote: "It seems to me that I ought to beg your pardon for complaining to you so often about the work. I hope your new assistant, if you have one, may find joy in cataloguing stars and that she may have no other ambition than to do that well."[71] In her next letter Wagner asked whether it would be possible for her to return to work, because, while she did not like the work, she preferred not to leave the impression that she had been fired. She felt that she could be content if her duties were different:

If I can use the telescope evenings and work four hours a day on the catalogues, I will be satisfied. I should like to have some opportunity of showing you and Professor Pickering that I can do some good work. If the burden of poverty is not so heavy I shall be happier.[72]

She was not invited back, but Pickering sent her a letter of recommendation at her request.

Wagner wrote a final letter to the observatory in February, saying that she had been unsuccessful in finding a job teaching high school, but was giving private lessons and planning on studying zoology at the University of Minnesota. She wrote:

It breaks my heart though when I think of the wooden post that I watched with so much interest from my prison window. I should like to know whether the telescope has been mounted or not. If I had not been so very poor, I would have been delighted with the prospect for I like to observe and I have a good eye, but Astronomy must be left in the hands of the wealthy, while I turn to something that will give me a living.[73]

Vassar alumnae records show that Wagner received a B.A. from the University of Minnesota in 1897 and taught mathematics and science in high schools. In 1902 she opened an inn in Poughkeepsie, New York, which she owned and managed for twenty years. In a letter to Mary Whitney, Pickering gave his final view of the incident:

> Miss Wagner's work was very satisfactory to Professor Searle, to whose department it was mainly confined, but she found the monotony of ordinary computing too much for her, although care had been taken before she came to explain to her that there was no prospect of pecuniary returns for work of a more interesting character.[74]

Wagner may have had personal reasons for her dissatisfaction as well, but her complaints show limitations in the available jobs that others may have felt but not said.

Mary Wagner's letters show that a woman with good scientific training from a women's college might have trouble finding what she considered a suitably challenging job. The fact that women received recognition as scientists yet were at least partially confined to routine jobs shows the complex nature of women's work in astronomy. At most observatories women found themselves limited to jobs requiring painstaking attention to detail. While some women found this unacceptable, others found ways of making important contributions to science with the material they were given.

THE FAMOUS WOMEN AT HARVARD

The most notable achievements of women astronomers in the United States before 1920 were made by four women at Harvard College Observatory: Williamina Fleming, Antonia Maury, Annie Cannon, and Henrietta Leavitt. The opportunities available to women at Harvard to do original work may have been larger than at other observatories, but more important, Harvard had its own approach to astronomy which gave more room for women's work. Pickering stated his philosophy in his introduction to the *Henry Draper Catalogue* prepared by Annie Cannon. He wrote:

> In the development of any department of Astronomy, the first step is to accumulate the facts on which its progress will depend. This has been the special field of the Harvard Observatory. An attempt is made to plan each investigation on a scale that it will not be necessary to repeat it shortly, for a larger number of stars. Speculations unsupported by fact have little value, and it is seldom necessary in such investigations as are carried on here, to form a theory in

order to learn what facts are needed. An observer also is likely to be prejudiced if he has already formed a theory to which he thinks the facts should conform.[75]

Pickering found women ideal employees for such large-scale work requiring little theoretical sophistication. He was right in considering such investigations important, and the women who did them were praised by grateful astronomers. However, this approach did not allow the women to pursue independent theoretical work, as some had the ability and desire to do.

WILLIAMINA FLEMING

Williamina Paton Stevens Fleming (1857–1911) was the first of the Harvard women to become famous, and her background and education were different from the other notable women. Fleming was born in 1857 in Dundee, Scotland, the daughter of a prosperous craftsman. She attended the public schools there and worked as a student teacher. With her husband, James Orr Fleming, whom she had married the previous year, she immigrated to Boston in 1878. A year after coming to the United States her husband deserted her, and she found a job as a maid (or housekeeper) in the house of Harvard director Edward C. Pickering. In 1879 she had a son, Edward Pickering Fleming.

A legend has grown up about how Williamina Fleming came to work at the observatory. As Dorrit Hoffleit (who did not know her) tells the story she heard, Pickering adopted the system of objective prism photography of spectra, and many beautiful plates were being taken. He wanted to use the material to set up a spectral classification system, so he assigned some male assistants to study the plates. However, the work did not progress satisfactorily. One day Pickering became so annoyed with the lack of progress that he went out of the building in a huff, vowing that his Scottish maid could do a better job. And so he hired her.[76] Other evidence casts doubt on parts of this story. All Pickering himself wrote about Fleming's hiring in his official memorial was: "Mrs. Fleming began work at the Harvard Observatory in 1881. Her duties were at first of the simplest character, copying and ordinary computing."[77] When the Draper Memorial, which first made possible substantial work on spectra at Harvard, was founded in 1886, the first assistant to study spectra was Nettie A. Farrar. After a few months she left to get married, and Pickering wrote to Anna Draper that Farrar was "instructing Mrs. Fleming who has assisted me, and who will I think take her place satisfactorily."[78]

Fleming's most important scientific work was on the Draper Memorial. Pickering assigned her to study the spectra on the plates and develop an empirical classification system, which she would then use to classify stars for a catalogue. The usual system of cataloguing spectra at that time had been developed by Father Angelo Secchi of Rome on the basis of visual

The Women's Workroom at Harvard. Williamina Fleming is standing; to the left of her in the rear is Antonia Maury; the seated figure in front is Evelyn Leland. Courtesy of Harvard College Observatory.

observations. Secchi divided spectra into four classes, depending on which sets of spectral lines were brightest. Fleming found, however, that in the photographs of spectra she could distinguish more varieties, so she developed a system of twenty-two classes. In the *Draper Catalogue*, volume 27 of the *Harvard Annals*, Fleming classified 10,498 stars according to her system. Her classifications are dependable; in other words, each class has its identifying features. However, the order and division of the classes were limited by the fact that Fleming worked from plates with low dispersion spectra, usually less than an inch long, and the plates were occasionally of poor quality. A number of Fleming's classes later turned out to be defects in the photograph, and these extra classes made it difficult for her to arrange the spectra in a consistent order.

In her examination of the plates, first in her cataloguing work and then as Curator of Astronomical Photographs, Fleming found many unusual and interesting objects. She discovered ten novae, more than three hundred variable stars, and fifty-nine gaseous nebulae.[79] Pickering published numerous notices in the *Astronomical Journal* about stars with unusual spectra that Fleming had discovered. He took responsibility for the work, but always gave her credit for doing the actual studies. She wrote a paper for a 1898 conference at Harvard (which resulted in the founding of the Astronomical and Astrophysical Society). A notice of the meeting described the presentation of her paper:

> The contribution of Mrs. M. Fleming, read by Mr. Pickering, Director, on "Stars of the Fifth Type in the Magellanic Clouds" contained some statements in reference to the stars having spectra consisting mainly of bright lines, designated as Fifth Type. . . . In conclusion Professor Pickering said that Mrs. Fleming had omitted to mention that of these seventy-nine stars nearly all had been discovered by herself, whereupon Mrs. Fleming was compelled by a spontaneous burst of applause to come forward and supplement the paper by responding to the questions elicited by it.[80]

Fleming published two papers in the prestigious *Astrophysical Journal* in 1895 under her own name, but most of her results were published by Pickering in Harvard Circulars and the Harvard *Annals*.

In later years Fleming's duties became more administrative, but she was honored by many astronomical societies for her earlier work. She held the position of Curator of Astronomical Photographs at Harvard from 1899 to 1911, the first woman ever to receive a corporation appointment from Harvard. She sternly supervised and checked the work of the women assistants, examined the plates as they came in, corrected proofs for the Harvard *Annals*, and acted as Pickering's secretary.[81] She was made an honorary member (because women were not allowed to be regular members) of the Royal Astronomical Society in England in 1906, only the fifth woman member. She also belonged to the Astronomical and Astrophysical Society of America and the Society Astronomique de France.[82]

ANTONIA MAURY

Pickering assigned the next major project of the Henry Draper Memorial to Henry Draper's niece, Antonia Caetana de Paiva Pereira Maury (1866–1952). Daughter of a minister and naturalist, Mytton Maury, Antonia received a B.A. from Vassar in 1887. She started work at Harvard in 1888 and continued to do at least occasional research at Harvard until 1935, mostly as a volunteer.

Maury investigated the spectra of bright northern stars as part of the Draper Memorial from 1888 to 1896. Pickering wanted to improve Fleming's classification system by a study of very detailed spectra made with two or three prisms in front of the telescope so that the light of one bright star formed a large spectrum on the plate. He assigned this work to Maury. On these spectra the relative intensity of hundreds of lines could be studied to better understand the relationship between one spectral type and another. Being of independent character, Maury abandoned Fleming's classification system almost entirely. Maury developed her own system, taking into account both the arrangement of spectral lines and their width and sharpness.[83] The two-dimensional system she developed was the source of Ejnar Hertzsprung's discovery that stars of a given temperature (and therefore of the same spectral type by Fleming's and Cannon's systems) differed in size and luminosity. Maury had included the width and sharpness of the lines in her system because it seemed to give a more natural classification; it turned out that the differences in width and sharpness she had noticed resulted from differences in the size and luminosity of stars.

The importance of Maury's contribution was not recognized at Harvard. Maury's classification system was awkward to use because it used roman numerals for the main series and had many obscure subscripts. More important, the determination of the second classification, the width and sharpness of the lines, required the study of more than one plate for each star because an apparent fuzziness could be due to poor focusing or sky conditions. Pickering would not permit extra resources and effort to catalog a distinction whose importance he did not see.

In addition, Maury's system was not adopted because she did not stay at Harvard to apply it. She left the observatory in 1892 without completing her study because she was unable to work with Pickering. She taught school for two years, while Pickering anxiously tried to arrange for her to complete her work or agree to turn it over to someone else. Maury wrote to Pickering in 1892 about closing up her work:

> I am both willing and anxious to leave it in a satisfactory condition, both for my own credit and in honor of my uncle. I do not think it is fair to myself that I should pass the work into other hands until it can stand as work done by me. I do not mean that I need necessarily complete all the details of the classification, but that I should make a full statement of all the important results of the investigation. I worked out the theory at the cost of much thought and elaborate comparison and I think that I should have full credit for my theory of the relations of the star spectra and also for my theories in regard to Beta Lyrae. Would it not be fair that I should, at whatever time the results are published receive credit for whatever I leave in writing in regard to these matters.[84]

Pickering wrote back that "It is the regular practice of this Observatory to make full acknowledgement in its publications of the credit due to authors of particular portions of them."[85] Maury wanted more than the standard acknowledgment; her work was the first volume of the *Annals* to have a woman's name on the title page (Fleming had received only an acknowledgment for the first *Draper Catalogue*). Pickering finally had Maury sign an agreement to finish or turn over the work by the beginning of 1893.[86] Maury was still not finished in December of 1894, and still unable to work comfortably under Pickering. After one incident her father wrote the following letter:

> I was greatly concerned last night, when my daughter came home, to find from her that you had spoken to her in a tone and manner which greatly hurt her feelings. I cannot believe that you were conscious of it, for I am quite sure that you would not willingly have spoken as you did. Of course my daughter is inexperienced in preparing matter for the press, and is far from unwilling to listen to any suggestions. She is glad to receive them. But you may judge of the impression which you have produced, when I tell you that I am convinced, were it not for other considerations, she would never enter the Observatory again as your assistant and seeing how she feels I should not be willing that she should. I scarcely need say to you that though she is working in the Observatory, she is a lady and has the feelings and rights of one. I am sure you cannot wish to injure either.[87]

Despite these problems the study was finally published in volume 28 of the Harvard *Annals* in 1897.

In part, the problem arose from conflicting approaches to science. Pickering did not appreciate the detailed, theoretical classification system Maury wished to develop. In an undated letter, probably written after her study of spectra had been finished, Maury wrote:

> But although I several times before have taken offense at things you have said to me I have always decided in the end that the only trouble was that I, being naturally unsystematic was not able to understand what you wanted and that you also, not having examined minutely into all the details, did not see that the natural relations I was in search of could not easily be arrived at by any cast iron system.[88]

Her search for a natural classification system conflicted with Pickering's philosophy that "it is seldom necessary in such investigations as are carried on here, to form a theory in order to learn what facts are needed."[89] Dorrit Hoffleit, who knew Maury in the 1930s, said that Pickering "did not think as highly of her as she merited" and that "I am convinced that Miss Maury was an original thinker."[90] Cecilia Payne-Gaposchkin, who came to Harvard in 1923, described Maury:

> Of course, I only knew her when she was old. I was very fond of her, but she just talked and talked and talked and talked. You couldn't do any work because she wanted to talk so much. It was just she needed an outlet; she needed to discuss. Nobody had ever listened to her, nobody had ever responded to her scientific questionings, I think.[91]

Although Maury's contributions were not appreciated at Harvard, they did have a significant influence on scientists elsewhere. The astrophysicist Ejnar Hertzsprung recognized the importance of Maury's two-dimensional classification system, and wrote to Pickering to protest the omission of Maury's second dimension, which she had indicated with the letters a to c, from a later catalogue (in his own uncertain English):

> But in one respect I have been disappointed, and I allow me directly to say a few words on that point. On my opinion the separation by Antonia C. Maury of the c- and ac- stars is the most important advancement in stellar classification since the trials by Vogel and Secchi. But in the new catalogue the spectra of some of them as Alpha Cygni and Delta Cephei are not even mentioned as peculiar. It is hardly exaggerated to say that the spectral classification now adopted is of similar value as a botany which divide the flowers according to their size and color. To neglect the c- properties in classifying stellar spectra, I think, is nearly the same thing as if the zoologist, who had detected the deciding differences between a whale and a fish, would continue in classifying them together.[92]

Having recognized that the differences that Maury had seen were crucial, Hertzsprung used that information as an essential step in his contribution to the discovery of the Hertzsprung-Russell diagram, which forms the cornerstone of modern stellar astrophysics.[93]

Maury's other major scientific investigation was a study of the spectroscopic binary Beta Lyrae. In 1889 Pickering had discovered the first spectroscopic binary, a star which looks single through the telescope but which shows itself to be double by a periodic doubling of the spectral lines (when one star is moving toward the observer and the other is moving away the doppler shift causes their spectral lines to separate). Maury discovered the second such star, Beta Aurigae, shortly thereafter. She studied the particularly complex spectroscopic binary Beta Lyrae for many years. She did not work steadily at Harvard on this research; she spent time studying spectra visually at Vassar (in 1896), teaching high school, giving lectures, and pursuing her interest in ornithology, but over the years she investigated this star in great depth. Beta Lyrae is a close double star, so close that the two stars are nearly in contact. The spectrum of the system also shows variations of longer term than the period of the stars, and an irregular variation which Maury could not explain.[94] Maury made major contributions to the understanding of this star and visited Harvard annually to check on its behavior until 1948.[95] But as Cecilia Payne-Gaposchkin said of Maury in 1977: "She had a flair for picking out tremendous problems. If you look at the literature for Beta Lyrae you will see that nobody has ever solved the problem."[96]

Women at the 1920 meeting of the American Astronomical Society at
Smith College. Left to right: Margaretta Palmer (Yale), Sarah Whiting
(Wellesley), Harriet Bigelow (Smith), Mary Hopkins (Smith), Anne
Young (Mount Holyoke), Helen Swartz (South Norwalk, Conn.), Annie
Cannon (Harvard), Antonia Maury (Harvard), Caroline Furness (Vas-
sar), Leah Allen (Wellesley), D. Peck (Smith?), Vera Gushee (Smith).
William Henry, courtesy of *Sky and Telescope*.

ANNIE JUMP CANNON

The spectroscopic work that Maury abandoned was continued by Annie
Jump Cannon (1863–1941), who had already started work on the southern
stars before Maury finished the northern. Daughter of a prosperous ship-
builder and state senator, she graduated from Wellesley in 1884. Partially
deaf from an early age, she spent the next ten years at home in Dover,
Delaware, as a dutiful daughter, though she wrote in her journal:

> I am sometimes very dissatisfied with life here. I do want to accomplish some-
> thing so badly. There are so many things that I could do if only I had the
> money. And when I think that I might be teaching and making the money,
> and still all the time improving myself it makes me feel unhappy and as if I
> were not doing all I can.[97]

In 1894, after the death of her mother, Cannon returned to Wellesley as
an assistant in the physics department, and in 1895 she studied astronomy
as a special student at Radcliffe.[98] In 1896 she started work at Harvard
College Observatory, where she remained until 1941.

Cannon's investigation of the bright southern stars was published in

Physics class at Wellesley, 1895–96. Annie Jump Cannon is third from the left; Sarah Frances Whiting is fourth from the left. Courtesy of Wellesley College Archives.

the same volume of the *Annals* as Maury's classification of the spectra of bright northern stars. In this third major Draper study the stars were classified by a third spectral classification system, developed by Cannon from her detailed study of spectra photographed at the Harvard station in Arequipa, Peru. Pickering explained: "In all three cases, it was deemed best that the observer should place together all stars having similar spectra and thus form an arbitrary classification rather than be hampered by any preconceived theoretical ideas, or by the previous study of visual spectra by other astronomers."[99] He explained that the different classification systems could be freely translated. Cannon rearranged Fleming's classes to the sequence which Maury had discovered, eliminated the spurious classes, added numbers for finer gradations, and dropped Maury's second dimension referring to the sharpness of the lines except when the star was unusual enough to be marked peculiar. The sequence of classes, which Fleming had labeled alphabetically, became O, B, A, F, G, K, M.

Having developed a satisfactory classification system, Cannon next set about applying it on a large scale. The first volume of her *Henry Draper Catalogue* was published in 1918 as volume 91 of the Harvard *Annals*. The entire catalogue contained 225,300 stars, and when it was finished she started work on an extension. The stars were catalogued according to the system she developed for the bright southern stars. For the large *Henry Draper Catalogue*, she worked from low-dispersion objective prism plates, with hundreds of spectra on each plate. Cannon would examine the plate with a magnifying lens and call out her identifications to an assistant, who would write them down. In the forty-five years Cannon work at the observatory, she spent a large part of her time classifying stars, and she learned to identify spectra almost instantaneously. Her record books reveal that she could classify spectra at a rate of more than three stars a minute.[100] All the classifications in the catalogue were done by Cannon, which resulted in a consistency that was one of the reasons for its great value. In a reexamination of the problem of classification of spectra in 1935 by Russell, Payne-Gaposchkin, and Menzel, the authors survey the physical processes which result in the differences in lines which are the criteria for the different spectral types and conclude: "Multifarious as these criteria are, they express the most conspicuous features from type to type. It is doubtful whether more outstanding bases for classification could be selected."[101]

Much of Cannon's scientific work was classification, but her work was not just routine. Cecilia Payne-Gaposchkin wrote: "Miss Cannon was not given to theorizing; it is probable that she never published a controversial word or a speculative thought. That was the strength of her scientific work—her classification was dispassionate and unbiassed."[102] This style, together with her reportedly charming personality, explains why she worked more smoothly with Pickering than did Maury. While Cannon did not make significant theoretical investigations, her work was not limited to routine classification. Margaret Mayall, one of Cannon's chief assistants, said that while Cannon's main work was somewhat routine, "her perception of the spectra and interpretation of them" were amazing and she had great skill at "seeing unusual things. She had wonderful eyes and she could see things that very few people would recognize until she pointed it out."[103] Because of this ability she made many significant discoveries of new and unusual stars.

Like Fleming, Cannon was honored by the astronomical community. She belonged to, among others, the American Philosophical Society, the International Astronomical Union, and the American Astronomical Society (as the Astronomical and Astrophysical Society was renamed), of which she served as treasurer. She was also an honorary member of the Royal Astronomical Society. At Harvard she held the position of Curator of Astronomical Photographs after Fleming's retirement, but the Harvard administration was reluctant to give her a corporation appointment. Harvard President Lowell wrote: "I always felt that Mrs. Fleming's position was somewhat anomalous, and it would be better not to make a practice of treating her successors in the same way."[104] As early as 1911 the Visiting Committee to the Observatory wrote of Cannon: "It is an anomaly that,

H.D. 159,257–162,147   ANNALS OF HARVARD COLLEGE OBSERVATORY
H.D. 320,086–320,725   CHART 182

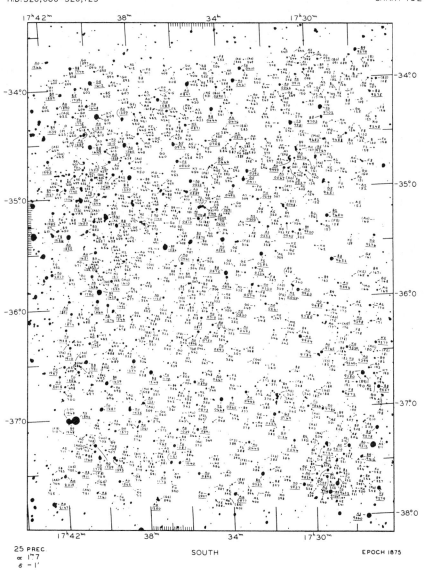

The second volume of the Henry Draper Extension was published in the form of star charts with the spectral type and HD number marked on the chart. This page gives an idea of the complexity of the task. *Annals,* vol. 112, p. 202.

A group of staff members at the Harvard College Observatory circa 1925. Left to right: (front) Irene Crossman, Mary B. Howe, Harvia H. Wilson, Margaret L. Walton (Mayall), Antonia C. Maury; (middle) Lillian L. Hodgdon, Annie J. Cannon, Evelyn F. Leland, Ida E. Woods, Mabel A. Gill, Florence Cushman; (rear) Margaret Harwood, Cecilia H. Payne (Gaposchkin), Arville D. Walker, Edith F. Gill. Courtesy of Harvard College Observatory.

though she is recognized the world over as the greatest living expert in this line of work, and her services to the Observatory are so important, yet she holds no official position in the university."[105] In 1925 Cannon became the first woman to receive an honorary degree of Doctor of Science from Oxford University in England. It was not until 1938, however, that Harvard gave her a corporation appointment as William Cranch Bond Astronomer.

HENRIETTA LEAVITT

The last of the famous women astronomers who did their work at Harvard before 1920 was Henrietta Swan Leavitt (1868–1921). The daughter of a minister, Leavitt attended Radcliffe College and received her degree in 1892. She worked as a volunteer at Harvard Observatory in 1895–96 and did some traveling, and then spent a number of years at home in Wisconsin

Henrietta Leavitt. Courtesy of Harvard College Observatory.

because of an illness that left her partially deaf. In 1902 she wrote to Pickering about continuing her work, and he persuaded her to return to Cambridge by offering her 30 cents an hour.[106] She worked at Harvard almost continuously from 1902 until her death in 1921. She showed great ability in her assigned work, but Pickering gave her little opportunity to use her talents for theoretical studies.

Leavitt's chief research was in photographic photometry, the problem of determining the magnitude (brightness) of a star from a photographic image. This is difficult because of the complexity of the photographic process itself (the darkness of the star image on the negative is not linearly proportional to the brightness of the star), and because each telescope and type of plate give different results on account of the differing colors of stars. Leavitt first determined a sequence of standard comparison stars in the region around the North Pole. Once a sequence was determined, magnitudes could be estimated by comparing one star with another and the problem of the actual relationship between brightness and the density of the image on the photographic plate could be avoided. The results of this research were published in volume 71 of the Harvard *Annals*. Later, Leavitt studied color indices, the difference in magnitude of a star depending on the color sensitivity of the photographic plates.[107] Pickering also gave her the difficult job of determining the corrections necessary for various telescopes, films, and filters.

Leavitt's investigations of variable stars were of much greater theoretical importance. She studied photographs of the Magellanic Clouds taken in Peru and discovered 1,777 new variable stars.[108] She determined the period of a few of these and discovered a relationship between their period and their brightness. Because the stars were all in the Small Magellanic Cloud, she was able to draw a crucial conclusion: "Since the variables are probably nearly the same distance from the Earth, their periods are apparently associated with their actual emission of light, as determined by their mass, density, and surface brightness."[109] She did not pursue this discovery further, and it was left to Hertzsprung to calibrate the curve of this period-luminosity relationship so that it could be used to determine the actual distance of stars from the earth. Pickering's replacement as director of the Harvard College Observatory, Harlow Shapley, used the relationship in his determination of the size of the galaxy, his most important contribution to astronomy.

If Leavitt had been free to choose her own research projects, she might have investigated the consequences of the period-luminosity relationship she had discovered. Leavitt certainly had scientific talent; Margaret Harwood, who worked with Leavitt, said that she had the best mind at the observatory.[110] Cecilia Payne-Gaposchkin described Leavitt: "I think she was the most brilliant of all the women [at Harvard]. I think she was certainly someone who would have considered that duty was foremost. She wouldn't even have complained, as Miss Maury apparently did."[111] Payne-Gaposchkin wrote in her autobiography of the complications of Leavitt's scientific work comparing color indices and magnitudes on plates from different telescopes:

> It may have been a wise decision to assign the problems of photographic photometry to Miss Leavitt, the ablest of the many women who have played their part in the work of Harvard College Observatory. But it was also a harsh decision, which condemned a brilliant scientist to uncongenial work, and probably set back the study of variable stars for several decades. . . . Pickering was wise in his day to invest his finest workers, and the time and resources of his Observatory, in laying the foundations of photographic photometry. Thirty years later it would be superseded.[112]

It is impossible to say what someone would have done if conditions had been otherwise, but Leavitt was certainly restricted by her position as an employee expected to do assigned work, so that she could not choose the scientific problem she wished to pursue.

The stories of these four women astronomers show that the women at Harvard were limited by the same approach that had resulted in the hiring of large numbers of women. Because Pickering believed the accumulation

of data to be more important than theoretical research, there were opportunities at Harvard for women whose talents ran toward patient and skillful cataloguing. Those women who might have had new ideas and made
theoretical contributions, however, were frustrated by the work they were
assigned and the lack of interest in their original investigation. Cecilia
Payne-Gaposchkin said:

> I remember Miss Maury saying to me, rather sadly, "I always wanted to learn
> the calculus but Professor Pickering didn't wish it." . . . I think Pickering hired
> people to do a specific job and didn't want them wasting their time doing
> anything else.[113]

Harvard College Observatory under Pickering was dedicated to one particular style of astronomy, and women were both limited and benefited by
this.

WOMEN ENTER THE MAINSTREAM

In 1925 Cecilia Payne-Gaposchkin received the first Ph.D. awarded to a
student at Harvard College Observatory; this marked the beginning of a
trend away from separate women's work in astronomy. Astronomy had
gained a reputation as being hospitable to women, and it continued to
attract a larger proportion of women than did the other physical sciences.
In the period after 1925 women faced frequent discrimination, but they
gradually moved beyond the research areas that were defined as women's
work. This chapter has been directed primarily to an examination of the
phenomenon of separate women's work astronomy, but it is worth looking
briefly at developments during the second quarter of the twentieth century.

When doctoral degrees became part of the career pattern for women
astronomers, their opportunities widened somewhat. A few women had
earned doctorates before 1920, but it was only later that the doctorate became standard. From 1923 to 1930 fifteen women earned Ph.D.'s in astronomy, making up more than 25 percent of the earned degrees in the
field. This high level of participation continued for the next two decades,
but in the 1950s and 1960s women earned only about 10 percent of the
doctoral degrees in astronomy.[114] Women astronomers with Ph.D.'s found
that many jobs were still not open to them because they were women, but
they commanded more respect and better positions than the traditional
computing jobs.

To some extent these women continued to work in those branches of
astronomy in which women had originally been encouraged. Helen Sawyer
Hogg, who earned a Ph.D. at Harvard College Observatory in 1931, studied
variable stars. She worked with her husband, first at the Dominion Astro-

physical Observatory in British Columbia, where she was not eligible for a job because her husband held one, then at the University of Toronto's David Dunlap Observatory. She eventually received the rank of professor in 1957, the same year she was elected president of the Royal Astronomical Society of Canada. Her major work was searching for and cataloging variable stars in globular clusters. These dense clusters of stars were formed early in the history of the galaxy, and studying their variable stars was an important way of learning about them.[115]

Dorrit E. Hoffleit also worked on variable stars, first as an assistant at Harvard. Harlow Shapley, director of the observatory, encouraged her to pursue the Ph.D., which she received in 1938. After working for the government and for Harvard, she became a research associate at Yale University in 1956 and also director of the Maria Mitchell Observatory, a small research institution on Nantucket. Much of her research was on variable stars and spectra, although she also worked on meteors and galactic structure.[116]

Women tended to stay in the astronomical fields traditionally assigned to women—variable stars and stellar spectra. They made solid contributions to the slow growth of knowledge in these fields, but had little opportunity to make major breakthroughs. Those patterns changed substantially only in the 1950s. Vera Rubin was perhaps the first important woman to make major contributions in other fields. She earned a Ph.D. from Georgetown in 1954 and studied galaxies and galactic dynamics, and also spectroscopy. She became a staff member of the Carnegie Institution Division of Terrestrial Magnetism in 1965.

One woman played an outstanding role in the development of astronomy during the interwar period. Cecilia Payne-Gaposchkin (1900–1979) received her undergraduate degree from Newnham College of Cambridge University and then went to study at Harvard, because she had been told that opportunities were greater for women astronomers in the United States than in Britain. She arranged to get a Ph.D. from Radcliffe College for her work at Harvard College Observatory.[117] Her dissertation was described in a history of twentieth century astronomy as "undoubtedly the most brilliant Ph.D. thesis ever written in astronomy."[118] She found a way to accurately relate spectral classes of stars to their actual temperature by taking into account the different abundances of the different ionized states of a particular element as shown by relative line strengths in their spectra. This work used the spectral classifications of Annie Cannon and the more detailed observations of Antonia Maury, but it was neither a new set of observations nor a new classification scheme. Rather, it was a major contribution to the theoretical understanding of the physics of stars, showing how the ionization theory developed by the physicist Meghnad Saha could be used to analyze the variety of spectra observed.

Payne-Gaposchkin continued in a distinguished research career, during which she published over 150 papers and 4 books. Her major research was on the spectra of extremely large, luminous stars and various types of variable stars. She also spent some time continuing Henrietta Leavitt's work on stellar photometry, establishing a standard scale for the magnitudes and colors of stars. Payne married Sergei Gaposchkin in 1934 and they collaborated on studies of variable stars. One project was the analysis of all known bright variable stars using the Harvard photographic plate collection, about 2,000 stars in all. She applied this comprehensive data to fit variable stars into theories of the structure of the galaxy and the evolution of stars. Her obituary in *Physics Today* makes the curiously stereotyped claim that "Most astronomers will agree with her self-assessment, namely that she was not one to fashion new theories, but contributed by collecting, turning over in her own hands, comparing and classifying the data of astronomy."[119] Yet her concern with how the data fit into the theories stands in sharp contrast with the simple data collection of most of the earlier generation of women astronomers.

Payne-Gaposchkin spent her entire career at Harvard despite the fact that official recognition came very slowly. She was given a corporation appointment and the title "astronomer" only in 1938. In 1956 the observatory finally gave her a professorship, the first full professorship not specifically limited to women given to a woman at Harvard.[120]

The change from women's work to a less differentiated role for women in astronomy had both benefits and costs. An exceptional scientist like Cecilia Payne-Gaposchkin could work on more important research problems. Yet without an accepted area of women's work, women were probably more isolated and faced more job discrimination. Perhaps this was part of the reason for the decline in the representation of women in astronomy in the 1950s and 1960s.

CONCLUSION

Before the 1960s or 1970s astronomers accepted women as employees at least in part for economic reasons. Women performed tedious work with great patience for low wages. The character of astronomy at the time resulted in unusual opportunities for women. The semi-independent status of observatories also helped make the hiring of women possible. The other sciences that had a large percentage of women—botany, zoology, and anthropology—were associated with museums, which had a semi-independent status similar to observatories and a similar need to process large amounts of data.[121]

Talented women at the observatories were accepted as colleagues because they performed, and performed well, a certain kind of work that was considered women's work. Probably men (and women) developed their assumptions about what was women's work from the sort of work they saw the ordinary women employees doing. Women executed projects set up for them by others; the talents for which they received praise were not originality or scientific thinking but patience and keen perception. Women with the appropriate talents found great opportunities open to them, but those whose talents lay elsewhere found that these opportunities were restrictions. Williamina Fleming and Annie Cannon were rightly praised for their catalogues and their discoveries of unusual stars, but Antonia Maury and Henrietta Leavitt received far less recognition for their contributions to theoretical understanding. It was an age in which theoretical work was not yet central to American astronomy, but others became famous for completing the theories for which these women laid the groundwork.

The women at the women's colleges undertook what was in essence the same sort of women's work, and for this they too were accepted by the men as colleagues. The work of the professors at the women's colleges required the same talents, patience and keen perception, as the work of the notable women at the observatories. The professors at the women's colleges observed the positions of comets and asteroids and the brightness of variable stars, all small parts of large compilations of data planned by someone else. These women were dedicated to research, yet they were no more successful in being full participants in science, by modern definitions, than most of the women at the observatories. They were limited by the burden of heavy teaching loads and by research problems they were advised to work on by their male colleagues. This advice was no doubt given with the best intentions, but the research suggested conformed to the stereotypes of women's work in astronomy.

On the whole, unusual opportunities were open to women in astronomy in this period, and they did make important contributions to science. That women were accepted as scientists at all, even if in limited ways, shows that American scientists in the late nineteenth and early twentieth centuries were somewhat open-minded about intellectual and professional opportunities for women. Women astronomers were prevented from doing the same sort of original scientific work that the best male scientists were doing, because of the expectations and institutional framework that surrounded them. Their work has, however, lasted; in fact Annie Cannon's catalogue of the spectral types of stars is of more value today as a collection of data than any number of creative and original theories that are now outmoded. The combination of availability of scientific training in the women's colleges and of jobs for women in the large observatories resulted in a large amount of valuable work done by women in astronomy, even if those very opportunities limited the types of contribution women could make.

NOTES

1. David H. DeVorkin, "A Sense of Community in Astrophysics: Adopting a System of Spectral Classification," *Isis* 72 (1981): 29–49.

2. Solon I. Bailey, *The History and Work of Harvard Observatory, 1839 to 1927* (New York: McGraw Hill, 1931), p. 116.

3. For further information on the growth and support of astronomical institutions in the nineteenth century see Howard S. Miller, *Dollars for Research: Science and Its Patrons in Nineteenth Century America* (Seattle: University of Washington Press, 1970), particularly chapter 5.

4. Deborah Jean Warner, "Women Astronomers," *Natural History* (May 1979): 12–26.

5. Margaret Rossiter, *Women Scientists in America: Struggles and Strategies to 1940* (Baltimore: Johns Hopkins University Press, 1982), p.53.

6. Statistics from Pamela E. Mack, "Women Astronomers in the United States, 1875–1920," A.B. thesis, Harvard University, 1977.

7. Amateurs still make important contributions to astronomy because of the large number of interesting stars that need to be observed. In the nineteenth century, before the rise of big observatories, some of the most important astronomers were amateurs.

8. Bessie Z. Jones and Lyle G. Boyd, *The Harvard College Observatory: The First Four Directorships, 1839–1919* (Cambridge, MA: Belknap Press of Harvard University Press, 1971), p.385.

9. Helen Wright, *Sweeper in the Sky: The Life of Maria Mitchell, First Woman Astronomer in America* (New York: Macmillan Co., 1950), p. 71.

10. Helen Lefkowitz Horowitz, *Alma Mater: Design and Experience in the Women's Colleges from Their Nineteenth-Century Beginnings to the 1930s* (Boston: Beacon Press, 1984), pp. 37–40.

11. Maria Mitchell, "Annual Report, June 12, 1868" (in the form of a hand-written letter), Vassar College Archives, Poughkeepsie, N.Y.

12. Maria Mitchell, "Annual Report, June 15, 1869," Vassar College Archives, Poughkeepsie, N.Y.

13. Maria Mitchell, "Annual Report, 1887–88," Vassar College Archives, Poughkeepsie, N.Y.

14. Phebe Mitchell Kendall, ed., *Maria Mitchell: Life, Letters, and Journals* (Boston, Lee and Shepard Publishers, 1896), p. 179.

15. Maria Mitchell, "The Collegiate Education of Girls," paper read in October 1880 at the Congress of the American Association for the Advancement of Women and published in Anna C. Brackett, ed., *Women and the Higher Education* (New York: Harper and Brothers, 1893), p. 69. For analysis of Mitchell's role in the women's movement see Sally Gregory Kohlstedt, "Maria Mitchell and the Advancement of Women in Science," in Pnina Abir-Am and Dorinda Outram, *Uneasy Careers and Intimate Lives: Women in Science, 1700–1945* (New Brunswick: Rutgers University Press, 1986), pp. 129–46.

16. Gertrude Mead to Mary Whitney, Sept. 29, 1870, Vassar College Archives, Poughkeepsie, N.Y.

17. Caroline E. Furness, "Mary W. Whitney," *Popular Astronomy* 30 (1922): 597–608 and 31 (1923): 25–35, see especially pp. 601–602.

18. Mary Whitney, "Annual Report, 1906–7," Vassar College Archives, Poughkeepsie, N.Y.

19. Mary Whitney, "Annual Report for 1896–7," Vassar College Archives, Poughkeepsie, N.Y.

20. Furness, "Mary W. Whitney."

21. Ibid., p. 26.

22. Caroline Furness, *Observations of Variable Stars Made During the Years 1901–12 Under the Direction of Mary W. Whitney,* Publications of the Vassar College Observatory, No. 3 (Poughkeepsie, N.Y.: 1913), pp. 216–17.

23. Mary Whitney, Introduction to *Catalogue of Stars Within One Degree of the North Pole,* Publications of the Vassar College Observatory, No. 1, by Caroline E. Furness (Poughkeepsie, N.Y.: 1900), pp. iii–iv.

24. Furness, "Mary W. Whitney," pp. 26–27

25. Frank Schlesinger to Caroline Furness, June 9, 1903, Vassar College Archives, Poughkeepsie, N.Y.

26. George E. Hale to Mary Whitney, June 12, 1901, Vassar College Archives, Poughkeepsie, N.Y.

27. Simon Newcomb to Mary Whitney, November 27, 1905, Vassar College Archives, Poughkeepsie, N.Y.

28. George A. Campbell to Vassar Department of Astronomy, May 25, 1912, Vassar College Archives, Poughkeepsie, N.Y.

29. The meridian circle is a small telescope in a special mounting designed to make accurate measurements of the positions of stars.

30. The dates are 1868–1903 according to the *Dictionary of Scientific Biography,* s.v. "Charles Augustus Young" by Richard Berendzen and Richard Hart, but Mount Holyoke records show 1869–1908.

31. Quoted in a memorial for Elizabeth M. Bardwell, Mount Holyoke Archives, South Hadley, Massachusetts.

32. Anne Sewell Young, "The Leonids," *Popular Astronomy* 4 (1896–97): 498–99; "Elliptic Elements of Comet g1896," *Astronomical Journal* 17 (1896–97):192. In the 1890s Carleton College had a full graduate program in astronomy, which trained a number of women. See Mark Greene, "A Science Not Earthbound: A Brief History of Astronomy at Carleton College" [pamphlet] (Northfield, Minn.: Carleton College, 1988).

33. Alice H. Farnesworth, "Astronomy at Mount Holyoke" (short typewritten paper dated July 3, 1950), Mount Holyoke Archives, South Hadley, Massachusetts.

34. At least two other candidates were considered: Winifred Edgerton, a graduate of Wellesley and student at Columbia, and Anna Winlock, a computer at Harvard. The position was as director of the observatory but not professor of astronomy; Mary Byrd mentioned in a letter "It is, I understand, contrary to the traditions of this institution to give any woman the title and pay of professor." Mary Byrd to E. C. Pickering, November 24, 1888, Harvard College Observatory Correspondence, Harvard Archives, Cambridge, Mass. (hereafter abbreviated HCO Corr.), UA V 630.17, group 1, folder Bua-By.

35. Mary Byrd, "Report of Smith College Observatory for the year 1891–2" (handwritten), Smith College Archives, Northampton, Massachusetts.

36. Byrd, "Report of Smith College Observatory for the year 1892–3," Smith College Archives, Northampton, Massachusetts.

37. Mary E. Byrd to E. C. Pickering, June 2, 1905, HCO Corr. UA V 630.17, group 2, folder Bua-By.

38. Mary Byrd and Harriet W. Bigelow, "Observations of Comets," *Astronomische Nachrichten* 169 (1905): 191.

39. Byrd did not discuss the reasons for her retirement in public, but after her death a friend recalled that "In accepting large sums from the various educational

foundations, she felt that colleges and universities limited in a measure their freedom of expression on economic problems." Article by Louise Barber Hobbit quoted in Richard Serena, "Mary Emma Byrd, 1849–1934, Professor of Astronomy and Author," June 1962, typescript, Smith College Archives. Byrd retired to her home in Kansas, where she spent most of the rest of her life, writing occasional articles on the teaching of astronomy. In a letter to a friend written fourteen years after she left Smith, she wrote:

> It is almost 40 years to the month since I met my first class in astronomy, at Wabash High School, Indiana. Never was the outlook darker in all that time for practical elementary astronomy. Ah, I have worked hard, given up so much to do my best to bring in better ways of teaching, and now the conviction comes home that I have failed. I should like to write just one more article, a scathing review of the unutterable indifference of astronomers, most of them, to the teaching of the elements of their science. I could do it too, I could say things in a way to make even the astronomers in the big observatories 'sit up and take notice' but I don't suppose I shall. It would mean hard work and probably would not do much good. Then too I am too kindhearted really to enjoy more than the first few sentences.

Mary E. Byrd to Mary Murray Hopkins, October 12, 1920, Smith College Archives, Northampton, Massachusetts.

40. Harriet W. Bigelow, "Declinations of Certain Circumpolar Stars," *Astronomical Journal* 24 (1904–5): 102. Comet Observations, vol. 25 (1905–8): 183; vol. 26 (1908–11): 68; vol. 27 (1911–13): 46, 108; vol. 28 (1913–15): 41; vol. 29 (1915–16):139.

41. The teaching role of observatory staff at Radcliffe was unusual; astronomy at Harvard College was taught by professors of mathematics and astronomy who were not connected with the observatory.

42. Sarah F. Whiting, "History of the Physics Department of Wellesley College from 1878 to 1912" (typewritten paper), p. 23, Wellesley College Archives. Wellesley, Massachusetts.

43. Cecilia Payne-Gaposchkin, interview held in Cambridge, Massachusetts, March 7, 1977.

44. Pickering, *A Plan for Securing Observations of the Variable Stars* (Cambridge, Massachusetts: John Wilson, 1882) p. 4.

45. Mary E. Byrd, "Anna Winlock," *Popular Astronomy* 12 (1904): 254–58.

46. The evidence is not conclusive, but Mrs. R. T. Rogers appears to have been a relative who kept house for William Rogers. When he was sick, she wrote notes to the observatory from his address to say that he would not be in.

47. William A. Rogers to C. W. Eliot, Nov. 23, 1875, HCO Corr. UA V 630.17, group 2, folder Ri-Rom, Harvard Archives, Cambridge, Massachusetts.

48. C. W. Eliot to Arthur Searle, Nov. 27, 1875, HCO Corr. UA V 630.17, group 1, folder Harvard President #2.

49. Howard Plotkin, "Edward C. Pickering and the Endowment of Scientific Research in America, 1877–1918," *Isis* 69 (1978): 44–57.

50. Margaret Harwood, interview held in Cambridge, Massachusetts, Dec. 1, 1976.

51. For a more detailed description of the founding of the Draper Memorial see Jones and Boyd, *Harvard College Observatory*, pp. 211–245.

52. The women employed for the Draper Memorial observed stars on objective prism plates. Henry Draper's photographs of stellar spectra had been made with a spectroscope attached to the eyepiece of the telescope, allowing him to make large photographs of the spectrum of a single star. For the first Draper Memorial catalogue, however, Pickering decided to use a prism placed in front of the objective

lens of the telescope (an older method) to produce a photograph showing many stars, each spread out to form a spectrum, on a single plate. The spectra were small, usually less than an inch long, but under a magnifying lens they showed sufficient detail for classification.

53. Helen L. Reed, "Women's Work at the Harvard Observatory," *New England Magazine* 6 (1892):174–75.

54. Mabel C. Stevens to E. C. Pickering, September 8, 1887, HCO Corr. UA V 630.17, group 1, folder Sn-Sten.

55. Annie E. Masters to E. C. Pickering, June 23, 1884, HCO Corr. UA V 630.17, group 1, folder Mart-Mat.

56. Florence Cushman to E. C. Pickering, January 31, 1888, HCO Corr. UA V 630.17. group 1, folder Cre-Cun.

57. Mary H. Cooke to E. C. Pickering, February 1, no year, HCO Corr. UA V 630.17, group 1, folder Cla-Cly.

58. Imogen W. Eddy to E. C. Pickering, July 11, no year, HCO Corr. UA V 630.17, group 2, folder Ea-Ed.

59. Women started to move into clerical work in the 1860s and in the late nineteenth century more women than men graduated from high school, but even in 1890 only 1.1 percent of the women in the labor force were office workers (Margery Davis, "Women's Place is at the Typewriter; the Feminization of the Clerical Labor Force," *Radical America* 8 [1974]: 7). Women considered astronomical jobs highly desirable because of good working conditions and comparatively pleasant duties, and because the jobs did not require a college education or an upper-class background. In a society where women had few job opportunities except as servants or as millworkers in the textile mills, or teaching if they were better educated, it is not surprising that the observatory always had a long list of applicants.

60. Arthur Searle to N. O. Smith, August 23, 1889, HCO Corr. UA V 630.14, letter book A9, p. 8.

61. For comparison, room and board could be found in 1875 for $6 a week, which totals $312 a year (William A. Rogers to C. W. Eliot, November 23, 1875, HCO Corr. 630.17, group 2, folder Ri-Rom). Domestic and personal service employed 40 percent of women workers in 1900, and manufacturing employed another 25 percent (Bureau of the Census, *Statistics of Women at Work*, based on the 12th census, in 1900 [Washington DC: Government Printing Office, 1907]). Members of the highest paid age bracket of female cotton millworkers in Massachusetts, 40–49 years of age, earned an average of 15.4 cents an hour according to a survey made in 1907–9 (Bureau of Labor Statistics, *Summary of the Report on Condition of Woman and Child Wage Earners in the United States* [Washington D.C.: Government Printing Office, 1916] p.25).

62. Mrs. Fleming, who was then supervising the work of eight other women, earned a salary of $1,200 a year in 1892. C. W. Eliot to E. C. Pickering, March 1, 1892, HCO Corr. UA V 630.17, group 1, folder Harvard President #2.

63. W. P. Fleming to Miss Hume, May 11, 1906, HCO Corr. UA V 630.14, letter book A17, p. 654.

64. Joseph Winlock to C. W. Eliot, November 10, 1887, HCO Corr. UA V 630.14, letter book A2, p. 475. For comparison, a male salesman or principal clerk in a dry goods store or tea store usually earned $15 a week, $780 a year, in 1887, according to C. W. Eliot to E. C. Pickering, December 1, 1887, HCO Corr. UA V 630.17, group 1, folder Harvard President #2.

65. In 1910, twenty-one women worked at the observatory; in 1920 there were sixteen (the reason for this decrease is unclear). The women hired after 1900 were

better educated than before, probably in part because college education was becoming more accessible to women. The previous experience of eight of the twenty women hired between 1900 and 1920 is known. Of these, four definitely graduated from college, one more probably did, two definitely did not have a college education, and the last one probably did not.

66. Mary S. Wagner to E. C. Pickering, May 15, 1893, HCO Corr. UA V 630.17, group 1, folder Wad-Way #2.

67. E. C. Pickering to M. S. Wagner, May 18, 1893, HCO Corr. UA V 630.14, letter book A11, p. 635.

68. M. S. Wagner to E. C. Pickering, May 22, 1893, HCO Corr. UA V 630.17, group 1, folder Wad-Way #2.

69. Arthur Searle to M. S. Wagner, May 23, 1893, HCO Corr. UA V 630.14, letter book A11, p. 646.

70. Wagner to A. Searle, May 26, 1892, HCO Corr. UA V 630.17, group 1, folder Wad-Way #2.

71. M. S. Wagner to A. Searle, December 11, 1893, HCO Corr. UA V 630.17, group 1, folder Wad-Way #2.

72. M. S. Wagner to A. Searle, December 19, 1893, HCO Corr. UA V 630.17, group 1, folder Wad-Way, #2. At least a few years later there were no strict rules about women using the telescopes. Pickering wrote to Annie Cannon, then at Wellesley: "One or more telescopes will also be available for the observations on variable stars which you wish to make." E. C. Pickering to A. J. Cannon, January 15, 1896, HCO Corr. UA V 630.14, letter book A13, p. 189.

73. M. S. Wagner to A. Searle, February 14, 1894, HCO Corr. UA V 630.17, Group 1, folder Wad-Way #2.

74. E. C. Pickering to Mary W. Whitney, March 3, 1896, HCO Corr. UA V 630.14, letter book A13, p. 242.

75. Annie J. Cannon and Edward C. Pickering, *The Henry Draper Catalogue*, Annals of the Harvard College Observatory, vol. 91 (1918): iii.

76. Dorrit Hoffleit, interview held at Yale University, New Haven, Connecticut, November 11, 1976.

77. E. C. Pickering, Official Memorial for Williamina Fleming, in a scrapbook of notices on her death, HUG 1396–5, Harvard University Archives, Cambridge, Massachusetts.

78. E. C. Pickering to Anna P. Draper, December 31, 1886, quoted in Lyle G. Boyd, "Mrs. Henry Draper and the Harvard College Observatory: 1883–1887." *Harvard Library Bulletin* 17 (1969): 95.

79. Annie J. Cannon, "Williamina Paton Fleming," *Astrophysical Journal* 33 (1911): 314.

80. Harriet Richardson Donaghe, "Photographic Flashes from Harvard Observatory," *Popular Astronomy* 6 (1898): 450.

81. The description of her supervision as stern comes from Cecilia Payne-Gaposchkin, interview held in Cambridge, Massachusetts, March 7, 1977. Interestingly, the common perception that Fleming worked more as a supervisor of the women's work than as a scientist does not reflect Fleming's view of herself. She wrote in her diary that she wished she was assigned to more scientific work and less administrative work and complained that she deserved a salary more comparable to those of the male assistants. See Rossiter, *Women Scientists in America*, pp. 56–57.

82. Cannon, "Williamina Paton Fleming," p. 316.

That women were welcomed fairly sincerely into the American Astronomical society is suggested by a story told by Furness:

In the winter of 1902 a pleasant incident happened which gave a final status to the position of women in the American Astronomical Society. The annual meeting was held in Washington in connection with the AAAS and it was planned to have a dinner. Notification blanks were sent out, and after some hesitation the invitation was refused with an intimation that perhaps the presence of women was not desired, that being the custom in several of the other scientific societies. A prompt response came from Professor Newcomb the president, which settled the question permanently.

"I am much disappointed to notice that although you hope to be here at our meeting, you do not propose to join in the dinner. Possibly you may be under a mis-apprehension, supposing that the dinner is only for the men of the society. Permit me, therefore to assure you that all members are equal, and that we should like very much to have our lady members with us."

Furness, "Mary W. Whitney," p. 31.

83. Antonia Maury and E. C. Pickering, *Spectra of Bright Stars Photographed with the 11-Inch Draper Telescope as Part of the Henry Draper Memorial*, Annals of the Harvard College Observatory, vol. 28, part 1 (1897), p. 4.

84. Antonia C. Maury to E. C. Pickering, May 7, 1892, HCO Corr. UA V 630.17.7, folder Antonia Maury.

85. E. C. Pickering to Antonia Maury, May 11, 1892, HCO Corr. UA V 630.14, letter book A11, p. 137.

86. The statement appears in the letter book as a letter from Pickering to Maury agreed to and signed at the bottom by Maury. E. C. Pickering to Antonia Maury, April 3, 1893, HCO Corr. UA V 630.14, letter book A11, p. 569.

87. Mytton Maury to E. C. Pickering, December 19, 1894, HCO Corr. UA V 630.17, group 1, folder Mau-Maz.

88. Antonia Maury to E. C. Pickering, December 21, no year. HCO Corr. 630.17.5, Group 1, folder Antonia Maury.

89. Introduction to Annie J. Cannon and Edward C. Pickering, *The Henry Draper Catalog*, Annals of the Harvard College Observatory, vol. 91, (1918), iii.

90. Dorrit Hoffleit, interview held at Yale University, New Haven, Connecticut, November 11, 1976.

91. Cecilia Payne-Gaposchkin, interview held in Cambridge, Massachusetts, March 7, 1977.

92. Ejnar Hertzsprung to E. C. Pickering, July 22, 1908, HCO Corr. UA V 630.17.8, folder Hertzsprung.

93. The other half of the discovery of the Hertzsprung-Russell diagram was made independently by Henry Norris Russell. DeVorkin, "A Sense of Community in Astrophysics."

94. One major paper on the subject was: Antonia Maury, "The Spectral Changes of Beta Lyrae," *Annals* of the Harvard College Observatory, vol. 84, p. 207.

95. Dorrit Hoffleit, "Antonia Maury," *Sky and Telescope* 11 (1952): 106.

96. Cecilia Payne-Gaposchkin, interview held in Cambridge, Massachusetts, March 7, 1977.

97. Annie J. Cannon, Diary, entry for September 21, 1885, private collection of Margaret Mayall.

98. Leon Campbell, "Annie Jump Cannon," *Popular Astronomy* 49 (1941): 345.

99. Annie J. Cannon and E. C. Pickering, *Spectra of Bright Southern Stars Photographed with the 13-inch Boyden Telescope as Part of the Henry Draper Memorial*, Annals of the Harvard College Observatory, vol. 28, part 2 (1901), p. 131.

100. Owen Gingerich, "Laboratory Exercises in Astronomy—Spectral Classification," *Sky and Telescope* 28 (1964): 82.

101. H. N. Russell, Cecilia Payne-Gaposchkin, and D. H. Menzel, "The Classification of Stellar Spectra," *Astrophysical Journal* 81 (1935): 107.

102. Cecilia Payne Gaposchkin, "Miss Cannon and Stellar Spectroscopy," *The Telescope* 8 (1941): 63.

103. Margaret Mayall, interview held in Cambridge, Massachusetts, December 8, 1976.

104. A. Lawrence Lowell to E. C. Pickering, October 11, 1911, HCO Corr. UA V 630.17.5, group 2, folder Harvard President's Office.

105. Report of the Committee to Visit the Astronomical Observatory of Harvard College, quoted in "Minor Notes," *Popular Astronomy* 20 (1912): 684.

106. Henrietta S. Leavitt to E. C. Pickering, May 21, 1902, HCO Corr. UA V 630.17.5, group 2, folder H. S. Leavitt.

107. See, for example, Henrietta Leavitt to E. C. Pickering, April 17, 1913, HCO Corr. UA V 630.17.5, group 2, folder H. S. Leavitt.

108. While she was in Peru, Annie Cannon wrote in her diary: "Magellanic Cloud (Great) so bright. It always makes me think of poor Henrietta. How she loved the 'Clouds.' " Annie Cannon, Diary, entry for April 20, 1922, private collection of Margaret Mayall.

109. Edward C. Pickering, "Periods of 25 Variable Stars in the Small Magellanic Cloud," Harvard College Observatory Circular no. 173.

110. Margaret Harwood, interview held in Cambridge, Massachusetts, December 1, 1976.

111. Cecilia Payne-Gaposchkin, interview held in Cambridge, Massachusetts, March 7, 1977.

112. Katherine Haramundanis, ed., *Cecilia Payne-Gaposchkin: An Autobiography and Other Recollections* (Cambridge: Cambridge University Press, 1984), pp. 146–47.

113. Cecilia Payne-Gaposchkin, interviewed by Owen Gingerich in Cambridge, Massachusetts, on March 5, 1968. Courtesy of the Niels Bohr Library of the American Institute of Physics.

114. Debby J. Warner, "Women Astronomers," *Natural History* (May 1979): 12–26. In contrast women earned 8 percent of the Ph.D.s in all physical sciences in the 1920s, 7 percent in the '30s, 5 percent in the '40s, 4 percent in the '50s, and 5 percent in the '60s. Betty M. Vetter, "Data on Women in Scientific Research," in Janet Welsh Brown, Michele L. Aldrich, and Paula Quick Hall, *Report on the Participation of Women in Scientific Research* (Washington, D.C.: American Association for the Advancement of Science, March 1978).

115. Helen S. Hogg, "Variable Stars in Globular Clusters," *Royal Astronomical Society of Canada Journal* 67 (1973): 8–18.

116. Dorrit Hoffleit, interview held at Yale University, New Haven, Connecticut, November 11, 1976.

117. Peggy A. Kidwell, "An Historical Introduction to 'The Dyer's Hand,' " in Katherine Haramundanis, ed., *Cecilia Payne-Gaposchkin*. See also Peggy A. Kidwell, "Cecilia Payne-Gaposchkin: Astronomy in the Family," in Abir-Am and Outram, *Uneasy Careers and Intimate Lives*.

118. Otto Struve and Velta Zebergs, *Astronomy of the 20th Century* (New York: Macmillan, 1962), p. 220.

119. Elske V. P. Smith, "Cecilia Payne-Gaposchkin," *Physics Today* 33 (June 1980): 64–65.

120. Kidwell, "Historical Introduction . . . "

121. Rossiter, *Women Scientists in America*.

This paper started out as a B.A. thesis, in which form it benefited greatly from criticism, advice, and encouragement provided by Owen Gingerich. The Center for Astrophysics, which includes the Harvard College Observatory, provided a grant to support some of the research.

This project has gained from the criticism and interest of more people than I can properly acknowledge. I would particularly like to thank Barbara Welther, Martha Liller, Bruce Rosen, David DeVorkin, and G. Kass-Simon. I am also very grateful to all the people I interviewed and to the staffs of the various archives who made possible my research and provided me with photographs.

JUDY GREEN AND JEANNE LADUKE

Contributors to American Mathematics

An Overview and Selection

D. E. Smith and J. Ginsburg, in *A History of Mathematics in America before 1900*, note that the *Mathematische Annalen* contains fifteen articles by Americans in the years 1893 to 1897. They list the authors of fourteen of these articles—all men.[1] The fifteenth was by Mary Frances Winston, an American student at Göttingen. The 1895 paper by Mary Winston[2] was not her dissertation, which she completed in June of the following year, but was based on a talk she had given in the mathematics seminar at Göttingen. She had in fact been the first woman to present such a talk, and did so less than three months after arriving in Germany. In December 1893 she wrote to her family of that presentation that "it went off reasonably well. . . . I do not think that anyone will draw the conclusion from it that women cannot learn Mathematics."

Unfortunately that conclusion has, and always has had, a certain amount of currency. Although there is a literature on women in mathematics, the bulk of it leaves the reader with the impression that the subject is exhausted by the study of seven women whose lives span more than fifteen hundred years.[3] The first, Hypatia (c. 370–415), a leading Neoplatonist of Alexandria, is said to have written commentaries on the *Arithmetica* of Diophantus and the *Conics* of Apollonius. Although her work has not been preserved, her life and violent death at the hands of a fanatical Christian mob are the subject of an 1853 novel by Charles Kingsley. In the eighteenth century the Marquise du Châtelet (1706–1749) and Maria Gaetana Agnesi (1718–1799) both contributed to the propagation of major new ideas in analysis. The Marquise du Châtelet, author of philosophical and literary works, made the first French translation of Newton's *Principia*. The translation first appeared posthumously in 1756; a version of it was issued in 1966 as a facsimile edition. Agnesi was the author of an early comprehensive treatise on algebra and the calculus (1748), which was translated into several languages. She was appointed to the chair of mathematics and natural philosophy at Bologna in 1750 but never taught there. She devoted herself to religious work after 1752.

The three major nineteenth-century women in mathematics were respectively French, English, and Russian. Sophie Germain (1776–1831), who worked both in mathematical physics and in number theory, was awarded a major prize by the French Academy of Sciences in 1816 for her memoir on the theory of elastic surfaces. Mary Somerville (1780–1872) was a generalist in the mathematical sciences. Although best known as a translator into English of Laplace's *Mécanique Céleste*, she also produced original works in a number of different areas. One of the first two colleges for women at Oxford was named after her. Sofia Kovalevskaia (1850–1891) is recognized for her work in partial differential equations. She was Professor of Mathematics at the University of Stockholm and is generally believed to be the first woman to receive a Ph.D. in mathematics, having received her degree from Göttingen in 1874. In 1888 she was awarded the Prix Bordin by the French Academy of Sciences for her work on the problem of a heavy rigid body rotating around a fixed point.

Emmy Noether (1882–1935) was one of the outstanding mathematicians of the twentieth century. She is best known among physicists for her theorem on the relationship between symmetries and conservation laws and among mathematicians for her development of abstract algebra. Efforts to assess the many aspects of her work continue and were highlighted by symposia organized in her honor during 1982, the centenary of her birth.[4] After spending most of her professional life at Göttingen, she came to the United States as a refugee from Nazism in 1933, just two years before her death.

There have been many works chronicling the contributions of these and a few other women. The first article of this type in English in a mathematical journal was "Women as Mathematicians and Astronomers" by R. C. Archibald, which appeared in 1918.[5] Even before then, in 1897, A. Rebière, in the second edition of *Les Femmes dans la Science*, included substantial entries on all the aforementioned women except Noether, each with a bibliography and references.[6] Other articles and books have appeared over the years.[7] The purpose of this chapter is not to chronicle again the contributions of this same small group of women mathematicians. Rather, we intend to show that in the United States, as long as women have been able to receive an undergraduate education, many of them have chosen, often in the face of considerable opposition, to pursue advanced studies in mathematics. We will also discuss some of the many different ways in which women have contributed to the American mathematical community.

We have identified the American women who received Ph.D.'s in mathematics before 1940, and we have studied their careers and contributions.[8] By 1939, 225 Ph.D.'s in mathematics had been awarded to women by American schools and another 4 had been awarded to American women by European universities. Ten of these degrees were conferred before 1900, 9 at American schools and 1 in Europe. The 219 degrees conferred after the turn of the century include 3 awarded by European schools. The 216 women who received Ph.D.'s from schools in the United States during the first four decades of this century constitute 14.3 percent of all the Ph.D.'s in mathematics earned at such schools during this period. This percentage is slightly higher than the 14.1 percent of all U.S. Ph.D.'s that were earned by women in all fields in the same period.[9] As was the case in other fields, the percentage of Ph.D.'s in mathematics awarded to American women declined dramatically starting in the 1940s, so that from 1950 to 1969 it was less than 6 percent.[10] Although the percentages increased in the 1970s, it was not until the 1980s that women doctorates in mathematics attained the relative proportions of the first four decades of the century.[11]

We focus below on some aspects of the lives and work of nine women, each of whom contributed significantly to mathematics or was in some way a pioneer in the American mathematical community. In addition, we will hint at some of the overall features that emerge as we study this community in the period before World War II. Because we wish to establish enough of a mathematical and historical context to convey a sense of the intellectual achievements of the women we discuss, we have had to select just a few of the many possible candidates for inclusion. We decided at the outset not to discuss the works of any women who are still living even though they may have received their Ph.D.'s before 1940. Thus, many who have made important contributions are not included in this discussion.

Of the nine to whom we direct attention, three (Charlotte Angas Scott, Anna Johnson Pell Wheeler, and Olive Clio Hazlett) were such distinguished mathematicians that their selection was inevitable. Another (Mildred Leonora Sanderson) showed at least comparable promise but died young. The remaining five (Christine Ladd-Franklin, Winifred Edgerton Merrill, Mary Winston Newson, Mary Emily Sinclair, and Mayme Irwin Logsdon) earned their degrees over a period of forty years and include the first two American women to earn Ph.D.'s in mathematics. While we cannot generalize from the study of these nine women, their educational and career patterns and their intellectual achievements do illustrate aspects of the rich and varied history of women in American mathematics.

BEFORE 1900

By the early 1880s undergraduate education for women was widely available. The Morrill Act of 1862 had created the land-grant colleges, the opening of women's colleges in the East had begun with that of Vassar in 1865, and

Senior mathematics class at Wellesley, 1886, taught by Helen A. Shafer, president of Wellesley 1888–1894. Helen Fitz Pendleton, president of Wellesley 1911–1936, is seated on Miss Shafer's far left. Helen A. Merrill, Ph.D. Yale 1903, is seated on Miss Shafer's right. Courtesy of Wellesley College Archives.

private coeducational institutions were proliferating. Excellent preparation in mathematics was available, particularly at Wellesley, where two of the first six presidents, Helen A. Shafer and Ellen Fitz Pendleton, had been members of the mathematics faculty, and where the early mathematics department was not only large but almost entirely female. Wellesley graduated far more women who earned the Ph.D. in mathematics before 1940 than any other college, coeducational or women's.[12]

The situation with respect to graduate education of women was entirely different. The establishment of The Johns Hopkins University in 1876 as the first research-oriented university in the country is especially significant in the history of American mathematics.[13] It was based on the model of German universities, which did not normally admit women, and, like so many other colleges and universities founded at that time, was funded by a private fortune. Daniel Coit Gilman, the president of Johns Hopkins, appointed J. J. Sylvester, a distinguished English mathematician, as the first Professor of Mathematics.

On 27 March 1878 Christine Ladd (1847–1930) wrote Sylvester to say she wanted to listen to his lectures and asked whether Johns Hopkins would refuse to allow her because of her sex.[14] Ladd, born in Connecticut, was one of the first graduates of Vassar College, having graduated in 1869. After her graduation she taught science in secondary school and continued to study mathematics. She attended lectures at Harvard and published several papers in a newly founded American journal, the *Analyst*. One of these was a report on the contents of *Crelle's Journal*,[15] in which she commented:

> It is not greatly to the credit of the mathematicians of the vicinity that Crelle's Journal lies on the shelves of the Boston Public Library with uncut leaves; unless, indeed, we are to suppose that all who give themselves to mathematics are so rich that they can afford to take the Journal for themselves. Certainly none who hope to extend the boundaries of the science can afford to do without their Crelle.[16]

Sylvester, having read some of her published work, wrote to President Gilman in January 1878 supporting her admission and noting her mathematical attainments and promise.[17] That same year Christine Ladd was formally admitted to Johns Hopkins and completed all her work, including a dissertation "On the Algebra of Logic,"[18] by 1882.

Ladd's dissertation made contributions to the relatively new field of symbolic logic. This field is often said to have begun with the publication in 1847 of George Boole's *The Mathematical Analysis of Logic, being an essay towards a calculus of deductive reasoning*,[19] in which logical propositions are expressed symbolically and mathematical operations on symbols are systematically developed. By 1882 a number of improvements on Boole's original algebra had been introduced; however, Boolean algebra had not yet attained what is now considered its definitive form. Well accepted at the time were the interpretations of the operations of addition, multiplication, and negation, now called union, intersection, and complementation when referring to sets, and disjunction, conjunction, and negation when referring to propositions. The logicians of the early 1880s did not use symbols to denote quantification, universal (every x), or particular (some x). Instead they used relations, called copulae, to denote certain quantified statements. For example, some used an inclusion copula, $<$, to denote "every x is a y" by $x < y$.

Ladd, in her dissertation, provided a very detailed analysis of a mutual exclusion copula, $\overline{\vee}$, and a partial inclusion copula, $\vee$, so that "no x is a y" is denoted by $x \overline{\vee} y$ and "some x is a y" is denoted by $x \vee y$. Her principal achievement in logic was her use of the system she developed to solve the problem of the syllogism. A syllogism is a particular type of formal argument that consists of two premises and a conclusion. The problem is to find a way to recognize when the syllogism is valid, that is, when the conclusion follows from the premises. Nineteen forms of the syllogism were considered valid by the scholastic logicians of the Middle Ages. The best known of these, Barbara, takes the form

Every m is a p; every s is an m; therefore every s is a p.

Others are Baroko and Bramantip:

Every p is an m; not every s is an m; therefore not every s is a p;

Every p is an m; every m is an s; therefore some s is a p.

Recognizing a valid form of the syllogism is equivalent to recognizing an antilogism, that is, recognizing the inconsistency of the triad of propositions consisting of the two premises and the denial of the conclusion. For instance, the validity of Baroko is equivalent to the inconsistency of the triad

Every p is an m; not every s is an m; every s is a p.

Ladd observed that if the triad is written out using her two copulae then it is an antilogism if and only if the following three conditions hold:

1. The three statements include exactly two formed with $\bar{v}$ and one with v.
2. Every pair of statements has one term in common.
3. Only the term common to the two $\bar{v}$ statements appears in both the positive and negative forms.

Ladd-Franklin's contribution, described by E. W. Beth as "the first adequate treatment of classical syllogism,"[20] is described in detail by Lewis and Langford.[21]

The symbolic forms of the triads associated with the syllogisms listed earlier are:

Barbara: $(m\bar{v} - p)(s\bar{v} - m)(s \vee - p)$

Baroko: $(p\bar{v} - m)(s \vee - m)(s\bar{v} - p)$

Bramantip: $(p\bar{v} - m)(m\bar{v} - s)(s\bar{v}p)$

the first two of which satisfy Ladd-Franklin's rule. The rule denies the validity of Bramantip as well as three other forms of the syllogism considered valid by scholastic logicians but rejected by modern mathematical logicians. Each of these four is now considered invalid because some one of the properties involved in the syllogism may not be possessed by anything at all.. For instance, if the second premise of Bramantip (every *m* is an *s*) is true and if nothing at all has the property *p*, then the first premise (every *p* is an *m*) is true as well, but the conclusion (some *s* is a *p*) is false.

Although Ladd's work satisfied the requirements for the Ph.D., and her dissertation was published in a collection of works of the students of her advisor, Charles Sanders Peirce, the trustees of Johns Hopkins refused to award Christine Ladd a Ph.D. because she was a woman. Vassar College, however, awarded her an LL.D. in 1887, the only honorary degree that college has ever bestowed. Upon finishing her graduate studies Ladd married Fabian Franklin, a member of the mathematics faculty at Johns Hopkins. Although she never lost interest in logic, most of her subsequent publications were in optics and psychology. She was a lecturer in logic and psychology at Johns Hopkins and later at Columbia and was active in the movement to open and provide support for graduate education for women. She was starred for psychology in Cattell's *American Men of Science*[22] and was included in *Notable American Women 1607–1950*.[23] In 1926, at the fiftieth anniversary of the inauguration of its first president, Johns Hopkins University finally awarded Ladd-Franklin the degree she had earned forty-four years before. She was then seventy-eight years old.

The first Ph.D. in mathematics that actually was awarded to a woman by an American university was conferred by Columbia University on Winifred Haring Edgerton (1862–1951) in 1886. She was, in fact, the first woman to earn any degree at Columbia in any subject. An 1883 graduate of Wellesley, she had "startled the academic world by applying for admission to Columbia,"[24] and "after much controversy she was finally admitted as an exceptional case and one which was 'to set no precedent.' "[25] Her son, Hamilton Merrill, wrote that she "said that a condition of her admission was to dust the astronomical [instruments] and so comport herself as not to disturb the men students."[26] Another Wellesley student later reported that some of Edgerton's prospective classmates in a course in celestial mechanics were sufficiently disturbed by the idea of studying with a woman that they asked their professor to use the most difficult text possible; the one he chose was the one from which Edgerton had studied at Wellesley.[27] Although in 1886 the trustees had not yet voted to grant degrees to women, Edgerton "received the degree of Doctor of Philosophy *cum laude*, 'in consideration of the extraordinary excellence of the scientific work' done by her."[28] In 1888, a year after her marriage, Winifred Edgerton Merrill helped to found Barnard College. In describing the founding of Barnard, Annie Nathan Meyer reported that Merrill was a member of the original board

Portrait of Winifred Edgerton Merrill, oil on canvas, by Helena E. Ogden Campbell, presented to Columbia University in 1933. Courtesy of Columbia University in the City of New York, Office of Art Properties.

of trustees until her husband objected to her constant attendance at meetings held at the offices of two male lawyers.[29]

In 1918 Merrill wrote:

> In my early education I was greatly impressed by a book by Benjamin Peirce, entitled "Ideality in the Physical Sciences", and it was the influence of this book upon my mentality that some years later led me to choose as the subject of the thesis for my Doctor's degree in Mathematics, at Columbia University, "The Unification of the Several Systems of Mathematical Co-ordinates."[30] Consciously and unconsciously throughout a busy life, as mother of four children and for thirteen years principal of a girls' school, I continued this mental search for co-ordinating elements in life-experiences, in art-forms, in the complexities of educational problems, always searching for a better understanding of the nature of things through some underlying unifying principle.[31]

Her portrait hangs in Columbia's Philosophy Hall; under it is the inscription: "She Opened the Door."

Beginning around 1890 opportunities for the graduate education of women in mathematics gradually increased. By the turn of the century nine more Ph.D.'s in mathematics had been conferred upon American women. The eight in the United States included three granted by Cornell, three by

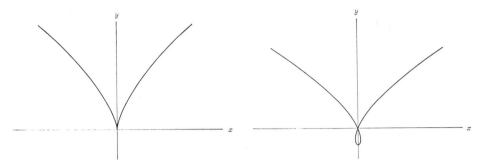

Figure 1.   Figure 2.

Yale, and two by Bryn Mawr. The ninth was awarded by the university at
Göttingen.[32] When Bryn Mawr College was founded in 1885, M. Carey
Thomas, the college's first dean and later its president, hired Charlotte
Angas Scott (1858–1931) of England, one of very few women in the world
at the time with a doctorate in mathematics, to head the mathematics de-
partment. Scott had been a student at Girton College, Cambridge, from
1876 to 1880 and was unofficially the first woman to place among the
Wranglers on the mathematical Tripos.[33] She was a Lecturer at Girton from
1880 to 1884 and received the B.Sc. in 1882 and the D.Sc. in 1885 from the
University of London.

An algebraic geometer of real distinction, Scott wrote many papers on
singularities and intersections of plane algebraic curves. A plane algebraic
curve can be thought of as the locus of a polynomial equation $f(x, y) = 0$;
a singular point is a point at which not only f but also both partial derivatives of f vanish. In two of her early papers in the *American Journal of
Mathematics*,[34] Scott set herself the task of elucidating the "geometrical
reality" underlying the algebraic decomposition of singularities. For example, the curve $x^2 - y^3 = 0$, which has a cusp at the origin (fig. 1), is a
special case of the more general curve $x^2 - a^2y^2 - y^3 = 0$, where a is a
real number. When a is not 0 the cusp is replaced by its penultimate form,
a node with a loop (fig. 2). A node, the crossing of two branches of the
curve having distinct tangent lines, is the simplest singularity a curve can
have. Scott showed that algebraic results on the decomposition of higher
(more complicated) singularities into nodes and cusps can be interpreted
as showing the existence of penultimate forms, with clusters of simpler
singularities replacing the more complicated ones.

An important problem connected with intersections of algebraic curves
is the determination of order. For example, the intersection of two curves
at a point is of order one if the point is not singular for either curve and
the curves cross each other with distinct tangents at the point. If two curves
are tangent at a point they have an intersection there of order at least two.

Further complications arise from the possibility that the point of intersection is singular for one or both of the curves. In her last paper Scott addressed this topic. In the introduction she wrote:

> While there is no obligation to reject the aid of some special algebraic process simply because it has no geometrical equivalent, yet the fundamental concepts of power series and convergence are in so different a category from those of algebraic geometry that a geometer may be allowed some reluctance to adopt them unless absolutely necessary.[35]

Scott showed that existing procedures for determining the order of intersection of two curves at a point, justified in the literature by convergence arguments, could, in fact, be justified by purely algebraic arguments. She went on to show that the entire theory of singularities of plane algebraic curves can be established by purely algebraic arguments, i.e., without reliance on convergence. Her work is little cited today, because, already at the time of her death in 1931, algebraic geometry was undergoing fundamental changes, brought on in part by the work of Emmy Noether in commutative algebra.

Scott was an excellent expositor. In an obituary, the English mathematician F. S. Macaulay described her as "an enthusiastic searcher and propounder of new ideas and an interpreter of the work of others, adding simplifications and extensions of her own."[36] In the first of her interpretive papers Scott gave her view of the accessibility of mathematics to students, a view that may help to explain why so many women studied mathematics at Bryn Mawr:

> In the following pages I attempt to show, as a matter of purely pedagogic interest, how simply and naturally Cayley's theory of the Absolute follows from a small number of very elementary geometrical conceptions, . . . my contention is not that every step in the rigorous proof can be presented under the guise of elementary mathematics, but that it is quite possible to develop the theory so as to be intelligible and interesting to average students at a much earlier stage than is customary.[37]

Scott was also the author of a highly respected textbook entitled *An Introductory Account of Certain Modern Ideas and Methods in Plane Analytical Geometry*, which appeared in 1894; later editions appeared in 1924 and 1961.[38] The analytical geometry of the title was not the subject that is taught today to high school seniors and college freshmen, but rather an introduction to projective algebraic geometry, with "Cartesian Geometry and the Differential Calculus" indicated as prerequisites in the preface. Scott often offered this subject at Bryn Mawr as a "post major" course, a course for graduates and undergraduates who had completed the major.

Charlotte Scott remained at Bryn Mawr for forty years. Between 1896 and 1925 she officially advised seven Ph.D. students.[39] She created a vigorous mathematics department, and it was to this department that Emmy Noether came when she left Germany in 1933. Charlotte Scott's influence on American mathematics was publicly acknowledged in many ways. She was starred in the first edition of *American Men of Science*, having been ranked by her peers as fourteenth among the top ninety-three mathematicians of the period.[40] She was elected to the first Council of the American Mathematical Society in 1894 and served again in 1895–1897 and 1899–1901. In 1906 Scott served as vice-president of that organization, being the first woman to hold that office and the only one to do so until 1976.[41] Charlotte Scott is the only woman listed as a mathematician in *Notable American Women 1607–1950*.[42]

At the end of the nineteenth century one of the major centers of mathematical research in the world was the university at Göttingen, and many American mathematicians and students of mathematics were drawn there. Among these was Mary Frances Winston (1869–1959). Mary Winston grew up in the Midwest and graduated from the University of Wisconsin in 1889; she taught school in Wisconsin and then, after two years, received a fellowship and studied with Scott at Bryn Mawr for a year. She returned to the Midwest and was a fellow by courtesy during the University of Chicago's inaugural year, 1892–93. Letters of that period indicate how eager she was to go to Germany to study.[43]

Among the major participants at the 1893 Congress of Mathematicians, held in conjunction with the World's Columbian Exposition at Chicago, was Felix Klein, the principal professor of mathematics at Göttingen. After the Congress Klein presented a series of lectures, known as the Evanston Colloquium, on recent mathematical developments in Germany. He also became acquainted with the members of the Colloquium, among whom was Mary Winston. At this colloquium Mary Winston met Christine Ladd-Franklin, who offered, and later provided, financial support to Winston so that she could go to Germany that fall.

With the committed support and influence of Felix Klein, three women—Mary Winston; Grace Chisholm, an English mathematics student; and Margaret Maltby, an American physics student—were admitted to Göttingen in the fall of 1893.[44] They were the first women officially admitted as regular students in a university administered by the Prussian government.[45] Mary Winston became the first American woman to receive a Ph.D. in mathematics from a foreign university, passing the examinations *magna cum laude* in 1896. During her final year in Göttingen she held a European fellowship from the Association of Collegiate Alumnae (now the AAUW).

Winston worked on problems that arise in connection with physical applications of the theory of analytic functions of a complex variable, a major theme in the work of Klein and his students. Specifically, Winston

described properties of solutions of the ordinary differential equations
known as the hypergeometric equation and Lamé's equation. The note in
the *Mathematische Annalen* to which we referred at the beginning of this
chapter was a remark on hypergeometric functions (solutions of the hypergeometric equation); it was later cited by Klein in his book on that
subject.[46] Her dissertation concerned solutions of Lamé's equation with
applications to the mechanics of the spherical pendulum and to the theory
of the top. Klein cited this work in his lectures, "The Mathematical Theory
of the Top," given at Princeton during its sesquicentennial year of 1896.[47]

Mary Winston returned to the United States in 1896 and attempted to
have her dissertation published in this country. She was not successful,
so her degree was not granted until 1897, after she had arranged for its
publication in Germany.[48] Winston taught at a high school in Missouri for
the year 1896–97 before being hired as Professor of Mathematics at Kansas
State Agricultural College (now Kansas State University). In 1900, after
three years at Kansas State, she married Henry Byron Newson, who was
then on the mathematics faculty at the University of Kansas. She had no
position at Kansas but did teach there some summers when her husband
was not teaching.

There is no evidence that Mary Winston Newson did any original mathematical research after her marriage. However, she made one more contribution to the literature of the subject. While at Göttingen she had studied
with, among others, David Hilbert, a dominant figure in the mathematics
of the late nineteenth and early twentieth centuries. In 1900, at the International Congress of Mathematicians in Paris, Hilbert delivered a now
famous speech in which he outlined what he considered to be the principal
mathematical problems facing mathematicians of the new century. It was
decided by the editors of the *Bulletin of the American Mathematical Society*
that an English translation of Hilbert's speech should be published. Henry
Seely White (later a professor at Vassar and the only president of the American Mathematical Society to spend most of his career at a women's college)
suggested to Hilbert that Mary Winston Newson be asked to prepare the
translation. Hilbert agreed, so her translation appeared in the *Bulletin* in
1902[49] and was reprinted in 1976 in the proceedings of a symposium devoted to developments in mathematics arising from the Hilbert problems.[50]

H. B. Newson died in 1910, at age forty-nine, leaving Mary Winston
Newson, at age forty-one, with children of nine years, six years, and three
months and with no pension. She was not hired then by Kansas, although
there were positions to be filled, nor was she hired a few years later when
she was actively seeking employment. Her son, Henry Winston Newson,
made these observations:

A possible reason that she was not considered for an opening at the University
of Kansas had also to do with the nepotism rules, since (immediately after my

father's death) my aunt Alice Winston received an appointment in the English Department. The long arm of nepotism was illustrated again in the early 1920's when my mother was (as I understand it) actually offered an appointment at the University of Texas at Austin. However this offer was withdrawn because her brother was a professor of economics! She told me once that she took the poor paying and not very prestigious job at Washburn College in order to be close enough to Lawrence (30 miles) to get there to see her children on weekends. I remember her saying that she turned down a more attractive opportunity which was farther away.[51]

Mary Winston Newson became an assistant professor at Washburn College in 1913 and became active in the field of mathematics education, chairing the Kansas Association of Teachers of Mathematics in 1915. Newson remained at Washburn until 1921, when she resigned after opposing the college president's position on the firing of a faculty member. She spent the rest of her career, from 1921 to 1942, at Eureka College in Illinois.

AFTER 1900

Soon after its opening in 1892, the University of Chicago supplanted Johns Hopkins University as the leading American producer of Ph.D.'s in mathematics. Under the direction of E. H. Moore, head of the mathematics department from its inception until 1931, Chicago exerted a major influence on the American mathematical community, in terms of both the size and quality of its graduate program in mathematics and the later eminence of many of its early students. From 1900 until 1939 the Chicago department conferred about 17 percent of all Ph.D.'s in mathematics in the United States.

Chicago led by far in the Ph.D. production of early American women mathematicians. Of the 229 American women who received Ph.D.'s in mathematics by 1939, the 46 who received their Ph.D.'s from Chicago[52] include virtually all the American-educated women whose work received substantial recognition from the mathematical community before World War II. Thirty of these 46 women were students of Leonard Eugene Dickson or Gilbert Ames Bliss. Dickson worked in algebra and number theory and Bliss in the calculus of variations; both received their Ph.D.'s from Chicago, Dickson in 1896 and Bliss in 1900. Twelve of Bliss's 52 doctoral students and 18 of Dickson's 67 were women.

In 1908 Mary Emily Sinclair (1878–1955) became the first woman to receive a Ph.D. in mathematics at Chicago.[53] Sinclair, the daughter of a professor of mathematics at Worcester Polytechnic Institute, graduated from Oberlin College in 1900, taught for a year in Connecticut, and then studied at Chicago. She received her M.A. there in 1903 and continued for

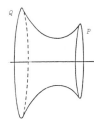

Figure 3. Adapted from Bliss 1925.

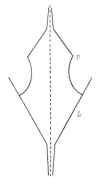

Figure 4. Adapted from Bliss 1925.

a year at Chicago before taking a position as Instructor of Mathematics at the University of Nebraska (1904–1907) and then at Oberlin (1907–1908).

Sinclair quickly published three papers and an abstract of other work on problems in the calculus of variations. These papers appeared in the *Annals of Mathematics* in 1907, 1908, and 1909.[54] Problems in the calculus of variations concern minimizing or maximizing quantities that are represented by definite integrals. Sinclair worked on the problem of determining a surface of revolution of minimum area. More precisely, given two points P and Q, the problem is to find a curve joining them such that the curve generates a surface of minimum area when rotated about an axis (fig. 3). This surface of revolution can be realized physically as a soap film stretched between two circles whose planes are parallel and whose centers lie on a common axis perpendicular to the planes.

A modification of the problem occurs when one of the endpoints varies along a curve rather than remaining fixed. In his monograph *Calculus of Variations*, Bliss credited Sinclair with having "studied in a very interesting way"[55] the case in which one endpoint, P, is fixed and the other varies along a straight line, L. He described her physical interpretation of the problem as follows:

If an inverted funnel is set in a larger funnel and moistened with soap solution, then when the smaller one is withdrawn in the direction of its axis a surface of revolution is formed [fig. 4]. . . . It is a catenoid one of whose bounding circles is always the greatest circle of the smaller funnel, while the other slides up or down the inner surface of the larger funnel as the distance between the two funnels is altered. At their intersection the soap film and the larger funnel meet at right angles. As the surface is elongated by enlarging the distance between the two funnels a certain point is reached when the surface becomes unstable and separates. . . . [56]

Sinclair determined mathematically the distance between the two funnels at the moment when the point P reaches the envelope of a one-parameter family of catenaries orthogonal to the line L "and verified by actual measurements that the film becomes unstable when that distance is reached. The agreement between her calculated and experimental results was surprisingly close."[57] Bliss also discussed other, more abstract aspects of Sinclair's work in his monograph.

Mary Sinclair returned to Oberlin as an associate professor in 1908, was promoted to professor in 1925, and was head of the department there from 1939 until her retirement in 1944. In the 1920s Sinclair read a number of papers in the calculus of variations before the American Mathematical Society and the Mathematical Association of America. One of her papers, describing a new mathematical model, was published by Schilling, the German manufacturer of the model.[58]

The second woman to receive a Ph.D. in mathematics from Chicago was Anna Johnson Pell (1883–1966), who received her degree in 1910. She was officially a student of E. H. Moore, although she had done much of the work for her dissertation before coming to the University of Chicago. Born in Iowa, she received her A.B. from the University of South Dakota in 1903 and her A.M. from Radcliffe College in 1905. She studied at Göttingen in 1906–07 and, in the summer of 1907, married Alexander Pell, a mathematician and one of her former teachers at South Dakota. She again studied at Göttingen during most of 1908 and enrolled at Chicago starting with the winter quarter of 1909.

In 1911 her husband suffered a minor paralytic stroke, so Anna Pell taught his classes at the Armour Institute of Technology (now Illinois Institute of Technology). She then taught at Mount Holyoke College for seven years. In 1918 she moved to Bryn Mawr College, where she spent most of the next thirty years. In 1921 Alexander Pell died, and four years later Anna Pell married Arthur Leslie Wheeler, a classicist, to whom she was married until his death in 1932. Although she lived in Princeton during most of her marriage to Wheeler, she remained on the faculty of Bryn Mawr, sometimes as Professor of Mathematics, at other times as Non-Resident Lecturer.

Pell Wheeler published twelve papers during her career, supervised the Ph.D. dissertations of seven students at Bryn Mawr,[59] and, in 1925, suc-

Anna Pell Wheeler, starred in the 1921 edition of *American Men of Science*. Courtesy of Grace Shover Quinn.

ceeded Charlotte Scott as the Alumnae Professor of Mathematics at Bryn Mawr. The mathematics department was under her direction and remained so as long as she was at Bryn Mawr, including the years 1933 through 1935 when Emmy Noether was there. Pell Wheeler retired in 1948.

In recent years a number of articles about Pell Wheeler have appeared, including an entry in *Notable American Women, The Modern Period*.[60] In an extensive discussion of her life and work, L. S. Grinstein and P. J. Campbell wrote:

> Pell Wheeler considered her work to be centered on "linear algebra of infinitely many variables," a branch of what is known today as functional analysis. Her interest in it derived from applications to differential and integral equations. Although she also pursued some purely algebraic results, her training was mainly in analysis, and she began her investigations at a time when functional analysis was emerging as a distinct area of mathematics.[61]

Pell Wheeler received more recognition from the mathematical community before World War II than perhaps any other American woman.[62] She was starred in *American Men of Science* in 1921; she was on the original board of trustees of the American Mathematical Society when it incorporated in 1923; she was a member of the Council of the AMS, 1924–1926; and in 1927, at a meeting of the AMS at the University of Wisconsin, she gave the prestigious Colloquium Lectures. According to Grinstein and Campbell, "The lectures, on the theory of quadratic forms in infinitely many variables and its applications, summarized and surveyed the broader scene in which her own work had contributed during the previous 20 years."[63] She was the only woman Colloquium Lecturer until 1980.[64]

The third woman to receive a Ph.D. in mathematics at Chicago was Marion Ballantyne White, a student of Bliss who received her degree in 1910, seventeen years after having received her bachelor's degree. She taught at the University of Kansas, at Michigan State Normal College, and at Carleton College until her retirement in 1937 from Carleton.

The next two women to receive Ph.D.'s in mathematics from Chicago were students of L. E. Dickson. Dickson played a major role in determining the direction of much of the American research in algebra in the first three decades of this century. He and his students worked primarily in the areas of finite groups, linear associative algebras (hypercomplex systems), and number theory.

Dickson's first female doctoral student was Mildred Leonora Sanderson (1889–1914). She was born in Waltham, Massachusetts, attended public schools there, and graduated from Mount Holyoke College with Senior Honors in mathematics in 1910. She spent the next year at Chicago as a fellowship holder from Mount Holyoke, completing her master's thesis under Dickson's direction in June 1911. A shortened version of that thesis appeared the same year in the *Annals of Mathematics*.[65] In it she treated a notion of congruence of polynomials and proved several generalizations of results in number theory based on this notion. In 1915 Dickson wrote, "This work might well have served for her doctor's thesis; but she was quite willing to undertake a new investigation in a wholly different field."[66]

The field in which Sanderson wrote her dissertation was the theory of modular invariants, a highly abstract and formalized subject introduced by Dickson in 1907.[67] He wrote later:

A simple theory of invariants for the modular forms and linear transformations employed in the theory of numbers should be of an importance commensurate with that of the theory of invariants in modern algebra and analytic projective geometry, and should have the advantage of introducing into the theory of numbers methods uniform with those of algebra and geometry.[68]

For those readers who are familiar with the language of linear algebra we define the terms needed to state Sanderson's main result. By a form in m variables $x_1, \ldots, x_m$ we mean a homogeneous polynomial (a polynomial each of whose terms has the same degree) in those variables with coefficients $a_1, \ldots, a_r$. Any non-singular linear transformation,

$$x_i = \sum_{j=1}^{m} c_{ij} x'_j, \; i = 1, \ldots, m,$$

also operates linearly on a set of forms by the rule

$$f'(x'_1, \ldots, x'_m) = f(x_1, \ldots, x_m).$$

An algebraic invariant of a group of non-singular linear transformations is a polynomial I in the coefficients of the forms with the property that

$$I(a'_1, \ldots, a'_r) = (det(c_{ij}))^n I(a_1, \ldots, a_r)$$

for some integer n and all transformations of the group (where det is an abbreviation for determinant). If we stipulate that the coefficients of the forms and the c_{ij} that define the linear transformations represent variables taking values in the field of complex numbers, we obtain classical algebraic invariant theory. If, instead, we think of the possible values as lying in the field Z_p of integers modulo p or in one of its finite extensions, $GF[p^n]$, we obtain the theory of modular invariants.

Connected to the notion of modular invariant is the important concept of formal modular invariant for which the congruence

$$I(a'_1, \ldots, a'_r) \equiv (det(c_{ij}))^n I(a_1, \ldots, a_r) \; (mod \; p)$$

holds on a formal level, where only the c_{ij} are regarded as elements of some $GF[p^n]$, while the coefficients of the forms are regarded as indeterminate, and reductions modulo p are allowed in the numerical coefficients that arise as a result of addition and multiplication. We can derive a modular invariant from any formal invariant by specifying that the coefficients of the forms also take values in $GF[p^n]$. Sanderson proved that every modular invariant arises in this way; i.e.,

To any modular invariant i of a system of forms under any group G of linear transformations with coefficients in the $GF[p^n]$, there corresponds a formal invariant I under G such that $I = i$ for all sets of values in the field of the coefficients of the system of forms.[69]

In 1915 Dickson wrote of Sanderson's dissertation:

This paper is a highly important contribution to this new field of work; its importance lies partly in the fact that it establishes a correspondence between modular and formal invariants. Her main theorem has already been frequently quoted on account of its fundamental character. Her proof is a remarkable piece of mathematics.[70]

The theorem was often cited as "Miss Sanderson's Theorem."

Mildred Sanderson presented her results at a meeting of the American Mathematical Society in March 1913 and received her Ph.D. in June of that year. In October her dissertation appeared in the *Transactions of the American Mathematical Society*,[71] and she began her career as an instructor at the University of Wisconsin. A year later she died after an illness of several months. In 1938, for the semicentennial celebration of the American Mathematical Society, E. T. Bell summarized the major interests and achievements of American algebraists. In his discussion of modular invariants he said, "Miss Sanderson's single contribution (1913) to modular invariants has been rated by competent judges as one of the classics of the subject."[72]

Sanderson and her work were well known to Olive Clio Hazlett (1890–1974), Dickson's second female doctoral student and the most prolific of all the American women working in mathematics before 1940. Hazlett was born in Cincinnati, moved to Boston at the age of nine, and attended public schools there before entering Radcliffe in the fall of 1909; three years later she received her bachelor's degree. From Radcliffe she went to Chicago, where she received an S.M. in 1913 and a Ph.D. in 1915.

Hazlett wrote both her master's thesis[73] and her Ph.D. dissertation[74] under the direction of Dickson in the area of linear associative algebras.[75] The development of these algebras was the result of a gradual enlargement of the concept of number from integers, to rational numbers, to real numbers, to complex numbers, to quaternions, to the more general "hypercomplex numbers." Technically, a linear associative algebra (a hypercomplex system) is a finite-dimensional vector space over a field with a multiplication that is associative and in which the distributive law holds. It is useful to think of a vector space as a set of mathematical objects that can be added together and multiplied by numbers. In an algebra the objects can also be multiplied together, although in general they cannot be divided by one another. One may consider as an example the algebra of real quaternions, namely, the collection of all elements

$$q = xu + yi + zj + wk$$

with the four generators (called basis elements in the present day terminology of vector spaces but called "units" in Hazlett's time) u, i, j, k satisfying

$$i^2 = j^2 = k^2 = -u, \ ij = -ji = k, \ ki = -ik = j, \ jk = -kj = i,$$

and with coefficients x, y, z, w real numbers. This algebra has a multiplicative identity u.

Hazlett observed in her first paper:

> Linear associative algebras of a small number of units, with coordinates in the field of ordinary complex numbers, have been completely tabulated, and their multiplication tables have been reduced to very simple forms. But if we had before us a linear associative algebra, the chances are that its multiplication table would not be in such a form that we could find out readily to what standard form it was equivalent. And so the question arises, "May we not find invariants which completely characterize these algebras?"[76]

In this paper, her master's thesis, she then described such algebraic invariants in the case of algebras with identity, of dimension two or three, over the complex numbers.

A common technique in attempting to understand any mathematical object is to decompose it into simpler objects. A theorem of Wedderburn, which extended an earlier result of E. Cartan, showed that any associative algebra over a field can be written as a sum of a semi-simple algebra and a nilpotent algebra. Since the structure of semi-simple associative algebras was well understood, Hazlett focused on nilpotent algebras in her Ph.D. dissertation and in a subsequent paper published the same year.[77] (A nilpotent algebra is one in which some power of every element is zero.) She classified all, not necessarily associative, nilpotent algebras of dimension four or less over the complex field. She also gave invariantive characterizations of certain other nilpotent algebras.

After receiving her Ph.D. from Chicago in 1915, Hazlett spent the year 1915–16 at Harvard as Alice Freeman Palmer Fellow of Wellesley College. There she continued her work on invariants of nilpotent algebras and obtained a position at Bryn Mawr starting in the fall of 1916. After two years at Bryn Mawr she moved to Mount Holyoke as an assistant professor. This was the same time that Anna J. Pell moved from Mount Holyoke to Bryn Mawr.

After more work on algebras, Hazlett began to work in the field of modular invariants and covariants. (A covariant of a system of forms with variables $x_1, \ldots, x_m$ and coefficients $a_1, \ldots, a_r$ is a function $I(x_1, \ldots, x_m; a_1, \ldots, a_r)$ with properties analogous to those of invariants.) Hazlett's first paper in this area answered affirmatively a question raised by Sanderson "as to whether all covariants of a system S can be expressed as polynomials in L [the universal covariant L is $x^{p^n}y - xy^{p^n}$] and the modular invariants of the system S enlarged by a linear form."[78] She wrote in this paper:

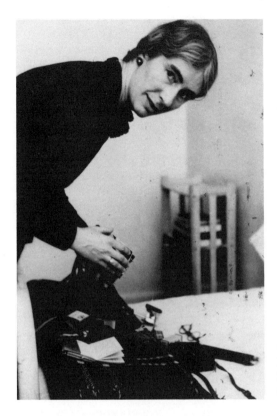

Olive C. Hazlett, starred in the 1927 edition of *American Men of Science*. Courtesy of Grace Shover Quinn.

The chief interest of this theorem seems to be in the light it throws on the very difficult and, thus far, unsolved problem of formal covariants, since the theory of modular covariants is a stepping stone from the theory of modular invariants to the theory of formal covariants.[79]

Four more of her papers in this area appeared within the next two years; she wrote two more, the last of which appeared in 1930. Meanwhile, American algebraists stopped working in the field of modular invariants. E. T. Bell remarked in 1938:

One of the most interesting phenomena of the past thirty years of American algebra is the sudden rise and equally sudden decline of the theory of modular invariants. Activity in this field is at present in abeyance, apparently for lack of an extension (if there is one) to covariants of M. L. Sanderson's theorem on correspondences between modular and formal invariants, and of a proof of the conjecture that all congruent covariants admit symbolic representation.[80]

Hazlett was at Mount Holyoke from 1918 to 1925, as an assistant professor until 1924 and then as an associate professor. During this time, in addition to her work on modular invariants, she presented a paper "On the Arithmetic of a General Associative Algebra"[81] at the International Mathematical Congress in Toronto in 1924. Dickson, in his 1923 book *Algebras and their Arithmetics*,[82] developed a generalization of the theory of algebraic numbers. Here he extended the definition of integer to that of integer in an associative algebra with identity over the rational field. Hazlett's Toronto paper modified and extended Dickson's definition of integer to one suitable in an algebra over an arbitrary field. She developed these ideas in two later papers.

On 11 March 1925, while she was at Mount Holyoke, Hazlett wrote to E. H. Moore that she had neither the time nor the library facilities to pursue her latest ideas in algebra.[83] So later that year, although an associate professor at Mount Holyoke, she accepted an assistant professorship at the University of Illinois. Throughout the 1920s and early 1930s she was very active in the mathematical community. She was a cooperating editor of the *Transactions of the American Mathematical Society* from 1923 to 1935 and was on the Council of the AMS from 1926 to 1928. She was starred in *American Men of Science* in 1927. In 1928 she was awarded a Guggenheim Fellowship and took a leave of absence from Illinois for study in Italy, Switzerland, and Germany. She began her year in Europe by presenting a paper, "Integers as Matrices,"[84] at the International Congress of Mathematicians in Bologna in September 1928. This paper was a continuation of the work described four years earlier in her Toronto talk. She requested and received an extension of her Guggenheim Fellowship and her leave for the year 1929–30; meanwhile, she was promoted to Associate Professor at Illinois.

The last three of Hazlett's seventeen research papers appeared in 1930, although she wrote reviews and remained mathematically active for several more years. She officially retired from the University of Illinois in 1959 as Associate Professor Emerita, having spent the last fourteen years of her appointment on disability leave.

Despite the large number of women who earned Ph.D.'s in mathematics at Chicago, only one woman, Mayme Irwin Logsdon (1881–1967), held a regular faculty position above the rank of instructor in the Chicago mathematics department before 1982.[85] Logsdon was born, attended public schools, and graduated from Hardin Collegiate Institute in Elizabethtown, Kentucky. She married in 1900 but was widowed early. In 1911 she entered the University of Chicago, received her B.S. in 1912, studied there in 1912–13 and again in the summer of 1914, and received an M.A. in 1915. Meanwhile, she went to Hastings College in 1913 as Professor of Mathematics and Dean of Women. After four years at Hastings and two years as an instructor at Northwestern University, she returned to the University of Chicago in 1919, as an honorary fellow, to resume her graduate work in mathematics. She followed Mildred Sanderson and Olive C. Hazlett as

Mayme I. Logsdon, on the faculty of the University of Chicago from 1921 to 1946. Courtesy of the University of Chicago Mathematics Department.

Dickson's third female doctoral student, writing her dissertation on "Equivalence and Reduction of Pairs of Hermitian Forms."[86] She was hired immediately by Chicago as an instructor and remained on the faculty there until her retirement twenty-five years later.

Logsdon's interests soon shifted to algebraic geometry, and she read a paper in this area at a meeting of the American Mathematical Society in December 1922. She studied in Rome in 1925–26 as an International Education Board Fellow. Logsdon frequently taught upper division courses in algebraic geometry, and she directed the Ph.D. dissertations of four students working in this area between 1933 and 1938.[87] In one case, a student wanted to work on "Knotted Varieties," an area without a specialist at Chicago, and Logsdon directed the work, largely because she was the faculty member who was willing to learn it along with the student.

Logsdon always maintained a keen interest in teaching. Her publications after 1925 consisted of abstracts, book reviews, two textbooks, and an expository paper. The two-volume *Elementary Mathematical Analysis*, published in 1932 and 1933,[88] was a text for a year's course in preparation for the calculus sequence. Her second book, *A Mathematician Explains*,[89] was published in 1935 (with a second edition in 1936) and was designed for college students in a general education course in mathematics and for persons whose interest in mathematics was nontechnical. In the book she described the historical development of elementary concepts of arithmetic, algebra, and geometry, and introduced analytic geometry and calculus. Her expository paper "Geometries"[90] appeared in 1938. Logsdon was promoted to Assistant Professor in 1925 and to Associate Professor in 1930. She remained at that rank until she retired in 1946. She moved to Florida after her retirement from Chicago and taught at the University of Miami for fifteen years before a second retirement in 1961.

CONCLUSION

The careers of the nine women whose lives and work we have discussed above illustrate both the difficulties that have confronted women who chose to make mathematics their profession and the success and recognition that some of them achieved despite these difficulties. Christine Ladd-Franklin should have been the first American woman to receive a Ph.D. in mathematics, but Winifred Edgerton Merrill was instead because Johns Hopkins would not grant the degree to a woman. Charlotte Scott created the only significant graduate program in mathematics in a women's college and was the first woman to achieve recognition within the American mathematical community. Mary Winston Newson was the first American woman to earn a Ph.D. in mathematics from a foreign university. Among these early women only Scott remained an active research mathematician throughout her career.

The five women we discussed in the period after 1900 all received their degrees from the University of Chicago. This largely reflects the importance of the Chicago department in graduate education in mathematics during the first third of the twentieth century. Mary Emily Sinclair remained active at Oberlin but published little after her first papers. Anna Pell Wheeler built upon Scott's work at Bryn Mawr while continuing to engage in research. Mildred Sanderson, possibly the most promising, died at age twenty-five after creating a classic in her field, while Olive C. Hazlett's prolific career was truncated by illness. Mayme I. Logsdon, who received her degree at age forty, became, until recent years, the only woman faculty member in the mathematics department at Chicago to attain a regular rank above that of Instructor.

In many ways the women we have discussed illustrate the great variety of careers and levels of research involvement of the American women who received Ph.D.'s in mathematics before 1940. Because we have chosen to restrict ourselves to women who are no longer living, we have omitted those women who are still making important contributions to various areas of the mathematical sciences. Although we have discussed some women who did little or no mathematical research after receiving their Ph.D.'s, we should emphasize that the incidence of women who continued to publish research articles after their Ph.D.'s is far too high for us to have mentioned all of them.

The women discussed in this paper are not isolated examples. They are not just the occasional prodigies who had rare opportunities to acquire the education necessary for real achievement. Rather, they are part of a mathematical community. In short, for as long as there has been a substantial American presence in the mathematical world, women have contributed to American mathematics far more than is generally recognized.

NOTES

1. David Eugene Smith and Jekuthiel Ginsburg, *A History of Mathematics in America before 1900*, The Carus Mathematical Monographs (Chicago: Mathematical Association of America, 1934; reprint, New York: Arno Press, 1980), p. 194.

2. M. Winston, "Eine Bemerkung zur Theorie der hypergeometrischen Function," *Mathematische Annalen* 46 (1895): 159–60.

3. In 1957, Edna E. Kramer wrote about twentieth-century European women mathematicians who have attained eminence (Edna E. Kramer, "Six More Female Mathematicians," *Scripta Mathematica* 23 (1957): 83–95). Since then, other works have appeared that attempt to correct the impression that there have been only seven women mathematicians. One example is Louise S. Grinstein and Paul J. Campbell, eds., *Women of Mathematics: A Biobibliographic Sourcebook* (Westport, Conn.: Greenwood Press, 1987), which contains articles about forty-three women mathematicians from Hypatia to the present.

4. One such symposium, sponsored by the Association for Women in Mathematics, was held at Bryn Mawr College, 17–19 March 1982. The papers delivered at that symposium are included in Bhama Srinivasan and Judith D. Sally, eds., *Emmy Noether in Bryn Mawr: Proceedings of a Symposium Sponsored by the Association for Women in Mathematics in Honor of Emmy Noether's 100th Birthday* (New York: Springer-Verlag, 1983).

5. R. C. Archibald, "Women as Mathematicians and Astronomers," *American Mathematical Monthly* 25 (1918): 136–39.

6. A. Rebière, *Les Femmes dans la Science*, 2d ed. (Paris: Librarie Nony et Cie, 1897). First edition published in 1894 with subtitle *Conférence faite au Cercle Saint-Simon le 24 février 1894*.

7. A section on "Women in Mathematics" appears in the annotated bibliography Joseph W. Dauben, ed., *The History of Mathematics from Antiquity to the Present: A Selective Bibliography* (New York: Garland Publishing, 1985), pp. 428–34, and includes earlier bibliographies. One of these appeared in a revised edition in 1983: Phyllis Zweig Chinn, *Women in Science and Mathematics Bibliography*, rev. ed. (Arcata, Calif.: Humboldt State University Foundation, 1983). The literature on women in mathematics has recently undergone a considerable expansion, and neither of these bibliographies can be considered up to date.

8. For a report that includes the statistics cited in this chapter see Judy Green and Jeanne LaDuke, "Women in the American Mathematical Community: The Pre-1940 Ph.D.'s," *Mathematical Intelligencer* 9 no. 1 (1987): 11–23. The authors are preparing a monograph that will include biographical and bibliographical material on all the women in their study.

9. The figures on women Ph.D.'s in all fields are adapted from Table 4 in Lindsey

R. Harmon, *A Century of Doctorates* (Washington: National Academy of Sciences, 1978), p. 17.

10. National Research Council, Commission on Human Resources, *Climbing the Academic Ladder: Doctoral Women Scientists in Academe* (Washington: National Academy of Sciences, 1979), p. 20.

11. Tables summarizing the sex, race, and citizenship of new doctorates appear in the October issues of *Notices of the American Mathematical Society* each year 1973 to 1979 and in the November issues in the years since 1980.

12. Seventeen such women graduated from Wellesley. Other women's colleges, many of whose undergraduates earned Ph.D.'s in mathematics, were Goucher with ten, Hunter with nine, and Mount Holyoke with eight. Among coeducational colleges, the Universities of Kansas and Wisconsin each conferred undergraduate degrees on six women who were awarded Ph.D.'s in mathematics before 1940.

13. For an introduction to nineteenth-century American mathematics see Judith V. Grabiner, "Mathematics in America: The First Hundred Years," in *The Bicentennial Tribute to American Mathematics*, ed. Dalton Tarwater (The Mathematical Association of America, 1977), pp. 9–24; also see Smith and Ginsburg, *A History of Mathematics*.

14. Daniel Coit Gilman Papers, file on Christine Ladd Franklin, The Johns Hopkins University Archives, Baltimore, Maryland.

15. The *Journal für die reine und angewandte Mathematik,* one of the most prestigious mathematical journals through most of the nineteenth century, was universally referred to as *Crelle's Journal* after its original publisher.

16. Christine Ladd, "Crelle's Journal," *Analyst* 2 (1875): 52.

17. Gilman Papers, file on Christine Ladd Franklin.

18. Christine Ladd, "On the Algebra of Logic," in *Studies in Logic by the Members of Johns Hopkins University,* ed. C. S. Peirce (Boston: Little, Brown, and Co., 1883), pp. 17–71.

19. George Boole, *The Mathematical Analysis of Logic* (Cambridge: Macmillan, Barclay and Macmillan, 1847; reprint, Oxford: Basil Blackwell, 1948).

20. Evert W. Beth, "Hundred Years of Symbolic Logic," *Dialectica* 1 (1947): 337.

21. Clarence Irving Lewis and Cooper Harold Langford, *Symbolic Logic,* 2d ed. (New York: Dover Publications, 1959), pp. 60–63.

22. J. McKeen Cattell, ed., *American Men of Science* (New York: Bowker, 1906), p. 113.

23. Dorothea Jameson Hurvich, "Ladd-Franklin, Christine," in *Notable American Women 1607–1950,* ed. Edward T. James, vol. 2 (Cambridge, Mass.: Belknap Press, 1971), pp. 354–56.

24. "Mrs. Merrill, 88, Columbia Pioneer," *New York Times,* 7 September 1951, p. 29. Obituary.

25. "Columbia Honors its First Woman Graduate," *New York Times,* 1 April 1933, p. 17.

26. Letter to Judy Green, June 1982.

27. Helen A. Merrill, "A History of the Department of Mathematics, Wellesley College, from the opening of the College in 1875," Wellesley College Archives, Wellesley, Massachusetts, 1944, p. 15.

28. J. H. Van Amringe, "Columbia University and the Education of Women," *Church Eclectic* 28 (1900): 35.

29. Annie Nathan Meyer, *Barnard Beginnings* (Boston and New York: Houghton Mifflin Co., 1935), pp. 146–47n.

30. The actual title of Edgerton's dissertation is *"Multiple Integrals:* 1) Their Geometrical interpretation in Cartesian Geometry; in Trilinears and Triplanars; in Tangentials; in Quaternions; and in Modern Geometry. 2) Their Analytical inter-

pretation in the Theory of Equations, using Determinants, Invariants and Covariants as instruments in the investigation." A copy of the dissertation is in the Rare Book and Manuscript Library, Columbia University.

31. Winifred Edgerton Merrill, *Musical Autograms* (New York: Schirmer, 1918), p. iii. Music by Robert Russell Bennett.

32. For a list of American women who received Ph.D.'s in mathematics through 1911 see Judy Green, "American Women in Mathematics—The First Ph.D.'s," *Newsletter of the Association for Women in Mathematics* 8 (April 1978): 13 and correction, ibid., 8 (July 1978): 9.

33. A student who scored first class honors on the mathematical Tripos (Cambridge's honors examinations) was called a Wrangler. Scott's exclusion from the 1880 Tripos list caused a furor; the following year women were formally allowed to take the Tripos and, if successful, be listed in the University Calendar.

34. C. A. Scott, "Higher Singularities of Plane Curves," *American Journal of Mathematics* 14 (1892): 301–25; idem, "The Nature and Effect of Singularities of Plane Algebraic Curves," *American Journal of Mathematics* 15 (1893): 221–43.

35. C. A. Scott, "On the Higher Singularities of Plane Algebraic Curves," *Proceedings of the Cambridge Philosophical Society* 23 (1926): 206.

36. F. S. Macaulay, "Dr. Charlotte Angas Scott," *Journal of the London Mathematical Society* 7 (1932): 232.

37. Charlotte Angas Scott, "On Cayley's Theory of the Absolute," *Bulletin of the American Mathematical Society* 3 (1897): 235.

38. Charlotte Angas Scott, *An Introductory Account of Certain Modern Ideas and Methods in Plane Analytical Geometry* (New York: Macmillan, 1894); 2d ed. with notes and corrections (New York: G. E. Stechert, 1924); 3d ed. with corrections and additions, *Projective Methods in Plane Analytical Geometry* (New York: Chelsea Publishing Co., 1961).

39. Scott's doctoral students were: Ruth Gentry, 1894; Isabel Maddison, 1896; Virginia Ragsdale, 1904; Louise Duffield Cummings, 1914; Mary Gertrude Haseman, 1916; Bird Margaret Turner, 1920; and Marguerite Lehr, 1925. Although Gentry's degree was conferred in 1894, the 1908 Bryn Mawr Register of Alumnae lists her degree as 1896, the year that her dissertation was published. Cummings, although formally advised by Scott, lists herself as a student of Henry Seely White of Vassar in Hans Strobbe, ed., *J. C. Poggendorff's Biographisch-Literarisches Handwörterbuch*, vol. 6, part 1 (Berlin: Verlag Chemie, 1936), p. 498. White and Cummings were colleagues at Vassar and, starting in 1915, jointly published papers in the general area of her dissertation topic.

40. Stephen Sargent Visher, *Scientists Starred 1903–1943 in "American Men of Science"* (Baltimore: The Johns Hopkins Press, 1947; reprint, New York: Arno Press, 1975), p. 132.

41. In 1976 Mary W. Gray was elected a vice-president of the AMS, and several other women have since held that office. Julia B. Robinson served as the AMS's first woman president, 1983–1984.

42. Edward T. James, ed., *Notable American Women 1607–1950*, vol. 3 (Cambridge, Mass.: Belknap Press, 1971), p. 719. The entry on Scott is by Marguerite Lehr and appears on pages 249–50 of volume 3.

43. The information concerning Mary Winston comes, in part, from communications we have had with her children and from letters she wrote her family while she was studying in Göttingen. The letters were transcribed and made available to us by her daughter, Caroline Beshers.

44. For information about Grace Chisholm see Ivor Grattan-Guinness, "A Mathematical Union: William Henry and Grace Chisholm Young," *Annals of Science*

29 (1972): 105–86; idem, "Mathematical Bibliography for W. H. and G. C. Young," *Historia Mathematica* 2 (1975): 43–58. For information about Margaret Maltby see Agnes Townsend Wiebusch, "Maltby, Margaret Eliza," in *Notable American Women 1607–1950*, ed. James, vol. 2, pp. 487–88.

45. Although Sofia Kovalevskaia received a Ph.D. from Göttingen in 1874, she was never officially a student there. Her degree was granted *in absentia*.

46. Felix Klein, *Vorlesungen über die hypergeometrische Funktion* (Berlin: Julius Springer, 1933), p. 326. The lectures on which this book is based were given in 1893–94 and were attended by Mary Winston.

47. Felix Klein, *The Mathematical Theory of the Top* (New York: Charles Scribner's Sons, 1897), p. 35n.

48. Mary F. Winston, *Über den Hermite'schen Fall der Lamé'schen Differential-gleichung* (Göttingen, 1897).

49. David Hilbert, "Mathematical Problems," *Bulletin of the American Mathematical Society* 8 (1902): 437–79.

50. Felix E. Browder, ed., *Mathematical Developments Arising from Hilbert Problems* (Providence, R. I.: American Mathematical Society, 1976), pp. 1–34.

51. Letter to Judy Green, 22 March 1978.

52. Other schools that granted many Ph.D.'s in mathematics to women prior to 1940 were Cornell with twenty-one, Bryn Mawr with nineteen, and Catholic with fifteen.

53. We believe Sinclair's dissertation advisor was Oskar Bolza, but we have been unable to document this explicitly.

54. Mary E. Sinclair, "On the Minimum Surface of Revolution in the Case of One Variable End Point," *Annals of Mathematics* (2) 8 (1907): 177–88; idem, "The Absolute Minimum in the Problem of the Surface of Revolution of Minimum Area," *Annals of Mathematics* (2) 9 (1908): 151–55; idem, "Concerning a Compound Discontinuous Solution in the Problem of the Surface of Revolution of Minimum Area," *Annals of Mathematics* (2) 10 (1909): 55–80.

55. Gilbert Ames Bliss, *Calculus of Variations*, The Carus Mathematical Monographs, The Mathematical Association of America (LaSalle, Illinois: The Open Court Publishing Company, 1925), p. 121.

56. Ibid., pp. 121–22.

57. Ibid., p. 122.

58. Mary E. Sinclair, *The Discriminantal Surface for the Quintic in the Normal Form* $u^5 + 10xu^3 + 5yu + \zeta = 0$ (Halle, Ger.: Schilling, n.d.).

59. Pell Wheeler's doctoral students were: Margaret Buchanan (Cole), 1922; Marion Cameron Gray, 1926; Laura Guggenbuhl, 1927; Rose Lucile Anderson, 1930; Olive Margaret Hughes, 1934; Vera Ames (Widder), 1938; and Dorothy Maharam (Stone), 1940. She was officially the advisor for Ruth Stauffer (McKee), 1935, and for Marion Greenebaum (Epstein), 1938, although Stauffer's dissertation was supervised by Emmy Noether before Noether's death in April 1935, and Greenebaum worked with H. W. Brinkmann at Swarthmore.

60. Louise S. Grinstein, "Wheeler, Anna Johnson Pell," in *Notable American Women, The Modern Period*, eds. Barbara Sicherman and Carol Hurd Green (Cambridge, Mass.: Belknap Press, 1980), pp. 725–26.

61. Louise S. Grinstein and Paul J. Campbell, "Anna Johnson Pell Wheeler: Her Life and Work," *Historia Mathematica* 9 (1982): 43–44.

62. The English born and educated Charlotte Scott is excluded from this comparison.

63. Grinstein and Campbell, "Anna Johnson Pell Wheeler," p. 47.

64. Julia B. Robinson gave the Colloquium Lectures in August 1980 at the 84th summer meeting of the American Mathematical Society.

65. Mildred Sanderson, "Generalizations in the Theory of Numbers and Theory of Linear Groups," *Annals of Mathematics* (2) 13 (1911): 36–39. The master's thesis was entitled "The General Linear Group with Respect to a Function and Composite Integer as Moduli."

66. L. E. Dickson, "A Tribute to Mildred Lenora [*sic*] Sanderson," *American Mathematical Monthly* 22 (1915): 264.

67. L. E. Dickson, "Invariants of Binary Forms under Modular Transformations," *Transactions of the American Mathematical Society* 8 (1907): 205–32.

68. Leonard Eugene Dickson, "On Invariants and the Theory of Numbers," in *The Madison Colloquium 1913*, American Mathematical Society Colloquium Lectures, vol. IV (New York: American Mathematical Society, 1914; reprint, New York: Dover Publications, 1961), p. 1.

69. Mildred Sanderson, "Formal Modular Invariants with Application to Binary Modular Covariants," *Transactions of the American Mathematical Society* 14 (1913): 490.

70. Dickson, "A Tribute," p. 264.

71. Sanderson, "Formal Modular Invariants," pp. 489–500.

72. E. T. Bell, "Fifty Years of Algebra in America, 1888–1938," in *Semicentennial Addresses of the American Mathematical Society*, American Mathematical Society Semicentennial Publications, vol. 2 (New York: American Mathematical Society, 1938; reprint, New York: Arno Press, 1980; reprint, Providence, R. I.: American Mathematical Society, 1988), p. 22.

73. Hazlett's master's thesis is "Invariants which Characterize Linear Associative Algebras of a Small Number of Units," 1913. The published version of it is Olive C. Hazlett, "Invariantive Characterization of Some Linear Associative Algebras," *Annals of Mathematics* (2) 16 (1914): 1–6.

74. O. C. Hazlett, "On the Classification and Invariantive Characterization of Nilpotent Algebras," *American Journal of Mathematics* 38 (1916): 109–38.

75. For a discussion of the context of Hazlett's work in this area see Jeanne LaDuke, "The Study of Linear Associative Algebras in the United States, 1870–1927," in *Emmy Noether in Bryn Mawr*, eds. Srinivasan and Sally, pp. 147–59.

76. Hazlett, "Invariantive Characterization," p. 1.

77. Olive C. Hazlett, "On the Rational, Integral Invariants of Nilpotent Algebras," *Annals of Mathematics* (2) 18 (1916): 81–98.

78. Olive C. Hazlett, "A Theorem on Modular Covariants," *Transactions of the American Mathematical Society* 21 (1920): 247.

79. Ibid., p. 249.

80. Bell, "Fifty Years," p. 22.

81. Olive C. Hazlett, "On the Arithmetic of a General Associative Algebra," in *Proceedings of the International Mathematical Congress held in Toronto, August 11–16, 1924*, ed. J. C. Fields, vol. 1 (Toronto: The University of Toronto Press, 1928), pp. 185–91.

82. Leonard Eugene Dickson, *Algebras and their Arithmetics* (Chicago: University of Chicago Press, 1923; reprint, New York: Dover Publications, 1960).

83. E. H. Moore Papers, The Department of Special Collections, Joseph Regenstein Library, The University of Chicago, Chicago, Illinois.

84. O. C. Hazlett, "Integers as Matrices," in *Atti del Congresso Internazionale dei Matematici Bologna 3–10 Settembre 1928* (VI), ed. Nicola Zanichelli, vol. 2 (Bologna: 1930), pp. 57–62.

85. Karen Uhlenbeck was appointed Professor of Mathematics at the University of Chicago in 1982.

86. Mayme I. Logsdon, "Equivalence and Reduction of Pairs of Hermitian Forms," *American Journal of Mathematics* 44 (1922): 247–60.

87. Logsdon's doctoral students were: Anna A. Stafford (Henriques), 1933; James Edward Case, 1936; and Clyde Harvey Graves and Frank Ayres, Jr., 1938.

88. Mayme I. Logsdon, *Elementary Mathematical Analysis*, 2 vols. (New York: McGraw Hill, 1932–33).

89. Mayme I. Logsdon, *A Mathematician Explains* (Chicago: University of Chicago Press, 1935); 2d ed. (Chicago: University of Chicago Press, 1936; Phoenix Science Series, 1961).

90. Mayme I. Logsdon, "Geometries," *American Mathematical Monthly* 45 (1938): 573–82.

Both authors have been Honorary Research Associates in the Division of Mathematics, National Museum of American History, Smithsonian Institution, since 1979. They wish to express their gratitude to the Institution for use of numerous resources and to the curator emeritus of the division, Dr. Uta C. Merzbach, for beneficial discussions and advice.

MARTHA MOORE TRESCOTT

Women in the Intellectual Development of Engineering

A Study in Persistence and Systems Thought

Technological change and engineering of various kinds have always been a part of human lives. The development of the modern engineering disciplines, linked as they are to the sciences and math, began to accelerate after the Renaissance. The construction of the Italian canals and hydraulic works in the fourteenth and fifteenth centuries, and the building of important bridges, roadbeds, and other civil engineering works in France within the next two centuries, gave much impetus to the rise of civil engineering as a scientific discipline. This branch of engineering was carried forward especially by French universities and by West Point in the United States. The coming of Newtonian mechanics and the development of steam engines, particularly in England, laid the foundation for much modern mechanical engineering. Upon it professional groups such as the American Society of Mechanical Engineers (1880) were founded. Developments in the professionalization of civil and mechanical engineering led to similar developments in the field of electricity and chemical technologies in the nineteenth and early twentieth centuries. Professional groups of engineers in a variety of specialties, such as refrigeration, heating and ventilation, computer applications, and bioengineering, have subsequently continued to proliferate.

Throughout human history, women have contributed to technological development in major ways. Indeed, before the modern era, women and technology were not seen to be antithetical as they are today. As the various scientific fields became professionalized, however, women technologists, theorists, and inventors were not generally welcomed or included in the sciences. Engineering was no exception in this trend. Yet, even in this atmosphere, some women did persist, and many have made significant contributions, to engineering and to the other sciences and technologies as well.

There is much more substance to the history of women in engineering than historians of women and of science and technology have generally believed. It is true that in terms of numbers which form a "critical mass" women engineers have been an insignificant factor until the 1970s, having constituted one percent or less of the total engineering population of the United States. However, in terms of women's impact upon the theory and practice of their respective engineering fields historically, they have often had an inordinately great effect. Many women have become successful engineers and have contributed to the formulation of entirely new systems of thought and design, often establishing new paradigms in their areas. As women in this "ultramasculine" field, engineers like Lillian Gilbreth and Edith Clarke have faced far greater hurdles than those which have had to be faced by either men or women who entered other professions. These women therefore had to be particularly persistent, determined, shrewd, and intelligent. The 1970s represent an unprecedented phenomenon in the history of engineering in terms of the number of women entering engineering fields; this was largely due to a combination of factors, including the influence of affirmative action, the women's movement, and high salaries.

Since engineering deals with systems, those who can think in the context of systems can make special contributions to the evolution of engineering thought. Early women engineers seem to have been particularly able systems thinkers. So people such as Lillian Gilbreth, Edith Clarke, Kate Gleason, and Emily Roebling probably had much more impact on their fields than their small numbers would imply. These early women engineers invariably report that they were assigned a variety of leftover, odd jobs (even a good bit of "dirty work") which no one else wanted. Since it was difficult for women to obtain engineering jobs, they frequently took what they could get. The variety of menial tasks which these women performed year after year must surely have been onerous. But such assignments also meant that these women had to master a wide variety of areas in their fields, thus becoming very well acquainted with an entire engineering system and work environment. This wide-ranging, basic knowledge then served to shape their overview of a system of work, thought, and design.[1]

Because of the great barriers they had to overcome, women engineers seem to have been, in general, more persistent than their male counterparts.[2] Despite this, and even though they were often highly visible during their careers due to their extreme underrepresentation, they have not been adequately recorded, remembered, and assessed. Even the contributions of the greatest intellects among them have been buried in various ways. And, although the record and significance of the work of women in other professions, and in science particularly, have also frequently been lost to history, ascertaining the history of women in engineering may be especially problematical.

For one thing, engineers and technologists have historically left fewer written documents than have other scientists, since engineers have been more oriented to invention and design than to writing. For another thing, while modern science, medicine, and other professions have typically been very male-dominated, engineering has been extremely so. Of all professions, only the ministry and engineering have been so underrepresented historically by women. Until the 1970s women represented far less than 1 percent of all practicing engineers; by the early eighties they constituted only 2.5 to 3.0 percent of the total engineering labor force. In 1960, 4.2 percent of all physicists, 8.6 percent of chemists, 26.4 percent of mathematicians, and 26.7 percent of all biological scientists were women.[3] Third, and related to the second point, women who have become engineers and (or) who have contributed to the evolution of engineering theory and practice may well have been educated or trained in other fields. Until fairly recently, many women who have excelled in engineering were not "really" engineers, having been educated primarily in physics, chemistry, mathematics, psychology, education, and other areas. To cover adequately the history of women engineers in general, one must include women in engineering who have earned engineering degrees and those who have not, those who are licensed engineers and those who are not, those working in engineering but with degrees in other areas, and even those who "dropped out" of engineering work and who are or have been ostensibly working in other areas but who still use their engineering training in these positions.[4]

The lack of role models, mentors, and counseling and other information about schooling and jobs has been severe, and women have often entered engineering after many years or after careers in other areas (such as secretarial work and teaching) and often very indirectly. Many of those who lacked fathers, brothers, and other male relatives who were engineers have said that they just "fell" into engineering by a somewhat haphazard process.

In treating specifically the history of the intellectual contributions of women to engineering, one encounters very great problems. Written documents, particularly primary sources, are difficult to locate, especially in sufficient wealth to permit detailed assessment. Engineers and technologists often were not prolific writers, and they may not have thought their letters and notes particularly worth saving. Since their thoughts often pertained to the more pragmatic areas of life and since engineers necessarily must consult with people and work in the "real" world—even academic engineers—their efforts may at times have been recorded as part of a team and, therefore, are difficult to assess individually. Furthermore, women engineers were conditioned, along with other women, to be modest and to allow their contributions to be subsumed under a man's—in some cases, a relative's—name. And, in general, credit has not been forthcoming for women engineers from male peers.

Finally, in the intellectual history of women in engineering, a great and disparate variety of fields must be surveyed. One must include not only the basic engineering fields, such as electrical, mechanical, chemical, and civil, but also industrial, metallurgical, aeronautical, ceramic, marine, biomedical, general, and many other engineering specialites as well. Because the task is so formidable, and this chapter covers such a variety of fields, only the case study approach is feasible. Therefore, only a few examples from different areas of engineering will be cited. In many cases, these women have helped to institute new paradigms in their fields.

There are many omissions here, not only because of lack of space but also because of the burial of the achievements of women engineers, even in more recent decades. With ever-greater numbers of degreed women engineers working in all areas, with more women engineering Ph.D.'s than ever before, women engineers still win very few awards and honors from professional engineering societies. For instance, in 1980 the National Society of Professional Engineers commented, "Although NSPE has some awards for which women engineers would be eligible, to date no woman has received any of our awards. I hope we change that soon, but—."[5] The following accounts show just how the work of women in engineering has often been obscured.

ELLEN SWALLOW RICHARDS AND THE ECOLOGY MOVEMENT

Ellen Swallow Richards (1842–1911), who was denied an earned doctoral degree in chemistry at the Massachusetts Institute of Technology in the 1870s, has been called "the woman who founded ecology."[6] If this is a somewhat grandiose claim, it nevertheless reveals that she, like various other women in engineering, was a systems thinker.

Ellen Richards, although not an engineer by training, contributed much to the establishment of forerunners of environmental and sanitary engineering. She was associated with the MIT Chemistry Department from the time she earned a bachelor's degree there in 1873. From 1873 to 1878, she taught there without title or salary. In 1878 she became an instructor in chemistry and mineralogy in the Woman's Laboratory, which she helped found. From 1884 to 1911 she was Instructor of Sanitary Chemistry at MIT. She performed numerous analyses of water and gas for the State of Massachusetts and others and also became a specialist in analysis of metals and minerals.[7]

Ellen Swallow Richards authored more than fifteen books and numer-

Ellen H. Swallow. From Caroline
L. Hunt, *The Life of Ellen H. Richards*, Boston 1912.

ous articles and reports, including *Home Sanitation, Cost of Living, Air, Water
and Food, Sanitation in Daily Life, Industrial Water Analysis*, and *Conservation
by Sanitation*.[8] It may not be a gross overstatement to call her the founder
of ecology. Certainly, she was a pioneer in the science and engineering of
the environment.

It is difficult to do justice to all the contributions Ellen Swallow Richards
made to the understanding of the total human environment. In promoting
concepts of environmental systems, she exhibited holistic thinking par excellence. With a thorough grounding in chemistry, she began her study of
the environment with analysis of one of its most crucial aspects—water.
And, as Robert Clarke has noted, "water analysis was a new branch of
science" in the 1870s.[9] Ellen Swallow Richards was first introduced to it
through one of her professors at MIT, William R. Nichols (who, ironically,
had not believed in admitting women to MIT). As his best student, she
performed extensive analyses on Massachusetts sewage and water supplies
under the auspices of the Massachusetts Board of Health in 1872, and
Clarke has stated that this study "made her a preeminent international
water scientist even before her graduation."[10] From the use of chemistry
for the study of water, she then began analyses of the earth's minerals

under another MIT professor and mineralogist and her future husband, Robert H. Richards. She distinguished herself also in this area, especially in her detection of small amounts of vandium in various ore samples, a rare and very difficult metal to detect at that time. Although still a student at MIT, she was gaining a worldwide reputation for her chemical genius in environmental analyses. Her husband has written:

> My wife's work under Professor Nichols for the Massachusetts State Board of Health was arduous and extensive. Hundreds of analyses of water were almost entirely done by her. Much of the good work that was done later by Dr. Drown, whereby the standards of purity for wells, by means of curves called isochlors, was established for the whole State of Massachusetts, had its foundation in her work at this period. . . .
>
> My friend, David Browne of Coppercliff, Ontario, was seeking information about his copper ore from the Coppercliff Mine. He sent samples to a number of assayers, and among others, to Mrs. Richards. All the others returned results in copper and, I dare say, they did not know that they were to look for anything else. She, on the other hand, gave him a percent of copper in the ore and also reported five percent of nickel. This, I believe, was the beginning of the great nickel industry of which the Coppercliff Mine was the center. David Browne always said that Mrs. Richards was the best analyst in the United States.[11]

Her husband went on to report that in 1879 she was the first woman to be elected to the American Institute of Mining and Metallurgical Engineers, an honor which pleased her very much. (Incidentally, he was, as Margaret W. Rossiter has noted, vice-president of AIMME at the time.)[12]

In a file of her letters at MIT is correspondence which shows that Ellen Swallow Richards worked with both the Mining and Metallurgical Laboratory there and the Chemical Laboratory in performing analyses, as well as with the state board. Also in these letters one can see her concern for sanitation and the environment. She wrote Dr. Noyes, MIT president, about the state of heating and ventilation in certain MIT buildings in 1907–09, saying that "one of the most serious problems of civilization is clean water and clean air, not only for ourselves but for the world."[13] Dr. Noyes responded that "it would seem that an investigation of it [ventilation of MIT buildings] in detail with the purpose of discovering the most serious evils and suggesting practicable remedies for them would form a suitable and valuable thesis for one or two students in sanitary engineering or in the heating and ventilating option of the mechanical engineering course."[14]

She became an authority in analysis of food and the human diet, another vital aspect of the environment, and also on the subject of safer and healthier buildings and their design. She became so committed to pure air and water that she almost completely redesigned and remodeled the house which she and Robert Richards occupied in the Boston area, placing particular emphasis on its heating and ventilation systems. Clarke describes what she did:

> She checked and adjusted the plumb and fit of pipes through the house, replaced most with modern-seal joints, put in traps and other precautions for waste water, discarded the old lead poisoning water lines. A hand pump in the kitchen pulled water up from the well into a storage tank on the second floor for bath and toilet. This was before the municipal mains were laid in her neighborhood. She redesigned an inefficient water heater in the basement, replacing its input pipe and burner so that water would heat faster with less fuel. She put a "water back" on the furnace, using the heat from it in the winter and the water heater itself in the summer. . . . These devices became industry standards, but neither Robert nor Ellen tried to patent them. . . . Working with new knowledge of air analysis gained at the Institute, the Richards designed and installed a mechanical system of ventilation and circulation, a radical innovation in homes of any structure of their day.[15]

This was environmental engineering to an impressive degree. One major significance of this undertaking, aside from the many engineering innovations involved, is that this house became a consumers' testing laboratory, and she housed and taught students and others there. Some of her most important instruction actually occurred in the the "Center for Right Living," as she called it.

Ellen Swallow Richards viewed her extensive work in the organization and education of women as a part of her environmental work. She felt that women, being the center of family life and therefore perhaps the most critical part of the human environment, needed to be educated about diet, adulteration of foods, proper ventilation of homes, and other features of home life to promote the health and safety of all. This viewpoint may appear, to a later age, to be advocating that women stay "in their place." However, no other woman in American history labored so tirelessly and successfully on behalf of women in science. To understand the extent of Swallow Richards's contributions, we must understand the context in which she lived and worked. There were virtually no opportunities for women to receive scientific education. The conditions under which most housewives lived were so unhealthy and unsanitary that they had to struggle to keep themselves and their families alive and even minimally free of disease.[16]

In 1876, the Woman's Laboratory at MIT, which Ellen Swallow Richards founded, introduced the first course in biology at MIT. She was very instrumental in establishing a life sciences curriculum there. This soon materialized into a full biology department. At that time, biology was not an acceptable area of instruction at schools of applied science. In 1892, at a public lecture in Boston, Ellen Swallow Richards called for the "christening of a new science"—"oekology."[17] The two main branches of this new

Women's Chemical Laboratory, Massachusetts Institute of Technology, opened 1876, torn down 1883. From Hunt, *The Life of Ellen H. Richards.*

science were the consumer-nutrition and the environment-education movements. As the twentieth century approached, Swallow Richards's interdisciplinary science of oekology fell victim to increasing specialization in the sciences. The study of environment became the province of the male-dominated life sciences (many of which she had helped to found and whose guiding lights she had taught). In this process environment became identified with plants and animals, and Swallow Richards's focus on the human environment was lost.[18]

The consumer-nutrition movement was carried forward more by women and by those untrained in science. Never a separatist in women's issues, and always identified in her own mind with the educated scientific community, Ellen Swallow Richards nevertheless found herself at the helm of this branch of oekology. It was not long before Swallow Richards's oekology was labeled "domestic science" and then "home economics." She was a founder of the American Home Economics Association and also of the New England Kitchen, one of the most innovative experiments ever undertaken in the daily, scientific preparation of wholesome foods for public consumption.

In addition, Swallow Richards continued to teach various courses at MIT, including sanitary engineering, a course which she introduced and first taught. In these classes, she educated "the men who went on to design and operate the world's first modern municipal sanitation facilities. 'Missionaries for a better world!' she lectured them."[19]

Ellen Swallow Richards at her desk in the laboratory. From Hunt, *The Life of Ellen H. Richards.*

Ellen Swallow Richards also helped found the Seaside Laboratory, which later became the Marine Biological Laboratory at Woods Hole, where the sciences of both oceanography and limnology were developed. But although Ellen Swallow Richards played a major role in the rise of these sciences, she has not been credited, as Robert Clarke recounts in some detail. Instead, men associated with the Massachusetts Board of Health and with MIT, many of whom she had taught and others with whom she had worked for years, were hailed as the "Father of Modern Sanitation" and the "Father of Public Health." But as Clarke comments, "if these were the 'fathers' of their individual fields, she was the 'mother' of them all."[20]

As a part of her early work in the consumer-nutrition movement, Swallow Richards had also been indirectly responsible for the design and introduction of the Aladdin Oven, which was a forerunner of today's ovens in home and industry. It was invented by the president of the Manufacturers Mutual Insurance Company, industrialist Edward Atkinson, whom she had met in conjunction with her survey on fire prevention in factories. This survey had, in turn, stemmed from an analysis she had undertaken of lubricating oils for factory machines and from the results of this analysis, which pointed to the combustible nature of many of these oils. She had then begun to develop noncombustible oils for machines to reduce the

The Water Laboratory, showing the famous Chlorine Map. From Hunt,
The Life of Ellen H. Richards.

frequency of industrial fires. This work led her and Atkinson to the design
of fire-resistant factories, which were "copied throughout industry"; much
of the systems engineering was reminiscent of that seen in the remodeling
of her own home. She ultimately became an authority on industrial and
urban fires and began to survey schools and other public buildings.[21]

Her work, then, also involved much safety engineering as well as sanitation engineering, and she set up standards in many of the areas in which
she worked. The development of isochlors, or the Normal Chlorine Map,
which serves as "an early warning system for inland water pollution," is
one of her lasting original contributions, along with the "world's first Water
Purity Tables and . . . the first state water quality standards in the United
States."[22]

Ellen was fortunate to be married to a fellow MIT scientist who openly
praised her work. Hers is a case where marriage did not seem to hamper
her work but perhaps enhanced it. From her student days until after her
death, her husband did not seem jealous of her and, in fact, seemed proud
of her and desirous of promoting her work among those he knew and in
his writings. (However, their marriage in 1875 may help to account for
MIT's unwillingness to pay her in the period 1875–1878.) Robert H. Richards also seemed to recognize her talents as a systems thinker. Perhaps
his influence is one reason she has been as well recorded and remembered
as she has been. Nonetheless, the extent of her work in founding many

of the ecological sciences and related fields in engineering, and in the consequent introduction of entirely new paradigms, has not generally been recognized.

LILLIAN MOLLER GILBRETH AND THE RISE OF MODERN INDUSTRIAL ENGINEERING

Dr. Lillian Moller Gilbreth (1878–1972) is probably the best-known woman engineer in history. Perhaps the illumination of the "burial" of the most impressive intellectual contributions of "America's first lady of engineering" will set the stage for later discussions of less visible, less popular, or less famous women engineers.[23] By far, more primary sources, more archival materials and more published works are available for her life and work than for any other woman engineer.[24]

Lillian M. Gilbreth has certainly been considered important in the history of engineering. Her contributions to the rise of modern industrial engineering have been acknowledged in various ways by engineers and by people outside the field of engineering. She "pioneered in the field of time-and-motion studies, showed companies how to improve management techniques and how to increase industrial efficiency and production by budgeting time and energy as well as money."[25]

In 1904, Lillian Moller married Frank Bunker Gilbreth, another pioneer in scientific management, who was especially noted for his very real genius in motion study. The Gilbreths worked together in many areas: in scientific management in their consulting firm (which advised many companies and became one of the most important firms in the United States); in research and writing (together they authored hundreds of documents); in lectures at various companies, universities, professional societies, and elsewhere; in conducting the Gilbreth summer schools on management topics; and in raising their twelve children (*Cheaper by the Dozen* is their story, written by a daughter and son).

Yet Lillian has received far less credit than has Frank from engineers and historians alike. She has been considered an adjunct to Frank, even though he died an untimely death in 1924, while she headed Gilbreth, Inc. for decades afterward. She has been considered primarily his assistant or his disciple, even though he never earned a college degree and she attained the Ph.D. Her own expertise lay in the realm of integrating psychology and considerations of mental processes with time-and-motion work, while it is recognized among their colleagues that "if Frank Gilbreth slighted any discipline in his consideration, it was psychology."[26] If she is cited by historians, discussions are often limited to her contributions to domestic engineering (e.g., design of kitchen and appliances).[27] Lillian outlived Frank by nearly fifty years and was vigorous and professionally active

Lillian and Frank Gilbreth and children. With permission from *Industrial Engineering,* vol. 4 (February 1972).

almost until the time of her death in 1972. During this time she not only headed Gilbreth, Inc. (in effect she had been at its helm during much of Frank's lifetime, too) but also became a full professor of management in the School of Mechanical Engineering at Purdue University in 1935. She had succeeded Frank as lecturer there in 1924. She became head of the Department of Personnel Relations at Newark School of Engineering in 1941, and visiting professor of management at the University of Wisconsin at Madison in 1955. She received many honorary degrees (both master's and Ph.D.) in engineering from the 1920s on.[28]

In short, she deserves to be recalled and viewed in her own right, and not merely listed together with Frank as "the engineer, inventor, psychologist, educator Gilbreth."[29] While the first two terms may well describe Frank, Lillian was not only an engineer but also the pyschologist and main educator of the couple. Despite the recognition and publicity of various kinds given Lillian Gilbreth in the past, the depth and breadth of her contribution to the establishment of modern industrial engineering have not been well understood or widely discussed by historians or by engineers. Her major contributions lie in two directions: 1) the incorporation of psychological considerations, as conceived in broad terms (problem solving and the behavior of individuals, and related topics such as incentives, the nature of the work environment, monotony, the transference of skill among jobs and industries, and so on), into time-and-motion thought and study; and 2) the establishment of industrial engineering curricula in engineering schools in this country and around the world.

With graduate studies in psychology at Brown (she obtained her Ph.D. in 1915), Lillian Gilbreth was perhaps the best-trained psychologist at the time who was also interested and working in time-and-motion study. Certainly neither Frederick W. Taylor, who is associated with innovations in time study and most often cited as the founder of scientific management, nor Frank Gilbreth had had any special training in psychology to enable them to analyze in a systematic and professional way areas dealing with psychology, work, and management. Since workers the world over often resisted the introduction of time-and-motion concepts quite strongly, it is doubtful that the work of either Taylor or Frank Gilbreth would have been wisely, if at all, accepted without the shrewd application of psychology to time-and-motion work. Lillian's insights and those of her students and other workers (both male and female) for whom she was a guide and authority in this area helped reduce workers' resistance. Such applications of psychology, along with Frank's work in physiological areas, helped establish the study of human factors in engineering design.

Of her many articles, books, reports, and lectures, her early book *The Psychology of Management*, which stemmed from her Ph.D. research, is the most important in the history of engineering thought. This book was termed by George Iles (who subsumed her work under Frank's name) as a "golden gift to industrial philosophy."[30] When the Society of Industrial Engineers made Lillian an honorary member in 1921 (she was the second person to be so honored; Herbert Hoover was also made an honorary member of SIE), it was commented that

> she was the first to recognize that management is a problem of psychology and her book, *The Psychology of Management*, was the first to show this fact to both the managers and the psychologists. This book had a very small sale for two years after its publication but the demand has continually increased, until today it is recognized as authoritative.[31]

In the literature of scientific management before World War I, there was little coverage of such topics as the psychology of work and management, and it was in this relative vacuum at the early date of 1914 that *The Psychology of Management* appeared. It is true that others such as Hugo Munsterberg had studied industrial psychology at about the same time that Lillian Gilbreth began her research in psychology and management. However, as Robert T. Livingston (professor of industrial engineering at Columbia) commented in 1960, "Munsterberg's writings went largely unrecognized" for a long while. Furthermore, no one before Lillian Gilbreth in *The Psychology of Management* had brought together the basic elements of management theory, which are 1) knowledge of individual behavior, 2) the theory of groups, 3) the theory of communication, and 4) a rational basis of decision-making.[32] Although not always using this modern terminology, Dr. Gilbreth dealt with all of these areas, some in more depth than others.

To the modern observer, "psychology" denotes a fairly well-prescribed area, but it is misleading to apply modern usage of that term to discuss the entire content of Dr. Gilbreth's book. Her subtitle, *The Function of the Mind in Determining, Teaching and Installing Methods of Least Waste*, more nearly captures the scope. Hers certainly is not primarily a concern with the field of "industrial psychology," as both contemporary and modern psychologists and others might have understood that field. Indeed, she was writing in the context of scientific management, which had previously focused mostly on the physiological rather than the mental and emotional characteristics of workers and managers. Her book analyzes in detail the "function of the mind," or problem solving, decision making, planning, communicating, measuring, and evaluating in various work and managerial environments.

Throughout *The Psychology of Management*, she is able to transform time-and-motion study into the rudiments of modern managerial practices. In setting forth the value of her approach, she draws on the literature available, including her husband's work, but it is clear that her analysis represents a new point of departure in management. Her insights into the process of disciplining the work force, for example, and her analysis of the inherent conflicts of interest if disciplinarian and foreman are one and the same person show simultaneously the stark contrast not only between her ideas and the approach of "traditional," older ways of management but also between her approach and that of Taylor and his male disciples. She indicates that the foreman or disciplinarian must consider not only a specific act committed by a worker but also the history of this worker's behavior, his or her physical condition, the relative effectiveness of different kinds of and settings for discipline for a particular person, the sensitivity of the person being disciplined, identification of any ringleader and the necessity of disciplining that person too, and so forth. And she then says that the words "disciplinarian" and "punishment" as employed by scientific management "are most unfortunate."

> The "Disciplinarian" would be far better called the "peacemaker," and the "punishment" by some such word as the "adjustment." . . . The aim is, not to put the man down, but to keep him up to his standard, as will be shown later in a chapter on Incentives.[33]

Her analysis here also demonstrates her empathy with how workers *feel*. There was little in the literature of scientific management at that time about sensitivity to workers' feelings, perhaps, ironically, since labor often felt very strongly opposed to the new management. In that area, she was very far ahead of her time, since the rise of sensitivity training in management has really only blossomed since the 1960s.

From her chapter on individuality it is easy to see her pioneering efforts in management theory, especially in "knowledge of individual behavior." And yet not even this most obvious contribution is noted in the American Society of Mechanical Engineers volume covering *Fifty Years Progress in Management, 1910–1960*. Munsterberg, Gillespie, Lecky, and other psychologists are cited but not Lillian Gilbreth (who, incidentally, was co-auther of the introductory, overview essay).[34]

In 1911 Lillian Gilbreth introduced the first mention of the psychology of management at any management meeting, at the Dartmouth College Conference on Scientific Management.[35] In 1924 Lillian commented on a paper by H. S. Person, "Industrial Psychology," delivered at the Taylor Society. In referring to the Dartmouth conference, she said, "It is now almost thirteen years since the importance of [the relationship between psychology and management] was stressed before those interested in scientific management. . . . For seven years before that time, steady progress had been made in correlating psychology and management, but from that time on the correlation was placed upon a scientific basis."[36] Recall that Lillian had received her Ph.D. in psychology in 1915 and had married Frank in 1904. (It is interesting that Frank's many publications are typically post-1904.) Since this work formed the basis for her dissertation,[37] which she was working on before 1915, it is clear that she had been among those studying the interfaces between psychology and management—both in industry and in the home—during the "seven years" before 1911. It is not an overstatement to term her a foremost pioneer in this area in the earliest days of such work.

In the Gilbreth Collection at Purdue, it is interesting to note a brief career sketch of her work, differentiating her contributions from Frank's. (She wrote this as part of her application for membership in the American Society of Mechanical Engineers (ASME), which had not exactly sought her as a member, despite her renown.)[38] Between 1904 and 1914, when Frank B. Gilbreth, Inc. "operated as a construction company," she said, "I was chiefly employed in the systems work, standardizing practice. The results were published in *Field System*, *Concrete System*, and *Bricklaying System*"—all of which show only Frank's authorship. She continued:

I was also engaged in the perfecting of the methods and devices for laying brick by the packet method, and in the design and construction of reinforced concrete work. This work had to do with the management as well as the operation end. . . .

1914–1924. In 1914 our company began to specialize in management work. I was placed in charge of the correlation of engineering and management psychology, and became an active member of the staff making visits to the plants systematized in order to lay out the method of attack on the problems, being responsible for getting the necessary material for the installation into shape, working up the data as they accumulated, and drafting the interim and final reports. I was also in charge of research and teaching, and of working up such mechanisms, forms and methods as were needed for our type of installation of scientific management, motion study, fatigue study and skill study. These had to do not only with the handling of men, but with the simplification and standardization of the machinery and tools, for the use of both the normal and the handicapped. During Mr. Gilbreth's frequent and prolonged absences both in this country and abroad, I was in responsible charge of all branches of the work. This was also the case while he was in the service, and while he was recovering from his long illness incurred therein.

1924. Since Mr. Gilbreth's death, June 14, 1924, I have been the head of our organization, which consisted of Consulting Engineers and does work in management, and have had responsible charge of the research, installation and the teaching, in this country and abroad.[39]

As both Lillian and Frank conceived the rise of scientific management, teaching was integral for implementing and disseminating its practice. It was new and unfamiliar and was sometimes resisted: it had to be taught within firms, in the universities, and to others. Lillian Gilbreth was a brilliant teacher as well as a researcher. She had a grasp of theory and practice, and knowledge of the evolution of the field, and could teach the methods of scientific management to workers and managers. Many of her colleagues noted she had great tact and diplomacy, which served her well in integrating developments in industry and the universities.

Her upper-level undergraduate and graduate courses in management at Purdue in the 1930s and the 1940s were "open only to graduate students and to seniors of outstanding ability" and covered "investigation of specific management problems in the fields of organization, time and motion study, industrial accounting, factory layout, economic selection and equipment, and similar topics." By the 1940s she was listed with the faculty of industrial engineering at Purdue, whereas earlier in the Purdue catalogues her courses had been listed with general engineering or with industrial management.[40]

She was an invited speaker on numerous college campuses throughout her life, and addressed not only engineering students, but also at times women students specifically. She authored papers encouraging women to go into industrial engineering and management in this country and abroad, and at the Gilbreth summer schools, which she conducted, at least half the participants from various countries (mostly European) were women.[41]

Indeed, Lillian's work with and on behalf of women—from the handicapped homemaker to the female worker in the factory and office, to the professional in management and engineering, and to women consumers in general—has not begun to be illuminated. Her efforts on behalf of women in various roles and jobs and in different social and economic classes are a forgotten chapter in women's history.

Certainly, not only her work with and for women but also her various contributions to industrial engineering and its precursors, spanning nearly seven decades, have been only partly and vaguely acknowledged. She, in fact, helped formulate much of the theoretical underpinnings of the field, but she has been too narrowly labeled a "psychologist."[42] The precursors of the modern notion of "the work of a professional manager," in terms of "planning, organizing, integrating and measuring," as in Harold Smiddy's conception, can be seen in Lillian's published works, including *The Psychology of Management*, and in the records of studies she did for various firms, held in the Gilbreth Library. Yet when Smiddy and others today view the evolution of ideas about the function of the mind in management, Lillian Gilbreth's pioneering work is typically not mentioned.[43]

After Frank's death in 1924, she continued to be a prolific writer and to participate in meetings of professional groups such as the ASME, presiding over sessions such as one on the "management researcher" in 1933.[44] Even before her husband died, the Gilbreths together authored well over fifty papers on scientific management topics, not including those written by each of them as sole author nor including their books and consulting reports. And while Lillian's name alone appears on a few of these papers, it is not difficult to suppose that she was the main author on at least such works as "The Place of the Psychologist in Industry," "The Individual in Modern Management," "Psychiatry and Management," "The Relation of Posture to Fatigue of Women in Industry," and others.[45] She may well have been the principal investigator and author on articles credited to them both in the areas of fatigue, standardization, and transference of skill, but that is difficult to determine. The fact is that her own originality was buried, not only because she was married to a man in the same general field but also because of Frank's wish that both their names appear on all they wrote, even though this did not always happen.[46] Yet, because her own expertise and that of her husband were so clearly differentiated in many areas, and because she outlived him long enough to establish authority "in her own right," it is possible, at least in part, to resurrect her unique contributions.[47]

It is important to remember the context of Lillian's work, which emphasized the human element in scientific management. Being among the first to be so concerned with the human factor meant that one undertook to explore a frontier. Many subject areas were legitimate aspects of her research, since the human element is pervasive in all areas of work. As she and others who came after her worked, an increasing number of avenues for investigation were opened and a certain definition of the field evolved. *The Psychology of Management* was a wedge, opening whole new areas to scientific management, which have later evolved into mainstream topics in industrial engineering. *The Psychology of Management* was a departure from classical scientific management and formed a basis for much modern management theory.

Because of her exceptional longevity, her creativity and productivity in her consulting work, her research and publications, her lectures, courses, and workshops all over the world, and, therefore, her prestige and popularity, Lillian Gilbreth may have contributed more in the first four decades of this century than any other single person to defining industrial engineering and its major areas of investigation and analysis. Yet, even though she did remain very active for a long time and was a prolific writer, it is alarming that history has ignored many of the most significant contributions of Lillian Moller Gilbreth, "Member No. 1" of the Society of Women Engineers and perhaps the foremost woman engineer in history.[48]

OTHER EARLY WOMEN ENGINEERS

Having thus set the stage with a detailed analysis of how the work of even the best-known woman engineer was partly hidden by history, we can now view briefly other major thinkers who have also been very important in the history of ideas in engineering and technological change and who, as with Lillian Gilbreth, have not generally been fully credited. Those chosen for analysis here are ones for whom certain primary sources have been located and of whose contributions we can be certain. Certain other early women who deserve mention but whose work must remain more obscure because of the lack of documentation will be covered in the next section. We will first discuss some of those who made their major impact before 1950, and will follow with mention of later women in engineering.

EDITH CLARKE

From the field of industrial engineering we now turn to electrical engineering and the work of Edith Clarke during the same period—the first four or five decades of this century.

Edith Clarke (1883–1959) was an outstanding mathematician and theorist in the field of electric power, having worked for General Electric for twenty-six years. As her friend and colleague Professor C. N. Weygandt commented, there were "relatively few good theoreticians in the power field . . . and Edith Clarke was one of the best . . . but there haven't been too many—not nearly as many as there have been doing theoretical work in communications and computers and electronics and things of that sort."[49] It is possible that because of the relative lack of such theorists, Edith Clarke's contributions achieved more recognition than if she had been working in another area such as electronics. However that may be, there was another woman, Bertha Lamme, who worked in the area of power engineering and was also a mathematician. Lamme worked for GE's major competitor, Westinghouse, and, as will be seen, her work has mostly been unrecognized.

Edith Clarke began her work with the electrical industry in 1920, after having received a master's degree in electrical engineering from MIT in 1919 (the first woman to be awarded such a degree at MIT). Yet finding a job proved difficult, even at General Electric, where she ultimately worked (even though Katherine B. Blodgett, a physicist, had "joined GE . . . in 1918").[50]

> Where better to start than at either Westinghouse Electric or General Electric corporations? But it wasn't that easy. "Neither company had an opening for a woman engineer," says Miss Clarke now [1947]. "I had letters of introduction, but letters of introduction don't do anybody any good. People are more polite because of them, but they don't get you a job."
>
> General Electric, however, did offer her a job directing the calculations in the turbine engine department. Again she found herself supervising a group of girls [sic] in much the same work she did at AT&T. She wasn't satisfied. After all, wasn't she an electrical engineer? After two years in Schenectady, she heard of an opening as instructor at the Constantinople Women's College, which was in Turkey. . . . She got the job.
>
> . . . When she got back to the U.S., she was offered another job with General Electric at Schenectady . . . this time as an electrical engineer. Her new work was that of analyzing problems of power transmission submitted by power companies throughout the nation.[51]

As Weygandt has commented, Edith Clarke was not a young woman by the time she finished her master's work at MIT. She had always wanted to be an engineer, although no one in her family had been one before her. Upon graduation from Vassar in 1908, where, of course, engineering was neither offered nor probably encouraged for women, she began three years of work as a teacher, first of math and physics in a girls' school and then of math at Marshall College in West Virginia. From there she decided to enter the School of Engineering at the University of Wisconsin, where she concentrated on civil engineering. She stayed at Wisconsin only one year and then left for employment with American Telephone and Telegraph Company in New York, where she supervised women who did computations for the research engineers there. This work was mostly in the area of electrical technology. She then enrolled in MIT in electrical engineering.[52]

While Clarke was at MIT, she apparently knew a Japanese student named Bekku, who was a cooperative student working in industry. For part of his sojourn at MIT, he was assigned to Edith Clarke. As Weygandt says she told the story to him:

> He, Bekku, showed up in her office one day, and she thought she'd better do something to keep him busy. So she showed him a problem she'd been working on for about a month. She had the solution, but she didn't give him the solution. She gave him the problem and told him he could busy himself on it. Well, the next day he came back with the solution, and it agreed almost exactly with hers. She said, "Well, how on earth did you get it so fast?" "Well, I used this system of symmetrical components." So then she got interested, and so this exchange student Bekku taught Edith Clarke symmetrical components, the first time she knew about it. And Bekku published a paper . . . I don't know what became of him, but Edith Clarke told me this, that was how she got interested in symmetrical components. And then, of course, she became one of the authorities, and she invented a new system called "modified"—she called them modified—symmetrical components. They're also called the alpha, beta, zero components and very frequently now they're called the Clarke components . . . , which is her invention really.[53]

The theory of symmetrical components, or coordinates, represents a mathematical means for engineers to study and solve problems of power losses and performance of electrical equipment caused by unbalanced loads on electric circuits "by means of equivalent balanced three-phase systems and their single-phase counterparts."[54] As Weygandt has explained, the theory of symmetrical components was originally a discovery of a man who worked for Westinghouse, Charles Fortescue, but Edith Clarke adapted Fortescue's work to engineering theory and practice for three-phase systems.

> And it really revolutionized the solution of unbalanced three-phase problems in the power field. Of course, symmetrical components itself is an interesting thing. Fortescue . . . wrote a paper in 1918 . . . on symmetrical components. But he was a mathematician, so he did it for "n" phases, which made everything very complicated. A mathematician thinks if he can do it for any number, "n," then he'd like to, rather than do it for three, although all our systems are three-phase systems. So nobody used symetrical components for about ten years, until 1928, when some other Westinghouse people, Wagner and Evans, set n = 3 in Fortescue's very complicated paper and found that it was very useful. It was brand new—1928 was when they started it, and when I was at MIT in 1932–33, they were teaching it to graduate students. . . . But when she [Edith Clarke] was there, they didn't have symmetrical components then at MIT.
> . . . They [the Clarke components] have various advantages. . . . Ordinarily, in solving a three-phase problem, you get a nine-element matrix. It is usually symmetrical, so that there are only six different parameters. But by using these transformations, you can make the matrix diagonal, so that all

these things become zero except the three terms down the main value—which is a great simplification. Well it's something that was well-known to people in advanced dynamics, mechanical problems, but it was new to electrical engineers and first developed by Fortescue. He picked one particular set. Now, actually, you could show that you can have infinitely many sets of components like that, which differ in various ways. And the advantages of the Edith Clarke ones is that the elements in the transformation matrix are real, as opposed to being complex. . . . In the modified symmetrical components of Edith Clarke's, all the coefficients in the transformation matrix are real, whereas in the Fortescue one they're complex. Now, with present-day digital computers, it doesn't make so much difference, but back when computation had to be reduced to a minimum because it was done by hand, then when you could use Edith Clarke's method with its real coefficients, there was much less computation.[55]

Dr. Weygandt's statement is quoted at length here not only because it represents a rare assessment by a male colleague who knew the woman in question, but also because it elaborates the context of her "invention." Also, even though Edith Clarke was certainly a first-rate mathematician, one can see in this account the great difference between her approach and that of Fortescue. She was thinking as an engineer, as she related to real-world problems which she encountered in her work for GE.

While still at GE, Clarke wrote what became the major textbook in the field, *Circuit Analysis of AC Power Systems, Symmetrical and Related Components* (1943). This was followed by a second volume in 1950. She was then a professor of electrical engineering at the University of Texas, where she taught a class, Symmetrical and Related Components, among other upper-level and graduate courses. She was the first woman to be elected a fellow of the American Institute of Electrical Engineers (now the IEEE, the Institute of Electrical and Electronics Engineers).[56]

In 1954, Edith Clarke was given the Achievement Award by the Society of Women Engineers for "her many original contributions to stability theory and circuit analysis." In the text of the award, her work in the development of calculators and charts for simplification of the calculation of transmission line performance, especially for sixty-cycle systems, is cited, along with her "systematic exposition of the circuit analysis method using components for the study of special circuit unbalance problems."[57]

Although her work in symmetrical components is perhaps best known, Weygandt, the Society of Women Engineers, and others stress that that was not her only contribution. While at GE, she also apparently worked on a machine called the "network analyzer." "GE was a leader in building these networks analyzers . . . , an obsolete machine because digital computers have taken over all the work they used to do. But they were essentially miniature models of the power system," and Edith Clarke "worked on the network analyzer."[58] Colleagues consider her contributions to range over "the whole idea of circuit theory as applied to power systems . . . putting it all on a sound mathematical basis was her contribution."[59]

During her lifetime, Edith Clarke was the subject of much publicity, especially in the Texas press after she joined the University of Texas faculty. It was frequently commented then that she was the first woman professor of electrical engineering to teach in a university in the United States.[60] Much as Lillian Gilbreth received attention for her "firsts," especially for her positions in academia, Edith Clarke also was recognized. And it is clear that, as in Lillian Gilbreth's case, Edith Clarke contributed to the evolving theory of her field in a most important way. Yet, if we limit our view of her contributions to the Clarke components, for which she is most remembered, we may, as with Lillian Gilbreth, overlook her larger grasp of her field and her work in integrating the power field.

KATE GLEASON

Mention should also be made here of Kate Gleason (1865–1933). Gleason attended Cornell in the 1880s but did not graduate. Although not an academic, Kate became "world famous for her original design of worm gears."[61] Kate's father, William, had founded Gleason Machine Tools Company, where she worked after leaving Cornell. The company grew into "the world's leader in specialized gears."[62] She was the "first woman to qualify for full membership in the American Society of Mechanical Engineers."[63] Kate, who never married, later entered other fields of endeavor; she became the first woman to be president of a national bank, the National Bank of East Rochester, which was "one of the first to construct low-cost housing in the United States."[64]

Alva Matthews, who has contributed much to engineering thought and practice, has studied Miss Gleason's life and indicates that the ASME selected her to represent them at various world conferences and that she was part of an invited engineering delegation which toured the Far East. Upon her death in 1933, Gleason's "multi-million dollar estate" helped found the Rochester Institute of Technology in her hometown.[65]

SOME MORE OBSCURE
EARLY WOMEN ENGINEERS

BERTHA LAMME

It is possible that those women who have contributed to the development of engineering design and theory especially excelled in systems thought, rather than in limited areas. If this is so, then we can imagine that the statement by one engineer that Lillian Gilbreth's influence "actually brought Frank around to his unique point of view in management circles" may well be true.[66] In a similar way, we can suppose that the ideas of a woman such as Bertha Lamme (1869–1954), for instance, did in fact influence both her husband and her brother in their work, perhaps especially her brother Benjamin in his contributions to the development of the system of power at Niagara Falls. The case study of the Lammes highlights those women whose contributions have been even more thoroughly buried than those we have already described.

It is important to include Bertha Lamme not only because she was the second woman to receive an engineering degree in the United States, which in itself is a substantial intellectual achievement, but also because Lamme represents the many other, later, degreed women engineers whose promising careers were curtailed by marriage to a man in the field who then became well-known for "*his*" engineering contributions. Margaret Rossiter has shown that the careers of male scientists were greatly enhanced by help from their wives in the same field, although the women received very little credit or compensation for their contributions.[67] Because these women did receive so little recognition, it is difficult to pinpoint the exact nature of their contributions. This holds true even for such a prolific researcher, writer, teacher, and public figure as Lillian Gilbreth. For those whose work has been even more obscured, such as Bertha Lamme, we can only speculate. However, it is important that we do so, for we know her work at Westinghouse involved designs and calculations similar to those needed for her husband's and her brother's impressive achievements in the field of electric power.

It is often claimed that Bertha Lamme was the first woman graduate of an engineering degree program. She was graduated from Ohio State University in 1893 with a degree in mechanical engineering "with an option in electricity" and subsequently went to work for Westinghouse in Pittsburgh.[68] Even though she was apparently the first female engineering graduate from OSU, she was not the first woman to graduate from an engineering degree program in the United States. Elizabeth Bragg was the first woman to obtain an engineering degree from an American university, the University of California at Berkeley in 1876 in civil engineering. There were also women graduates of other engineering schools, such as the University of Illinois, from the 1870s on. However, most of these women, if not all of them, majored in architecture and architectural engineering, some programs of which later became incorporated into schools of architecture.[69] However, Lamme was the first woman who graduated in a major field of engineering other than areas related to civil engineering.

Just after her graduation from Ohio State, Lamme entered employment at Westinghouse Electric and Manufacturing Company, where her brother, Benjamin Garver Lamme, was also working. Benjamin had also been gradua-

Bertha Lamme at her drafting ta-
ble. From *News in Engineering,*
Ohio State University, vol. 45
(January 1973).

ted in mechanical engineering at OSU several years before. Bertha worked
for Westinghouse for twelve years in the East Pittsburgh plant in the engineering department, where she apparently designed motors and generators. In 1905 Bertha Lamme married her supervisor, Russell S. Feicht,
who had graduated as a mechanical engineer from OSU in 1890.

Guenter S. Holzer has researched Bertha Lamme's life and work and
has searched archival materials held by the family, at OSU, and at Westinghouse. But he has said that information on her work at Westinghouse
is difficult to find and to assess, since her work there was "fused into a
team."[70] That Bertha Lamme contributed something to the work of Westinghouse and to that of her husband and her brother, both of whom lived
with her, can hardly be doubted. She obviously stayed interested in the
field after her marriage, even though she retired at that time to become a
wife and mother.

Russell Feicht had designed the 2,000-horsepower induction motors
displayed at the St. Louis World's Fair in 1904, and later retired from Westinghouse as its director of engineering. Benjamin rose to the position of
chief engineer at Westinghouse and had been in charge of the design of
the turbogenerators at Niagara Falls in the historic development of a.c.

power there during the 1890s and the early 1900s.[71] Benjamin at least mentioned Bertha in his own autobiography:

> Miss Bertha Lamme, my sister, entered the employ of the company at Mr. Schmid's request in the latter part of this period. Schmid became head of the Engineering Department at Westinghouse in 1895. She had taken an Engineering degree in the Ohio State University, more for the pleasure of it than anything else[!]; but sometime Mr. Schmid gave her a serious invitation to enter the employ of the company, and so she took up the work of calculation of machines and stayed until she married.[72]

Apparently Bertha was very skilled in mathematics, and there are pictures of her engaged in calculations, presumably on machine design.[73] It was commented in the *OSU Monthly* in 1916 that the Feichts' six-year-old daughter, Florence, "already displays mathematical ability," and indeed Florence went on to become a physicist for the U.S. Bureau of Mines in Pittsburgh.[74] It is more than likely that her mother's abilities and interests played an important part in Florence's choice of career.

One cannot say how much Bertha's skills in mathematics helped change theory or practice in the design of motors and generators at that time. It is clear from later experiences of women engineers in Westinghouse in similar design work that Bertha was probably not allowed access to the shop floor or the field, where she would have seen her designs tested and implemented (and design under these constraints is incredibly difficult).[75] Yet knowing how much husbands and male relatives in science, engineering, and other fields have been able to profit from work done by females in their families,[76] one suspects that the innovations in the state of the art attributed to both Bertha's husband and her brother may have been influenced by Bertha's ideas, calculations, and designs. Indeed, this seems very likely, since both men not only worked alongside her but also lived with her; Benjamin, who never married, lived with his sister and her husband for many years.

Marriage had different effects on the careers of Lillian Gilbreth and Bertha Lamme. With twelve children Lillian remained active in her field, while Bertha Lamme retired at the time of her marriage. Yet marriage in both cases served to bury real contributions of both women. The overwhelmingly male tenor of engineering meant that it was unlikely that any engineers would truly have acknowledged the abilities and intelligence of such women, nor would these men have assessed adequately, if at all, the real contributions of women to engineering theory and practice.

Emily Roebling

There were also women in civil engineering, as well as the other major engineering fields mentioned here. Alva Matthews has noted that Emily

Warren Roebling, wife of Colonel Washington Augustus Roebling, who was the son of John Augustus Roebling, was responsible for directing the building of the Brooklyn Bridge during 1872–1883 (the plans for the bridge were approved in 1869) after her husband became paralyzed, partly blind, deaf, and mute because of an attack of the "bends." She was the only person who could communicate with him, and he taught her "higher mathematics, the calculation of catenary curves, strength of materials, stress analysis, bridge specifications, and the intricacies of cable construction."[77] Alva Matthews notes that she was perhaps the first woman field engineer, or one of the first, and that she gave a most moving speech before the American Society of Civil Engineers in 1882, twelve years after construction on the Brooklyn Bridge began, to discuss why her husband should not be replaced as formal director of the construction—the first time a woman had addressed the ASCE. Alva Matthews says:

> There is no record that I can find of any more of her work. Surely to stay successfully in that situation her talents must have been considerable. However, except for the plaque on the bridge and the record of the dedication speeches in which she was mentioned, her name nowhere appears on the official records of the bridge.[78]

Emily Roebling, in charge of completing the bridge's construction day after day for years and well-steeped in relevant mathematics and theory, can be hailed as someone who contributed to the evolution of engineering practice in the field of civil engineering. The main "schools" for learning civil engineering in this country in the nineteenth century were West Point and the major construction projects—such as the Erie Canal and other canals, railroads, and bridges. The reputations of civil engineers were formed during such construction projects, and they were subsequently hired on the basis of such works; these projects were then as important, if not more so, than a formal engineering degree.[79] We can surmise that had Emily been a man she would have been given full credit for such a feat and all the innovations entailed. (After all, she was on site day after day, while her husband could only watch at times from his room with binoculars.) She would most certainly also have been made an honorary member of the ASCE and received other awards. In addition, she would have been sought after for other such jobs. It is interesting to speculate on her salary for this work, since her husband retained title as the official director of the work and since married women's income and property were then solely owned by their husbands. It is doubtful that income for the work went directly to her, or even that she realized much of it, since her husband's illness must have been a great expense. Indeed, being a wife, she would also have had charge of the bulk of his care, as well as the management of the bridge's construction.[80] Emily died in 1903, her husband in 1926.

JULIA BRAINERD HALL

Another woman whose chemical contributions were obscured by such an environment is Julia B. Hall (1859–1925), an older sister of Charles Martin Hall. He has been credited with the development of a process for the electrolytic reduction of alumina in a molten bath of cryolite to produce aluminum metal. The Hall process then formed the basis for the establishment of the Pittsburgh Reduction Company, the forerunner of ALCOA (the Aluminum Company of America).[81] What is not always recognized is that much of the invention and innovation which culminated in the formation of Pittsburgh Reduction in 1888 was due to Julia Brainerd Hall.

Like Charles, Julia had graduated from Oberlin; the date of her matriculation was 1881 and his, 1885. Like Charles, Julia took chemistry at Oberlin, also in her junior year and also under Professor Franklin F. Jewett. At Oberlin, Julia completed slightly more credits in science than Charles, even though she was officially enrolled in what was called the "literary course," an outgrowth of the earlier "ladies' course." While Charles received a degree for his four-year course of study, Julia received a diploma.[82] In childhood, Julia was Charles's closest companion and confidante, a role which she continued to play in the adulthood of this shy, reclusive man. Neither Charles nor Julia ever married. Upon graduation, because of her mother's illness (she died in 1885), Julia assumed the responsibilities of raising her two younger sisters and of directing other household tasks.

Charles used the Oberlin home as his base of operations, primarily before 1887; he set up his laboratory in the woodshed, next to the kitchen—Julia Hall's headquarters. Being in the home, Julia was often present in Charles's lab, helping out with the experiments and consulting with him on scientific and technical matters. She served also as a scientifically astute, well-educated, and competent eyewitness for Charles's experiments and for the letters and papers he wrote concerning the aluminum invention. She also acted in this capacity for certain of his other inventive ideas.[83] These and similar activities resulted in the issuance of a family of patents to Charles Hall on April 2, 1889 for the production of electrolytic aluminum. This inventive activity since 1882 was nonrandom and was planned by Julia and Charles Hall with an eye toward potential markets from the beginning of serious work on the invention. Just as with today's research and development teams, invention in the nineteenth century was also a team effort.

Besides assisting in the lab and offering her technical advice and expertise, Julia Hall faithfully and minutely recorded the steps in the invention process—that is, the results of a given day's work, with technical details and the date, and with evidence which could substantiate the date.

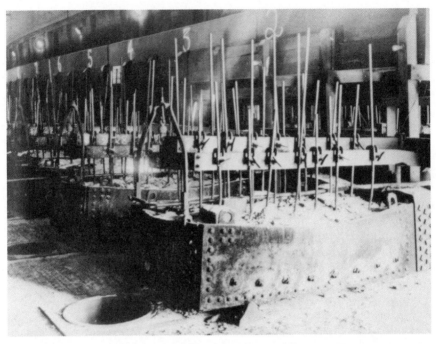

Aluminum reduction pots at Massena, New York, 1914. From Junius Edwards, *The Immortal Woodshed*.

Modern pot room, Aluminum Company of America, 1955. From Junius Edwards, *The Immortal Woodshed*.

She also contacted family and friends who might have leads on financial backing or finances themselves; she advised Charles on difficult questions such as to whom he might wish to sell his process. She acted as an information center, relaying information to and from Charles about people potentially interested in the process. She advised him that he should write to the *Scientific American* for their "free" advice on patents, a service apparently offered to inventors of the time by the journal. She dated his papers and letters pertaining to the invention and made sure that the copies he sent were clearly legible, and she acted as a censor of important names, dates, and other facts in letters Charles wrote her, in case the letters fell into the wrong hands, including relatives. She also advised Charles not to leave her letters lying around for the same reasons.

Finally, she composed a "History of C. M. Hall's Aluminum Invention," a six-page document, as her 1887 eyewitness account for the important Hall-Héroult patent interference case. Her testimony served to clinch Hall's victory in this case. Since Hall had filed his patent application on July 9, 1886, and Héroult in May, Hall had to establish without a doubt that he had reduced his invention to practice before May and in particular before April 23, 1886, the date on which Héroult's French patent had been granted. The witnessess were Charles himself, Julia, Charles's father, and two of Charles's professors, one of whom was Jewett. Only Julia could positively identify as an eyewitness the production of aluminum, verifying the identity of the product, produced on February 23, 1886. Julia was the first person to whom Charles Hall had fully disclosed his ideas about the invention, on February 10, 1886, a date which Julia fully documented in court.[84]

Yet Julia received little real recognition for her scientific, technical, informational, managerial, and entrepreneurial contributions here. Charles went on to become extremely wealthy, with $170,000 annual income from ALCOA stock alone at the time of his death in 1914. Julia at that time until her death in 1925 averaged about $8,000 income from her stock.[85] And not even Charles credited her in his 1911 acceptance speech for the cherished Perkin Medal of the American Chemical Society, when he told of family involvements in the invention and innovations leading to the establishment of the Pittsburgh Reduction Company.[86] Not even Julia's 1887 account was referenced as such by Hall's biographer, Junius D. Edwards, nor by Charles C. Carr, ALCOA company historian, although both authors evidently used it and both knew about it.[87] Nevertheless, it is apparent that the team of Charles and Julia Hall brought to the market cheap aluminum.

Julia Hall assisted, undoubtedly more than we can know, in the invention process, much as a member of a research-and-development team in a chemical corporation might do. Also, among her outstanding contributions was management of the entire Hall system of invention for the electrolytic process and management of much of the entreprenuerial activities which

Julia Brainerd Hall at age 22. From Junius Edwards, *The Immortal Woodshed*.

combined to form the Pittsburgh Reduction Company. These activities are similar to those many engineers have undertaken when inventing and in establishing firms such as Dow, Anaconda, and Union Carbide. And later, after the company had been established and was producing aluminum, Julia continued to act as a consultant to Charles, though less frequently than in the early days. In Julia's careful management of the system, with attention to many details she felt would be critical to the granting of the patents (and which did, in fact, turn out to be critical, even though Charles had not foreseen their significance in some cases), and in the establishment of the Pittsburgh Reduction Company, we see another instance of the broad systems approach. Julia was, in fact, a very good manager of the research, engineering, and business aspects of the system which led to the formation and development of one of our largest and most important firms.

We could continue almost indefinitely with citations of outstanding women who contributed in this early period to the intellectual advancement of professional engineering. These cases are sufficient, however, to illustrate not only the achievements women have made in some of the major fields of engineering, but also the various ways in which these women's contributions have been buried or their careers cut short.

SOME MODERN, LESS OBSCURE WOMEN ENGINEERS

Only a small space remains to say something of the women who came after these pioneers. At best, mention can be made of only a few among the many women who should be included here. The numbers, if not generally the percentages, of women graduating from engineering schools increased from the 1920s into the 1940s. There was still some tendency for women to concentrate in architecture, but they increasingly began to enter the spectrum of engineering fields—from chemical, marine, and aeronautical engineering, to engineering administration. The number of women majoring in mechanical engineering remained almost nil until the 1940s. Some of the women who became successful in engineering fields in this period are still living and have been interviewed by this author.

ELSIE EAVES

Certainly, Elsie Eaves (1890–   ), in the civil engineering and construction field and longtime manager of the Construction Economics Department of *Engineering News-Record*, must be mentioned. She was the first woman to be elected an honorary member of the American Society of Civil Engineers (1979); in addition, she was the first female to be elected to the ASCE as an associate member, a member, a fellow, and a life member. She was also the first woman to receive the Honorary Life Membership Award from the American Association of Cost Engineers. Among many other accomplishments, one of her most important contributions to the construction industry was that she

> organized and directed the measurement by *Engineering News-Record* of "Post War Planning" by the construction industry. This was used by the Committee for Economic Development and the American Society of Civil Engineers as the official progress report of the construction industry on work that could go ahead promptly at the close of World War II.
> She converted these "Post War Planning" statistics to develop the first continuous inventory of construction in the planning stage.[88]

One can see in her work evidence of extensive familiarity with her field and true systems thought. These are especially evident in her "basic research on indexes and cost trends" and in her innovation of a "nationwide system of collecting, analyzing, and reporting statistical and economics information on construction that has become a vital business tool for public and private owners, designers, and builders of projects and for the manufacturing, service, and supply organizations that serve the construction industry."[89] Her work formed the basis of industry standards and was cited in congressional hearings and in other legal and political contexts, such as arbitration proceedings. Elsie Eaves has received many engineering awards for her work in engineering economics and has authored dozens of papers and speeches, including several on women in engineering.

It is important to include in engineering history contributions such as those of Eaves for several reasons. First, engineering is a very diverse profession, and the jobs which engineers hold are myriad. One of the most important aspects of engineering is economics, and cost engineering is a part of engineering too often ignored by intellectual historians who focus on the more scientific or strictly technical achievements. Eaves's mind was no less active than the minds of those in engineering design, and her holistic thinking was no less impressive and significant for the advancement of engineering. Cost engineering is a crucial part of civil engineering and construction, and Eaves is widely recognized as an important civil engineer.

Second, women engineers have often excelled in advancing engineering literature, whether in heat transfer, as in the careers of Florence Buckland and Nancy D. Fitzroy, in groundwater and hydraulics, as with Helen J. Peters, in the rise of computer software, as with Grace Murray Hopper, in refrigeration and air conditioning, as with Margaret Ingels, and in many other and various fields mentioned here and in my book on women engineers. In addition to these contributions of technical research, women engineers have often been editors of engineering periodicals, as in the case of Eaves and Ivy Parker (to whom we turn next). Recognized as holistic intellects, managers, and writers, women engineers have often been in charge of engineering literature, including company manuals and reports. This may be because of discrimination ("women can be writers but not designers"), but it more likely relates to women's superior verbal skills. And these verbal abilities in many areas of engineering thought, even if not well compensated, have contributed significantly to the instigation and diffusion of advances in engineering.

IVY PARKER

Dr. Ivy Parker, (1907–   ), chemist and research engineer in the petroleum industry, should also be noted. She worked for Shell Oil in the 1930s and the early 1940s and later joined Plantation Pipe Line Company in Atlanta, Georgia. In 1944, Dr. Parker was selected as the first editor of *Corrosion*, the official publication of the National Association of Corrosion Engineers, and she retired from that position in 1965. She became a specialist in research on the causes and prevention of corrosion of pipelines and published and presented many papers on this subject. She was a systems thinker and made contributions on the development of pipeline technology, especially in the South as it expanded after World War II. She researched and developed data on "quality of product" flowing through the pipeline, "corrosion protection" (innovations in both water and oil-soluble inhibitors), "keeping the pipeline clean," filtration, and "one other big area . . . , tank painting and tank seals and developing internal coatings for tankage . . . and eventually going to some short pipelines in place of having filterings; just coating the inside of the pipeline." She said that "I suspect that I really accomplished more in the field of filtration," with the exception of her work on

> developing an understanding of how oil-soluable inhibitors work. . . . When it became possible to use the oil-soluable inhibitor and get rid of all this attention you had to give at each pump station for getting the water samples, at the same time electronic and electrical equipment was being developed so that stations could be operated by remote control. . . . Automation was coming along at the same time. All these things had to fit together. It is not so much a matter of making a specific contribution as keeping all phases—keeping the system—in tune.[90]

Again, we may well find a high proportion of systems thinkers among women engineers who made their mark in the first half of this century. Margaret Ingels, a mechanical engineer who specialized in research and design in refrigeration, Margaret Hutchinson, a chemical engineer who, like Ivy Parker, worked in the petroleum industry, and many others could be named here.[91]

Women engineers who have contributed to advances in the theory and state of the art in their respective engineering fields mostly after 1950 are listed in the Society of Women Engineers Citation Award Booklet. Systems thinking is seen in the work of Maria Telkes in the field of solar energy, Dorothy Simon in space engineering, Martha J. Thomas in her work on phosphors at Sylvania-GTE, Alva Matthews in engineering mechanics and especially in analysis of shock and wave propagation in various materials, Grace Murray Hopper in design of computer programming systems, Nancy Fitzroy in thermal engineering, Barbara Crawford Johnson in research and development work on manned space flight, Shella Widnall in fluid mechanics, and Ada Pressman in innovations in controls in power plants of various kinds, and in others who have received this award.[92] The few female members of the heavily male-dominated National Academy of Engineering should be noted also, including Betsy Ancker-Johnson, Ruth M. Davis, Mildred S. Dresselhaus, Jean Sammet, and Grace Murray Hopper.[93] Betsy Ancker-Johnson and Grace Hopper are also listed among the Fellows of the Institute of Electrical and Electronics Engineers, along with Jenny Rosenthal Bramley, Elizabeth Laverick, and Irene C. Peden (who also won an SWE Citation Award in 1973).[94] Betsy Ancker-Johnson, whose undergraduate and doctoral work (Tübingen University, Wellesley College) were in physics, was cited by the IEEE "for contributions to the understanding of plasmas in solids and to the development of government science policy."[95] She was made Assistant Secretary of Commerce for Science and Technology in 1973 and has worked in both industry and academia. In 1980, she was named vice-president for environmental activities of General Motors Corporation, making her one of two female vice-presidents at GM; the other, Dr. Marina von Newmann Whitman, was appointed vice-president and chief economist at the same time. In fact, they were the only two women vice-presidents in the whole automobile industry.[96]

JENNY ROSENTHAL BRAMLEY

Jenny Rosenthal Bramley (1910–   ) was cited by the IEEE "for achievement in spectroscopy, optics, mathematical techniques and their application for electron tubes, displays and light sources to engineering."[97] She said that this honor, conferred on "less than 2% of electrical and electronic engineers," best sums up her achievements. She invented the "microwave pumped high efficiency lamp," whose brightness in the "narrow spectral band . . . far exceeds that of high power xenon arc lamps." The basic ideas in this invention are now being applied to "the generation of very high efficiency lasers ranging in wavelength from vacuum ultraviolet to submillimeter." She served as head of the mathematics department at Monmouth Junior College in the 1950s, where she and her husband, Arthur Bramley, did some of their path-breaking work on "applications of electroluminescence to solid state display and storage devices." For this work, they obtained independent patents, which they licensed to IBM. In addition, she invented techniques of coding and decoding pictorial information, now being used in work, she said, "of which I . . . cannot be apprised. My very recent extension of the principle of alphanumerics appears to be of interest to the FBI." Finally, she has researched "hyperfine structure anomaly, which is still being cited in the literature more than 40 years after publication and which . . . is required reading of graduate students at New York University planning to do research in nuclear physics or hyperfine structure. The effect is referred to in the literature as the Breit-Rosenthal effect."[98]

MILDRED DRESSELHAUS

Dr. Mildred Dresselhaus (1930–   ) currently directs the Center for Materials Science and Engineering at MIT and is the Abby Rockefeller Mauze Professor of Electrical Engineering there. She has earned an international reputation "as a solid-state physicist and electrical engineer . . . , based on her work in superconductivity, the electronic properties of solids, semimetals, and magneto-optical phenomena."[99] She received her Ph.D. from the University of Chicago in 1958, having been a Fulbright Fellow at the Cavendish Laboratory of Cambridge University during 1951–52. Following postdoctoral work at Cornell, she spent seven years engaged in research at the Lincoln Laboratory at MIT and then was appointed professor of electrical engineering in 1968. When asked what she considered her most important discoveries, she commented, "Generally to show how magneto-optics can be used to study engineering bands in complicated systems, like semimetals."[100]

OTHERS

There are so many, many others who could be cited here, such as Thelma Estrin, a senior member of IEEE and pioneer in biomedical engineering and brain research. (Bioengineering is now attracting very capable men and women, such as a 1978 Ph.D., Dr. Janice Jenkins, now of Northwestern University.)[101] Also, Dr. Vera Pless, an electrical engineer and professor of mathematics and computer science at MIT, should be noted. Increasingly, younger women, such as Dr. Elizabeth L. White, research associate in civil engineering at Pennsylvania State University, who has researched bridge structure, channel hydraulics, flood hydrology, cements for geothermal wells, and other areas, are contributing to engineering research.[102] Then, too, there are those who have contributed much to the theory and practice of engineering education, such as Lois Broder Greenfield, who holds a Ph.D. (but no engineering degree), who is on the engineering faculty at the University of Wisconsin at Madison, and who has researched "problem-solving processes" in graduate study. She continues to do so within her work in engineering education.[103]

It is clear that there are many more women who are living and working today as engineers who deserve recognition by the engineering profession at large. Women in engineering have historically contributed intellectually to a wide spectrum of engineering fields and are still doing so in ever greater numbers, especially since increasing numbers of women are obtaining the Ph.D. in engineering.

It has been and remains difficult for women to become recognized as authorities in engineering. This will undoubtedly abate somewhat as the old myths and stereotypes about women and their supposedly inferior mental capacities, especially in technical fields, are being put solidly to rest. Such prejudice has been particularly unfortunate as applied not only to all women engineers but especially to those among them who have made substantive intellectual contributions to the theory and practice of engineering. Those women who combined the characteristics of pioneer, thinker, engineer, and women in our society—many breaking new ground—may well have represented a highly select group of systems thinkers. They therefore may be reassessed as having contributed much more to technological change and the advancement of engineering knowledge than one might have supposed from their relatively small numbers.

N O T E S

1. Martha M. Trescott, *New Images, New Paths: A History of Women Engineers in the United States, 1850–1970* (forthcoming).

2. This is a truism in the engineering literature and is also overwhelmingly a response on the returned questionnaires (see note 1, above). Also see, e.g., Ellis Rubinstein, "Profiles in Persistence," *IEEE Spectrum*, November 1973, pp. 52–64.

3. Carolyn Cummings Perruci, *The Female Engineer and Scientists: Factors Associated with the Pursuit of a Professional Career* (West Lafayette, IN: Purdue University, 1968), pp. 1–2.

4. This is the scope of the author's current research project on the history of women engineers in the United States.

5. Letter, Jean Robertson, Director of Information Services, National Society of Professional Engineers, to Martha M. Trescott, July 9, 1980.

6. Robert Clarke, *Ellen Swallow, the Woman Who Founded Ecology* (Chicago: Follett, 1973). See also Robert H. Richards, *Robert Hallowell Richards, His Mark* (Boston: Little, Brown, 1936) p. 153.

7. Richards, *Robert Hallowell Richards, His Mark*, p. 170.

8. Lists of published writings of Mrs. Richards in MIT Archives.

9. Clarke, *Ellen Swallow*, p. 38.

10. Ibid., p. 39.

11. Richards, *Robert Hallowell Richards, His Mark*, pp. 159–160.

12. Ibid., p. 161. Also Margaret W. Rossiter, *Women Scientists in America, Struggles and Strategies to 1940* (Baltimore: The Johns Hopkins University Press, 1982), p. 91.

13. Letter, Ellen H. Swallow Richards to Dr. Noyes, President of MIT, n.d. (in December 1907), MIT Archives, file on Ellen Richards.

14. Letter, Dr. Noyes to Ellen Swallow Richards, January 22, 1909 (they had had correspondence on this question during 1907–1909).

15. Clarke, *Ellen Swallow*, pp. 66–67.

16. Ibid., see especially chapter 10.

17. Ibid., see especially chapter 12.

18. Ibid., see especially chapters 13–15.

19. Ibid., p. 141.

20. Ibid., p. 149

21. Ibid., pp. 122–125.

22. Ibid., p. 147.

23. Anon., "Lillian Moller Gilbreth: Remarkable First Lady of Engineering," *Society of Women Engineers Newsletter*, XXV (November/December 1978), 1.

24. Gilbreth Collection, Purdue University, West Lafayette, Indiana. See acknowledgment above.

25. "Lillian Moller Gilbreth," *SWE Newsletter*, p. 1.

26. Discussion by William G. Caples, American Society of Mechanical Engineers, *The Frank Gilbreth Centennial* (New York: ASME, 1969), p. 72.

27. For example, see Siegfried Giedion, *Mechanization Takes Command, A Contribution to Anonymous History* (New York: Norton, 1969), pp. 121, 525, 615–616.

28. "Lillian Moller Gilbreth," *SWE Newsletter*, pp. 1–2, and Dr. Gilbreth's resume, located in the biography file, Schlesinger Library.

29. See the outline on the History of Scientific Management in NHZ 0830–23, Gilbreth Collection, Purdue. It is evident that Lillian, as with Julia B. Hall (see Martha M. Trescott, "Julia B. Hall and Aluminum," *Dynamos and Virgins Revisited:*

Women and Technological Change in History, ed. Trescott (Metuchen, NJ: Scarecrow Press, 1979), pp. 149–179), cooperated in allowing her work to be subsumed under Frank's name. This was probably due to her generally self-effacing nature and to her conditioning that men should get most of the credit while women should feel content and privileged to be considered their assistants. Also see, for example, the ASME *Fifty Years Progress* index, where all of the Gilbreths' work is listed together under "Gilbreth, Frank B., and Lillian M.," even though the text referred to may only mention one or the other (usually Frank).

30. George Iles, Introduction to Frank B. Gilbreth and L. M. Gilbreth, *Applied Motion Study, A Collection of Papers on the Efficient Method to Industrial Preparedness* (New York: Sturgis & Walton 1917), p. xi. Under Frank's name here are given his title, "Consulting Management Engineer," and a list of his professional society memberships, whereas with Lillian's initials only "Ph.D." is noted.

31. Anon., "Honorary Member No. 2," *Society of Industrial Engineers Bulletin*, III (May 1921), 2–3, Gilbreth Collection.

32. American Society of Mechanical Engineers, *Fifty Years Progress in Management, 1910–1960* (New York: ASME, 1960), p. 126.

33. Lillian M. Gilbreth, *The Psychology of Management: The Function of the Mind in Determining, Teaching and Installing Methods of Least Waste* (New York: Sturgis & Walton, 1914), p. 72.

34. ASME, *Gilbreth Centennial*, especially p. 126.

35. Item labeled "Dartmouth College Conference" (NHZ 0830–23, with type-written note by LMG, 2113–41, saying (in reference to her remarks at the Dartmouth Conference in 1911), "This is the first mention of the Psychology of Management [*sic*] at any management meeting." Gilbreth Collection, Purdue University.

36. Typescript of Lillian M. Gilbreth, "Discussion of Dr. Person's paper—'Industrial Psychology,' Taylor Society, Boston, Mass., April 25, 1924," p. 1, found in the Gilbreth Collection (NHZ 0830–11).

37. See especially Nancy Z. Reynolds, "Dr. Lillian Moller Gilbreth, 1878–1972," *Industrial Engineering*, February 1972, p. 30.

38. See letters to and from Mrs. Gilbreth on her election to ASME membership during 1925–1926 in the file NHZ 0830–1, Gilbreth Collection. Had she been a man in the field with such outstanding qualifications and publications, she would have undoubtedly been sought for membership and honored, instead of having to go to such lengths. This is another fascinating story.

39. Lillian M. Gilbreth, biographical memo, 1926, in NHZ 0830–1 Gilbreth Collection.

40. From the Bulletin of Purdue University, 1934/35–1949/50, as found in a search of these university catalogues by Keith Dowden.

41. See letter, W. H. Faunce to Lillian Gilbreth, May 7, 1928, Gilbreth Collection, NHZ 0830–4, and letter, Richard L. Anthony to Lillian Gilbreth, February 26, 1941. Also see letters, Arnaud C. Marts to Lillian Gilbreth, December 12, 1940 and January 8, 1941, and letter, Dorothy Dyer to Lillian Gilbreth, February 13, 1941, Gilbreth Collection, NHZ 0830–5. Also, "Stevens Honors Five Engineers," *The Christian Science Monitor* (n.d., n.p.), clipping from Gilbreth biography file, Schlesinger Library.

An inkling of the number of speeches and her extensive travel through the decades can be seen in the index to boxes of material on Lillian Gilbreth alone, Gilbreth Collection, Purdue.

Materials on the Gilbreth summer schools, particularly the one held in Baveno, Italy, June 10–25, 1927, Gilbreth Collection NHZ 0830–31.

Lillian M. Gilbreth, "Opportunities for Women in Industrial Engineering," Mimeo, October 20, 1924, pp. 2–3, Gilbreth Collection, NHZ 083–60. On women see also letter, Lillian Gilbreth to George C. Dent, September 19, 1924, and letter, George C. Dent to Lillian Gilbreth, October 1, 1924, and letter, George C. Dent to Lillian Gilbreth, October 1, 1924, containing a cover letter (sent with questionnaires) to all SIE women members, Gilbreth Collection, NHZ 0830–42. Letter, Dent to L. M. Gilbreth, October 1, 1924; also see in this same file the seven questionnaire responses (besides Mrs. Gilbreth's), and form letter to the "Women Members of the Society of Industrial Engineers" from George C. Dent, November 29, 1924, same file as noted above. Also see ASME, *Gilbreth Centennial*, p. 109, and ASME, *Fifty Years Progress*, pp. 136, 138–139. In addition, refer to Catherine Pilune, "The Industrial Engineer," mimeo, p. 1, Gilbreth Collection, NHZ 0830–112, and Gilbreth, "Opportunities for Women in Industrial Engineering," pp. 1–2. See also Lillian M. Gilbreth, "Industrial Engineering as a Career for Women," mimeo, n.d., Gilbreth Collection, NHZ 0830–67.

42. Dr. Gilbreth did treat such topics as psychologists typically do. See, for example, "Possible Psychopathic Types in Industry," a one-page typescript, n.d., NAPEGTG 0099., listing twelve types, attached to a one-page manuscript, in her handwriting, entitled "Psychiatry and Management." However, her main contributions should not be narrowly construed as those of a psychologist.

43. Harold F. Smiddy, "Management as a Profession," ASME, *Fifty Years Progress*, pp. 26–41.

44. Caples, in *The Frank Gilbreth Centennial*.

45. Ibid.

46. Letter, Frank B. Gilbreth to Irene M. Witte, as cited in the ASME, *Gilbreth Centennial*, p. 107.

47. Historians have, in fact, not covered either of the Gilbreths well. For some mention of them, see David F. Noble, *America by Design: Science, Technology, and the Rise of Corporate Capitalism* (New York: Knopf, 1977), especially pp. 274–275; Melvin Kranzberg and Carroll W. Pursell, Jr., ed., *Technology in Western Civilization* (New York: Oxford University Press, 1967): Richard S. Kirby, Sidney Withington, Arthur B. Darling, and Frederic G. Kilgour, *Engineering in History* (New York: McGraw-Hill, 1956); David S. Landes, *The Unbound Prometheus*, (Cambridge: Cambridge University Press, 1969); Eugene S. Ferguson, *Bibliography of the History of Technology* (Cambridge, MA: MIT Press, 1968), pp. 301–303 and also 116–117; and Alfred D. Chandler, *The Visible Hand: The Managerial Revolution in American Business* (Cambridge, MA: Belknap Press, 1977), p. 466.

48. "Lillian Moller Gilbreth," *SWE Newsletter*, p. 1.

49. Transcript of interview with Cornelius N. Weygandt by Martha M. Trescott, March 12, 1980, Moore School of Electrical Engineering, University of Pennsylvania, Philadelphia, p. 20.

50. Rossiter, *Women Scientists in America*, p. 257.

51. Dudley Early, "Miss Edith Clarke: Fate Placed Her on the Path to Fame," *Austin American Statesman*, October 19, 1948.

52. Ibid., p. A6. Early commented that Clarke took the "oblique approach" to an engineering career. However, due to lack of mentors, this was not atypical for the female pioneers.

53. Weygandt transcript, pp. 8–9.

54. G. O. Calabrese, *Symmetrical Components Applied to Electric Power Networks* (New York: Ronald Press, 1959), p. v. Incidentally, this author, who has worked with Fortescue, does not mention Edith Clarke in his book. See also C. L. Fortescue, "Method of Symmetrical Co-ordinate Applied to the Solution of Polyphase Networks," *Transactions*, AIEE, 37 (1918), pt. II, 1027–1140.

55. Weygandt transcript, pp. 8 and 10. Clarke worked intermittently for approximately three years with Weygandt when she visited the Moore School of Electrical Engineering. Cornelius N. Weygandt was then a professor at the school.

56. Anonymous, "Teaching Opens New World to Woman, 65," *Dallas News*, December 12, 1948. Also, "Woman Enters Male Field," *Dallas News*, May 4, 1947; "Texas Professor Is Paid Honor," *Marshall News Messenger* (Marshall, Texas), March 7, 1954; Mary Ann Beaumier, "UT's Lady Engineer Writes Second Book," *Daily Texan*, October 10, 1950, and many other newspaper clippings through 1957, as supplied by archivist William H. Richter. See also Edith Clarke, *Circuit Analysis of A.C. Power Systems*, Volume I (New York, 1943), and Volume II (New York, 1950).

57. Society of Women Engineers, *Achievement Award, 1952–1976*, New York, 1976, p. 4.

58. Weygandt transcript, pp. 13–14.

59. Ibid., p. 15.

60. See, e.g., "Miss Edith Clarke: Fate Placed Her on the Path to Fame," along with other clippings, as noted in note 56 above.

61. Alva Matthews, "Some Pioneers," mimeo, n.d., p. 5.

62. Ibid.

63. Eve Chappell, "Kate Gleason's Careers," *The Woman Citizen*, January 1926, p. 19.

64. Matthews, "Some Pioneers," p. 5.

65. Ibid.

66. Discussion by William Gomberg, ASME, *Gilbreth Centennial*, p. 77.

67. Margaret W. Rossiter, "Women Scientists in America Before 1920," *American Scientist*, 62 (1974), reprinted in Trescott, ed., *Dynamos and Virgins Revisited*, pp. 130–32; also see section on Julia B. Hall, below. Also see Rossiter, *Women Scientists in America*, pp. 140–143 and 208–209.

68. See, e.g., Timothy Collins, "1893 OSU Graduate First Woman Engineer," Ohio State University *Lantern*, March 27, 1973, and *OSU Monthly*, 35 (December 1943), p. 13 for her obituary, "Bertha Lamme Feicht." Also, a short biographical sketch in the OSU Archives. "Bertha Lamme Feicht," *OSU Monthly*, 35; "Mrs Feicht's Funeral Today," newspaper clipping, OSU Archives, along with various other obituary clippings, some unlabeled but all probably Ohio newspapers.

69. Data from my historical survey of women in engineering, not yet published. Also see Peggy L. Evanich, "American Women in Engineering," mimeo, Spring 1979, p. 2.

70. Collins, "1893 Graduate." Also, conversation by telephone with Dr. Holzer, July 10, 1980.

71. Collins, "1893 Graduate." Also see Benjamin G. Lamme, *Benjamin Garver Lamme, An Autobiography* (New York: G. P. Putnam's Sons, 1926).

72. Lamme, *Benjamin Garver Lamme*, p. 91.

73. See, e.g., clippings from "New in Engineering," January 1973, held in the OSU Archives.

74. Under the report for the class of 1890, on Bertha Lamme Feicht, *OSU Monthly*, 7 (May 1916), 35, and Collins, "1893 Graduate."

75. This information was given to me in a series of interviews with women engineers in Pittsburgh in March 1980, some of whom had worked for Westinghouse as well as other local firms.

76. See note 67 above.

77. Matthews, "Some Pioneers," p. 3.

78. Ibid., p. 4. Also see Richard S. Kirby et al., *Engineering in History* (New

York, 1956), p. 307, where Emily's name is not mentioned, but she is referred to as Washington's "wife."

79. Daniel H. Calhoun, *The American Civil Engineer: Origins and Conflicts* (Cambridge, Mass.: Technology Press, 1960).

80. Matthews, "Some Pioneers," p. 4.

81. Junius D. Edwards, *The Immortal Woodshed* (New York: Dodd, Mead and Company, 1955), and also by Edwards, "A Captain in Industry," 1957; Charles C. Carr, "Alcoa, An American Enterprise," 1952. For an abstract of the oral presentation before the Society for the History of Technology, October 19, 1975, see Deborah Shapley, "History of American Technology—A Fresh Bicentennial Look," *Science*, 190 (November 21, 1975): 763. Also, see Trescott, "Julia B. Hall and Aluminum," *Dynamos and Virgins Revisited*, pp. 149–179.

82. We are especially indebted to Mrs. Gertrude Jacobs of the Oberlin College Alumni Office for providing us with alumni necrology on the Hall family and also with transcripts of college courses for both Charles and Julia.

83. This is not only borne out in the letters but also in the documents used in court in the interference case. ALCOA furnished us with Julia Hall's written statement of 1887, "History of C. M. Hall's Aluminum Invention," and with the actual testimonies before the patent examiner.

84. See Charles M. Hall vs. P. L. V. Héroult, In Interference in the United States Patent Office, October 24, 1887, pp. 5–8 for Julia's testimony and pp. 5–6 for commentary on February 10, 1886.

85. Edwards, *The Immortal Woodshed*, p. 226. From ALCOA we obtained ALCOA stock ledgers for Julia B. Hall, 1909–1925, and for her sisters, Edie and Louie, 1909–1919 and 1909–1925, respectively, along with a table of "Dividends Paid on Common Stock," 1895–1943, all supplied by Mrs. Anna G. Lydon.

86. "The Perkin Medal, Remarks in Acknowledgment by Mr. Hall," *Industrial and Engineering Chemistry*, III (1911) 146–148.

87. Edwards had Julia Hall's account typed, along with the letters from Charles to Julia and others, and transmitted to company historian Charles Carr in 1936, according to the file from ALCOA's archives.

88. Elsie Eaves, "Professional Record, 1980," mimeo sent the author by Ms. Eaves.

89. Interview, Elsie Eaves by Martha M. Trescott, Port Washington, New York, March 9, 1980, p. 13 of the transcript and biographical sketch; "Elsie Eaves," *Civil Engineering* (October 1979), p. 89. In addition, Ms. Eaves has sent to me numerous clippings, articles, essays, and other items on her life and work.

90. Interview, Dr. Ivy Parker by Martha M. Trescott, Austin, Texas, June 1, 1980, pp. 3–10 transcript.

91. See, e.g., Anon., "Personalities in Industry, Margaret Ingels, M.E.," *Scientific American*, April 1941, p. 197; Anon., "Margaret Ingels of Carrier Honored as 'Pioneer' in Women's Careers," *Refrigerating Engineering*; and Anon., "Miss Chemical Engineer of 1955," *Chemical and Engineering News*, p. 3504.

92. SWE, *SWE Citation Award*, entire booklet. I have interviewed Dr. Matthews, Dr. Simon, and Barbara C. Johnson, and plan interviews with Dr. Telkes and Ada Pressman.

93. National Academy of Engineering, *Organization and Members, 1977–1978* (Washington, D.C., 1977), pp. 24, 25, 27, and 30.

94. IEEE, *Membership Directory*, 1975, pp. 13, 26, 74, 91, 117, and 128.

95. Ibid., p. 13.

96. Biographical sketch on "Betsy Ancker-Johnson," from the files of Dean Vivian Cardwell, College of Engineering, University of Illinois at Chicago Circle.

(Dean Cardwell has been interviewed by this author.) Also see "Two VP's at GM Are AAUW Fellows," *AAUW Newsletter* for Members-at-large, May 1980, p. 3.

97. IEEE, *Membership Directory*, p. 26.

98. Jenny Bramely, "Qualifications Statements," pp. 2–3, sent to me by Ms. Bramley.

99. Georgia Litwack, " 'If You Go into Science and Engineering, Go in to Succeed': A Conversation with Professor Mildred S. Dresselhaus, Solid-State Physicist," *Harvard Magazine*, January–February 1980, p. 49.

100. Ibid, p. 51. In addition, see the transcript of an interview with Dr. Dresselhaus, held in the MIT Archives; also, I have corresponded with both Dr. Dresselhaus and Ms. Litwack and have conversed with Dr. Dresselhaus, with an interview planned. See also Mildred S. Dresselhaus, "Some Personal Views on Engineering Education for Women," *IEEE Transactions*, vol. e-18 (February 1975), copy of reprint sent this author by Dr. Dresselhaus.

101. See, e.g., Thelma Estrin, "Ms. Biomedical Engineer: A New Professional Opportunity for Women," *IEEE Transactions on Education*, vol. e-18 (February 1975), pp. 11–14. An interview with Dr. Estrin is planned. Also, I interviewed Dr. Janice Jenkins during 1980.

102. Elizabeth L. White, "Narrative Autobiography," her curriculum vitae, and bibliography, as sent to me by Dr. White. An interview with Dr. White is planned. Also, see Sam Merrill, "Women in Engineering," *Cosmopolitan*, March 1976, pp. 162–172.

103. See interview, Lois B. Greenfield by Martha M. Trescott, College of Engineering, University of Wisconsin, Madison, April 23, 1980, transcript of interview. Also, see biographical sketch on "Lois Broder" in *Cadette Gazette*, 1980, p. 3; Benjamin S. Bloom and Lois J. Broder, *Problem-Solving Processes of College Students, An Exploratory Investigation* (Chicago: University of Chicago Press, 1950), and a paper on problem solving in engineering education delivered by Dr. Greenfield at the 1980 American Society of Engineering Education. Also, see Lois B. Greenfield, "Women in Engineering," a history of women in engineering at the University of Wisconsin, mimeo supplied by Dr. Greenfield in 1980. This paper is useful for information on certain women engineers mentioned in this essay, notably Lillian M. Gilbreth, Edith Clarke, Thelma Estrin, and others.

The author gratefully acknowledges the Rockefeller Foundation, which provides grant funds for her study of the history of women in engineering in the United States, 1850–1975, and also the College of Engineering at the University of Illinois, which has housed and nurtured this project. Also, she acknowledges the many female and male engineers who have taken time to be interviewed and who have otherwise shared their life experiences and insights. She is also grateful to many librarians and archivists who have helped, especially Keith Dowden, Chief Archivist, Gilbreth Collection, Purdue University, without whose help and access to that marvelous collection this essay would have suffered. Also, William H. Richter, Assistant Archivist, Barker Texas History Center, University of Texas at Austin, provided much help on Edith Clarke. In addition, Deborah Cozort, Archivist, Massachusetts Institute of Technology, has been very helpful for years, especially on Ellen Swallow Richards. Dorothy Ross, assistant in the Ohio State University Archives, helped greatly. The staff of the Schlesinger Library at Radcliffe has also helped. Also, Dr. Guenter S. Holzer, historian, St. Cloud, Minnesota, was helpful in providing information about Bertha Lamme whom he has researched. Thanks also to my husband, Thomas F. Lieb, for his help in proofreading the manuscript.

L . M . J O N E S

Intellectual Contributions of Women to Physics

The history of physics contains many very early developments, such as the "atomic theory" of the Greeks and the practical use of levers, wheels, etc. However, the field did not really get organized until the seventeenth, eighteenth, and nineteenth centuries. A revolution occurred when the meshing of experiments with new methods of calculation made it apparent that there are only a very few "laws of nature" on the elementary level; we expect that everything else can be predicted from these basics if we have the calculational strength.

Milestones in this progress were the development of the calculus to understand gravitation, work done largely by Newton (1642–1727); further mathematical work by Lagrange (1736–1813) on mechanical equations of motion; experiments showing the relation of electricity to magnetism by Faraday (1791–1867); and the discovery by Maxwell (1831–1878) of equations governing the electromagnetic field.

After this point, the field grew very rapidly. Since the start of the twentieth century, a genuine mathematical atomic theory has been developed and subjected to extensive testing. This is based on quantum mechanics, analogous to the classical mechanics studied by Newton and Lagrange but different in one basic point. A quantum field theory called quantum electrodynamics, which contains the application of quantum mechanics to the Maxwell equations, has been developed and verified in all details. We now use it as a model for quantum field theories of other interactions.

Women began to be important in the history of physics near the end of the nineteenth century, at exactly the time when the field was starting to take off. In this chapter we discuss the work of Emmy Noether (who contributed an important theorem to the field theory applications of the Lagrangian approach), of Marie Curie (one of the early pioneers in experiments involving radioactivity), and of Lise Meitner (a contributor to both the experimental and the theoretical side of nuclear physics), as well as a number of women who belong completely to twentieth-century science. As we will see, there is no "women's branch" of the field. Women have contributed to both experiment and theory, and to all the various subfields of the discipline.

Historically, however, the percentage of physicists who are female has been quite low. In many instances, women have been actively discouraged from becoming professional physicists; in some cases they have even been discouraged from taking physics courses. As we examine the work of important contributors in this chapter, we will see various instances of this discrimination and of the effort necessary to overcome it.

For many (but not all) of the women discussed in this chapter, the contribution of greatest importance was done in collaboration with someone else. This has given rise to a persistent myth that women are incapable of creative innovation in physics. This point of view is still expressed openly by certain persons. For instance, as late as the 1971 annual meeting of the American Physical Society, a session on Women in Physics elicited the statement from a well-known male physicist that "If I had been married to Pierre Curie, I would have been Marie Curie."[1]

Since women have been spectacularly underrepresented in physics,[2] it is not at all clear that the total of all women physicists in the last century really provides a good sample of what could be expected if more women attempted the subject. Rather than formulate a statistical analysis on this basis, therefore, we will focus on several outstanding individuals to determine their impact—both on the science of their era and on later work.

In cases of collaboration, any observer pretending to take an intellectual stance should attempt to separate the contributions of both collaborators to assess their relative input. This is, of course, only possible for present-day and ongoing work where one can question the individuals separately. In the more historical cases discussed here, where the culture and requirements of the times may have prevented one of the collaborators from speaking freely, the accumulated weight of their other work may well be used as a substitute.

The short sketches presented below of eminent women physicists tend to demonstrate that women awarded significant honors have made a number of contributions, with different sets of collaborators. In many cases, work which was not noted specially at its time of completion is quoted widely today because it is being used, whereas the prize-winning or attention-getting work has become frozen in textbooks and is no longer cited in research articles.

EMMY NOETHER

I have chosen to begin this chapter on physics with Emmy Noether (1882–1935), who is better known as a mathematician. There are two reasons for this choice: First, chronologically she appears as one of the earliest figures of twentieth-century physics. Her work on invariants used by physicists was published in 1918. Second, and more important for our purposes, her work was *theoretical* physics. Many people (both inside and outside scientific circles) have the notion that women perform well in routine laboratory work but that they are incapable of theoretical advances. It is for this reason that it is so instructive to begin with Noether; since she was always considered a mathematician, no one can define her contributions to physics as anything but theoretical.

The biography by Auguste Dick[3] gives a very complete account of Emmy Noether's background, early life, and struggles to become an academic mathematician. She had to wait for rules to change in order to be allowed to register as a student at the University of Erlangen; her thesis was accepted (summa cum laude) in 1908, when she was 26. Her early work on invariants is considered rather uninteresting by mathematicians; the study by Dick makes very short work of this and he clearly considers it to be mere preparation for her later contributions to abstract algebra.

In modern elementary particle physics, however, "Noether's theorem" plays a central role. To understand this role, a little background in theoretical physics is necessary; let me expand a bit to provide an overview.

The goal of theoretical physics is to write a set of equations which completely predicts the motion of objects. Typically these are differential equations, and their solution for any special case can be a difficult mathematical exercise. It is thus very useful to know that certain quantities are *invariant* (i.e., they do not change) while the motion is going on. These invariant quantities have a mathematical function (they are some of the parameters defining the solution), but their value to physicists is much greater than you might expect from their mathematical role alone. These quantities (energy, momentum, angular momentum) are the basic stuff of physics—the tools which physicists use when they think.

Students in the first physics course study Newton's Laws, the differential equations of motion in their "raw" form. In later years, however, they learn that these equations can be obtained from a function—the Lagrangian—which contains by itself all the dynamical content of the system. This is even true for physical quantities which vary from point to point—fields, like the electromagnetic field $\vec{E}(x)$: all of the behavior of the system is completely determined, once the Lagrangian is written down.

Emmy Noether's discovery relates conserved quantities to symmetries of the Lagrangian for a system of fields. If your world consists of some set of fields $\vec{A}_i(x)$, ordinarily it will look quite different after a "transformation" such as rotation of space through 169° about the z axis, or translation along the x axis. However, you may sometimes have a system whose Lagrangian is unchanged after any rotation around the z axis. This system will have as a conserved quantity L_z, the angular momentum about the z axis.

Noether's theorem states that "Any continuous symmetry group induces a conservation law for some physical quantity, which can be derived for any given system once the Lagrangian is known." More technically, if a Lagrangian is invariant under any continuous group of transformations, there exists a conserved current calculable from the Lagrangian, and the spatial integral of the fourth (time) component of this current is a conserved charge.

Special cases of this had been studied earlier by Lorentz, Weyl, Klein, and others. Noether's contribution was the general formulation of the problem in a manner which yields conserved charges for any continuous symmetry. The previous work had considered such special cases as translation invariance, and she makes contact in her paper with several examples already available in the literature.

All of modern elementary particle physics can be formulated in terms of a theory of quantum mechanical fields, and the solution of the field equations is in general extremely difficult (it is frequently impossible to find a solution in useful form). This is a situation tailor-made for application of Noether's theorem, since knowledge of the conserved quantities enables physicists to understand most of the motion. For this reason, all students in modern field theory classes know "Noether's theorem," and most have learned it with this name attached. Most of them do not know, however, that E. Noether was a woman.

Noether's paper with this theorem, "Invariante Variationsprobleme," appeared in the *Nachrichten Mathematisch- Physikalische Klasse* (1918) of the Göttingen Königlichen Gesellschaft der Wissenschaften. Noether was not a member of the Royal Society, and the paper was submitted by F. Klein, as was another contribution by Noether on a similar topic in the same volume. Due to prejudice against women, she never became a member, despite the spirited arguments of her supporters.[4] Although these supporters had immense stature in the field, the "old boy" network was resistant to the inclusion of a "girl." One very famous anecdote concerns Hilbert's attempt during World War I to get Noether a *Habilitation* (a junior faculty appointment) in Göttingen. He is supposed to have said, "After all, we are a university and not a bathing establishment."[5] This appeal did not succeed, however; it was not until later that she was able to get a university position.

There is no doubt that Noether was foremost a mathematician, and chiefly interested in the more abstract development of that subject. Her early papers on invariants show, however, a knowledge of the physical examples of interest at that time and a desire to abstract them which today would certainly be considered theoretical physics.

MARIE CURIE

Marie Curie (1867–1934) is by far the most famous woman physicist. There are well-deserved reasons for this: she is the only woman to have won two

Nobel prizes—in physics in 1903, in chemistry in 1911. Also, the epic biography by her daughter Ève,[6] shown in the United States in the form of a television series in 1978,[7] has made public many details of her struggles for scientific mastery.

Of the women discussed in this chapter, Marie Curie was born earliest, in 1867. Her family was very supportive of the educational aims of their daughters, and this support was a communal exercise—Marie worked for some time as a governess in Poland to help pay the expenses of her sister's studies in France. As a result her own studies were somewhat delayed; she obtained her license in physics in 1893 and the corresponding degree in mathematics in 1894, and did not obtain the doctorate until 1903. Apparently the climate of opinion in France made it relatively easier for a woman to study science there than in Germany; at least there seem to have been fewer "impossibilities" in attending lectures and obtaining degrees than in the career of Noether, for instance.

Marie Curie was fortunate indeed in her choice of thesis topic. Radioactivity was a very new subject at that time and no one had any idea of the vastness of the field. Inspired by the discovery of Becquerel that uranium gave off a mysterious type of penetrating radiation, Madame Curie examined a number of substances and found that thorium and its compounds behaved in the same way as uranium. Examining pitchblende, a uranium ore, she discovered radium and polonium. In 1910 she succeeded in isolating pure radium metal.

The work of the Curies in isolating these substances had a great impact on the general study of radioactivity by others. In 1899, Becquerel used a radium salt given him by the Curies to show the similarity between the rays emitted and cathode rays. On the other hand, he found that the radiation emitted by polonium was much less penetrating than that from radium. We now know that polonium emits only alpha rays, whereas radium emits beta rays also.

Rutherford, also in 1899, managed to demonstrate that the rays emitted from uranium were of two kinds: what we now know as alpha and beta rays. In 1900 Madame Curie suggested that the alpha rays might consist of particles ejected by the radioactive substance, a hypothesis which we now know to be true. Note that this was the same year in which Villard discovered gamma rays in the radiation from radium; thus not all the rays of interest at that time were in fact thought of as ejected particles.

Marie Curie was also instrumental in setting up the Curie laboratory in Paris. While the funds for the laboratory were given basically in honor of her husband, his death in a traffic accident meant that she had to assume the responsibility for the lab; she also succeeded him as professor at the Sorbonne. During World War I she did quite a bit of work in early medical applications of radioactivity; her daughter Irène also took part in these activities.

Like many of the early workers in radioactivity, Marie Curie had no idea of the potential dangers of these penetrating radiations to the human system. She died in 1934 of leukemia, almost certainly brought on by prolonged exposure to the concentrated radioactivity of the ore she was purifying.

In Madame Curie we have a very interesting example of a woman physicist. She got a late start, like many women of her time (if they got a start at all). However, once launched she was comparatively successful: her thesis was on a spectacular new topic and she was able to spend the rest of her career working on basically similar ideas. She wound up in a position of influence and honor. Unlike Noether and our next subject, Lise Meitner, both of whom were forced out of Germany because of their backgrounds, Marie Curie was not forced to flee her working basecamp due to political pressure. However, even such a fortunate person had terrific setbacks and has been the object of carping to this day. The French Academy of Science rejected her for membership (it was not until 1979 that they appointed their first woman); the remark at a meeting of the American Physical Society quoted in the introduction to this chapter indicates the general flavor of criticism.

LISE MEITNER AND IDA NODDACK

The next very well known woman in the history of radioactivity is Lise Meitner (1878–1968), whose life is described in the excellent biography by Deborah Crawford.[8] Born in 1878 (four years before Emmy Noether), Meitner experienced many of the same problems with discrimination in the German system as did Noether, although perhaps things were not quite as difficult in Austria, where she studied (at Vienna), as they were for Noether in Erlangen.

Meitner's long collaboration with Otto Hahn is described in the book by Crawford, as is much of the work of that time and place on radioactivity. Hahn and Meitner discovered thorium C (1908) and protactinium (1917), and made a number of studies of nuclear isomerism. Meitner is perhaps most famous for her 1939 letter with Frisch, "Disintegration of Uranium by Neutrons: A New Type of Nuclear Reaction,"[9] which discusses the possibility that various puzzling results of experiments (done by Hahn, Meitner herself, and Strassmann, and by Fermi independently) could be due to the fission of the uranium nucleus (uranium 92 plus neutron goes to barium 56 plus krypton 36). This was rapidly taken up in the United States and applied to the concept of the chain reaction.

In the biography by Crawford, it is pointed out that in 1933 Ida Noddack (see below) had suggested that it was possible that the Fermi experiments (initially considered to be creation of an element with a higher atomic number than uranium) might be due to the breakup of the nucleus—in

Lise Meitner in the lab with Otto Hahn. From *Otto Hahn: A Scientific Autobiography* (1966) courtesy of Charles Scribner's Sons, New York.

fact she had suggested this to Meitner. The idea was too new and strange to be accepted by anyone at that time, however.

Meitner is also given credit, along with C. D. Ellis, for uncovering the nature and properties of gamma and beta radiation.[10] Her basic article on gamma rays is found in *Zeitschrift für Physik*;[11] it follows an article in the same journal by Hahn and Meitner on beta rays.[12] Her monograph *Atomvorgange und ihre Sichtbarmachung*, a lecture given at the Munich Chemical Society in July 1925,[13] is a very clear exposition of the observation of alpha, beta, and gamma rays. Another example of Meitner's clear style is her short book with Max Delbrück, *Der Aufbau der Atomkerne*.

Let us look further at the idea of nuclear fission. Ida Noddack was somewhat younger than Meitner. Born in Germany in 1896, she received a Doctor of Engineering degree in 1921 from the University of Berlin-Charlottenburg. She worked with her husband, Dr. Walter Noddack, on a number of chemical investigations; they are credited with the discovery of the elements rhenium and masurium in 1925. Not only did she express to Lise Meitner the idea that the experiments by Hahn, Meitner, and Fermi might have a different explanation than the creation of a transuranic ele-

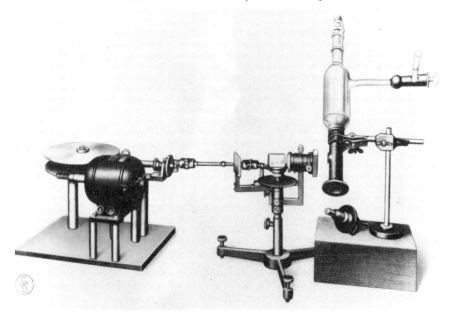

Apparatus used by Ida Noddack for discovering the element Rhenium. Courtesy of the American Institute of Physics, Center for History of Physics, Niels Bohr Library.

ment; she discussed this point of view in a paper, "Über das Element 93," published in 1934.[14] Her chief criticism, expressed in this paper, is that chemical analysis of the products should have been carried out to search for all elements of lower atomic numbers—not just those down to lead.

Ida Noddack survived World War II and returned to Germany to work at the geochemical institute in Bamberg. Her contribution to the concept of nuclear fission was basically ignored at the time it was made, and very few sources since then have brought it up. Most recently, a book by Leona Marshall Libby[15] mentions Noddack's contributions.

In 1944, Otto Hahn received the Nobel Prize for chemistry for the discovery, reported in 1939 by Hahn and Strassmann,[16] that barium was among the reaction products obtained by bombarding uranium with slow neutrons. This work was a continuation of work begun with Meitner and reported in 1938;[17] she left the project only because she was forced out of Germany by the policies of the Hitler regime.[18] It is not clear why Meitner was not included in the prize for her contributions to the understanding of fission. Perhaps it is a case of genuine oversight (there is quite a bit of oversight in science—look at Noddack's paper); more likely it was due to the politics of the Nobel awards. What is clear at this time, however, is that Meitner was really one of the basic people in both experimental and theoretical nuclear physics and chemistry in the years of their development.

IRÈNE JOLIOT-CURIE

After examining some of the immense struggles by the earliest women physicists, it is refreshing to see what happens if a person with genuine talent falls into a maximally supportive situation. Irène Curie (1897–1956) is a perfect example of this.

Born while her mother, Marie Curie, was working on her doctoral thesis, Irène was surrounded from the start by a world of scientific endeavor. Although her mother was very busy, and the children were mostly brought up by governesses, relatives, and friends, Irène worked at mathematical and scientific exercises set by Marie. On reading the correspondence between them,[19] one is struck by the continuous development of skills and the way in which this was thought of as natural. Letters from Irène to Marie (while Irène was on summer vacation and Marie at work) show that mastery of algebra and calculus was considered as normal as swimming, and both processes went on at the same time. Irène advanced steadily, although she certainly had normal reactions to the material. On July 22, 1914, at the age of seventeen, she wrote to her mother "Je maudis de bon coeur la formule de Taylor qui est la chose la plus laide que je connaisse" ("I heartily curse Taylor's formula, which is the ugliest thing I know").[20]

When World War I broke out, Irène desperately wanted to be of some use, and eventually persuaded her mother to allow her to help the medical teams by running X-ray machines (Marie had managed to set up a series of mobile X-ray machines to study the wounded). Irène worked with a nursing team in a different part of the front from that at which her mother worked. Thus it was as a mature and independent adult that she returned to the Curie laboratory after the war.

While working with radioactive samples in the lab, Irène Curie met and married Frédéric Joliot. Joliot, whose interest in science had previously led him to engineering school, did not have the sort of academic background that was typical for researchers in the lab, and he had to work hard to prove his worth. First, to satisfy Marie Curie, he had to obtain all the appropriate academic credentials. Since he was, in fact, a brilliant experimentalist, this was possible. He had a more difficult time fending off envious critics who thought he was trading on the Curie name.[21]

Study of the complete works of the Joliot-Curies[22] shows that typically in scientific papers Irène and Frédéric signed their own names (Irène Curie et Frederic Joliot, or vice versa). In papers with other workers, they also used their own names. However, he used Joliot-Curie in writing popular articles,[23] she used Joliot-Curie in her Nobel Prize address, and they are listed as M. et Mme. Joliot in their report to the seventh Solvay conference. Thus, the criticism that he used the Curie name to improve his position seems unwarranted; it is interesting to us here as an example of scientific backbiting aimed at the male member of a team rather than the female.

Irène's early work dealt with studies of alpha particle emission from polonium (her thesis), and with determination of half-lives and other properties of various other radioactive sources. She was thus well prepared for the series of experiments in which she and Frédéric studied numerous properties of various nuclear weak interactions. They made important contributions to knowledge about the neutron and positron, but their most original work was the discovery of "artificial radioactivity."

Prior to this discovery, studies of radioactive decay assumed that the only substances which could decay were those which people found naturally occurring in mineral deposits. The Joliot-Curies showed that this was not the case. By subjecting boron, magnesium, and aluminum to a beam of alpha particles, then removing the beam, they observed positron emission which decreased with an exponential time dependence similar to that of the naturally radioactive substances previously studied (*Comptes rendus des séances de l'Academie des Sciences* 198, p. 254 [1934]). The important difference with natural radioactivity was that the half-lives of the artifically produced isotopes were of the order of minutes rather than years.

This phenomenon was explained as follows: the initial bombardment creates new nuclei, which are unstable and decay rapidly. By working very quickly, the Joliot-Curies were able to chemically separate the small number of transformed nuclei (*Comptes rendus* 198 p. 559 [1934]) and to show that they had the properties expected of the hypothesized reaction products. Numerous results then followed, including a value for the mass of the neutron which was larger than the mass of the proton; this led them to suggest that the neutron might decay into a proton and an electron. (We know today this is correct, except that a neutrino is also given off.)

Irène and Frédéric were awarded the Nobel Prize in Chemistry in 1935 for this work. Although the distinction between radiochemistry and physics was never very clear during the period of early development, any physicist scanning the works of the Joliot-Curies will recognize fundamental concepts in nuclear physics under construction. It seems reasonable, therefore, to include Irène in this section on physics.

Long hours of work with radioactive elements during her early career probably contributed to the leukemia which caused Irène Joliot-Curie's death at the age of fifty-eight. Although scientists in general (and especially Frédéric Joliot) became sensitive to the hazards of radioactivity in the late twenties because of accidents in the lab involving large doses of radiation, Marie and Irène did not suspect that there could be hidden effects which would surface later. In this regard, Irène's early exposure to the laboratory practices of her parents was probably a disadvantage; in a letter to Marie dated August 5, 1927, Irène describes a Mme Cotelle in the lab who is seriously ill, but states that there can't be much in the air "étant donne que j'ai beaucoup travaillé la-dessus sans en être incommodée" ("since I've worked up there a lot without being inconvenienced by it").[24]

KATHARINE BURR BLODGETT

That all the contributions discussed in the previous pages relate to nuclear and particle physics may have given a lopsided view of physics, and of the efforts of women in the late nineteenth and early twentieth centuries. Women were also active in other areas of the field; the most well-known of these is perhaps Katharine Blodgett (1898–1979).

A very good biography of Katharine Burr Blodgett can be found in *American Women of Science* by Edna Yost.[25] This describes her early life and rather unusual education. After obtaining a bachelor's degree from Bryn Mawr and a master's degree from the University of Chicago, she went to work for General Electric Company in Schenectady, New York. There she worked as one of Irving Langmuir's assistants. Langmuir used his influence to get her accepted at the Cavendish Lab, and in 1926 she become the first woman to earn a Ph.D. in physics from Cambridge University.

Much of her work at General Electric involved the deposition of films on glass, and in 1938 she was announced as the inventor of nonreflecting glass. Yost's book treats this discovery in some detail, as do many texts. We should not dwell overlong on nonreflecting glass, however, for today the techniques that she developed for depositing the films that make the glass nonreflecting are more important in many other areas.

A search of recent Science Citation Indices reveals that two of her papers, published before 1938, are particular "hot sellers" today. The first of these, "Films Built by Depositing Successive Monomolecular Layers on a Solid Surface,"[26] was by Blodgett alone. In the citation index for the years 1980–84 this paper was quoted 122 times in papers on chemistry, biophysics, and device applications. The method is still used and useful in many different circumstances.

The other paper, by Blodgett and Langmuir, "Built-Up Films of Barium Stearate and Their Optical Properties,"[27] was quoted 92 times in the same five-year period—again in journals on solid-state physics, device applications, and biology. Several other papers of hers were also quoted during this time. Most of these cited papers were from the thirties, but a couple were from the fifties and sixties.

Any physicist living today will tell you that it is truly extraordinary for a paper forty-five years old to be quoted at all, much less for the technique to be so spectacularly useful that the paper is referenced more than ten times a year.

Although her work on films is most famous, Blodgett worked on various

General Electric group working on early weather modification experiments. From left: Irving Langmuir, Robert Smith-Johannsen, Katherine Blodgett, and Vincent Schaefer. Courtesy of General Electric Research and Development Center.

other projects during her long career at General Electric. Immediately after her return from Cambridge, she worked with Langmuir to improve the tungsten filaments in lamps. During World War II, she worked on the problem of de-icing aircraft wings, and devised a greatly improved smokescreen for troops.[28] Recognition of her work outside the company included honorary degrees, the Garvan Medal of the American Chemical Society (1951), and the Progress Medal of the Photographic Society of America (1972).

Tame stuff, perhaps, compared to Marie Curie's two Nobel Prizes. However, for an American woman born in 1898, the job at General Electric would have been unusual even had she never achieved scientific repute. Some perspective is provided in an article by Margaret Rossiter[29] on American women scientists before 1920 (Katharine Blodgett first went to work for Langmuir in 1918). At that time there were fifteen women with Ph.D.'s of the twenty-three identifiable women physicists. Of these twenty-three, eighteen had jobs as professors, and three were high school teachers—industrial jobs were clearly unusual.

MARIA GOEPPERT-MAYER

No one could possibly discuss the contributions of women to physics without coming to grips with an important social problem: the discrimination suffered by married women with respect to positions and pay. While it has frequently been difficult for women to gain recognition, even men with a strong belief in male superiority would agree that single women have to eat. As a result, once it was clear that they were emancipated from their father's household, single women physicists of the past could frequently obtain a job which provided them with a small amount of money to live on. Occasionally they even attained a "real" job. Married women of the same scientific calibre have had a much harder time. Maria Goeppert-Mayer (1906–1972) provides us with a classic example of the difficulties encountered by married women in American science.

Maria Goeppert entered the scientific arena in Göttingen some years after Emmy Noether, and the academic situation for women was much better. She obtained her Ph.D. in theoretical physics in 1930 at the age of twenty-four. (According to the Manpower Statistics Division of the American Institute of Physics, the average age of persons obtaining the Ph.D. in 1979 was twenty-nine.) In the same year, Maria Goeppert married and came to the United States with her husband, Joseph Mayer. Her entire postdoctoral career was spent in the United States.

A short sketch of her career is included along with her Nobel lecture in *Nobel Lectures in Physics 1963–1970*.[30] The tone of this two-page biography is so self-effacing, and indeed, apologetic, that one would deduce from it that she was a bumbling idiot helped at every turn by patient and kind collaborators. Perhaps this accounts for the impression of many physicists that her Nobel Prize-winning discovery of the role of spin-orbit forces in the shell model was entirely suggested by Fermi. Let us examine the record to see what her publications show.

First, her thesis "Über Elementarakte mit zwei Quantensprungen" ("On Elementary Processes with Two Quantum Jumps,")[31] was quoted fifteen times in 1978–1980 by papers on optics, atomic physics, and molecular physics. The work is still a classic.

The next stage of her career, in Baltimore, was complicated by the birth of two children and by university nepotism rules which kept her from having a real job. (She earned a few hundred a year—and an office in the attic—for helping a member of the physics department with his German correspondence.[32] The university was, however, happy to have her list "Dept. of Physics, Johns Hopkins U" on her papers!) According to the Nobel lecture biography, she "slowly developed into a chemical physicist," under the influence of her husband and of Karl Herzfeld. This all sounds rather meager, until you realize that one of the papers from this period,

Maria Goeppert-Mayer discussing with Robert Atkinson (left) and Enrico Fermi (center). Courtesy of the American Institute of Physics, Niels Bohr Library.

"Calculations of the Lower Excited Levels of Benzene," by Mayer and Sklar[33] is still quoted in chemical journals today, and that the text she wrote on statistical mechanics with her husband[34] remains a classic in the field.

Another pre–nuclear physics paper still quoted today was written at Columbia. "Rare Earth and Transuranic Elements"[35] is cited by both chemistry and solid-state publications on these elements.

The Mayers moved to Chicago in 1946. This marked Maria's "nuclear period," which was to result in her Nobel-winning discoveries. To provide some idea of the concepts involved in her work, and of their place in the history of physics, let me take a paragraph or two to discuss nuclear models.

Today every high school science student "knows" that atoms are like tiny versions of the solar system, with light electrons running in "orbits" around a very heavy nucleus. Students are also taught that the nucleus is made up of protons and neutrons, with the number of protons determined by the charge of the nucleus and the number of neutrons determined by its weight. This theory of the nucleus dates no further back than 1932, when the neutron was discovered. It is not surprising, therefore, that models for the actual motion of the protons and neutrons inside the nucleus should change rapidly over the decades immediately following 1932.

In discussing the work of Meitner, I mentioned the neutron capture experiments of the 1930s. The game these people played was to bombard a known nucleus with neutrons, in hopes that the nucleus (with Z protons and A-Z neutrons) would "swallow" the neutron and become an object of atomic weight A + 1. Then, if a beta decay occurred, one of the neutrons might turn into a proton (giving off a beta ray electron and a neutrino) and they would have a new nucleus of weight A + 1, and charge Z + 1.

In those times, they thought of the nucleus as basically indivisible; and Meitner did not come to the idea of nuclear fission until she was exposed to the "liquid droplet model," in which the protons and neutrons were imagined to swarm around within the nucleus more or less at random. As we all know, near the end of World War II the concept of nuclear fission became familiar to the man in the street; we should not be surprised, then, that nuclear physicists in the mid-1940s thought of the nucleus as a very disordered place.

In this context, it was very hard to justify the "magic numbers"—particular values of proton number Z or neutron number A-Z for which the nucleus seemed especially stable. Maria Mayer set out to understand these numbers, using a theory in which the properties of the nucleus were calculated in the same way as the properties of the electrons in their nice, well-defined orbits around the nucleus.

This use of a model which resulted in "shells" of neutrons and protons (like the electron shells in atomic theory) was heretical. Also, it didn't work very well—the shells that came out most simply from analogy with atomic physics fit only a few of the "magic numbers." But Goeppert-Mayer kept worrying about the problem, and eventually she came to realize that the magic numbers *could* be explained if one type of force, a so-called spin-orbit force, was large enough.

Her two most influential papers from this period are phenomenological—a great change from her purely theoretical thesis. These papers (by her alone), "On Closed Shells in Nuclei"[36] and "On Closed Shells in Nuclei II,"[37] established her priority to the spin-orbit coupling idea. This was cemented by the book she wrote with J. H. D. Jensen, who (together with O. Haxel and H. E. Suess) had independently discovered the same regularities. Their book, *Elementary Theory of Nuclear Shell Structure*, remains a widely quoted basic reference.

Clearly this work was basic and important. If one were to believe the biography in the Nobel Lectures ("In 1946 they went to Chicago. This was the first place where she was not considered a nuisance, but welcomed with open arms. She was suddenly a Professor in the Physics Department and in the Institute for Nuclear Studies. She was also employed by the Argonne National Laboratory with very little knowledge of Nuclear Physics"), Maria was in her element, in demand, and properly compensated at Chicago. Unfortunately, this was not the case.

Although she was listed as an associate professor and a member of the Institute for Nuclear Studies, she received no salary, due to the university's interpretation of their nepotism rule.[38] The university did not offer her a full professor's salary until 1959, when they tried to prevent her from going to La Jolla. She was, however, paid half-time by Argonne.

Nuclear physics was a subject she had to learn from scratch; but the Chicago of that era was an active center of research in the area and the atmosphere was right. With Fermi, Teller, and other well-known figures dominating the scene, Maria's reticence and persistence were interpreted by some as signs of stupidity. She never developed the charisma we often associate today with scientific greatness; one physicist who was a graduate student at Chicago recalls that the students considered her lectures poor, and that their complaints were a source of embarrassment to the department (which was nonetheless getting her services free).

The most persistent attack on Dr. Goeppert-Mayer's reputation comes, however, from those people who interpret her acknowledgment at the end of "Closed Shells in Nuclei II"—"Thanks are due to Enrico Fermi for the remark 'Is there any indication of spin-orbit coupling?' which was the origin of this paper"—to mean that Fermi had all the ideas for which she won the prize. Clearly this is unlikely, since she had previously undertaken a very thorough study of the shell model problem (as evidenced by "Closed Shells I"). The imputation is almost certainly due to the strong Fermi-cult among physicists who began their careers in the 1940s, and to the fact that she delayed submission of her most important paper for months due to some concerns about the work of others.[39] The lesson for young physicists in this is clear.

Unfortunately, while Maria Mayer's ability was finally recognized by her appointment to a professorship at La Jolla in 1960, she suffered a stroke soon thereafter and was unable to make much of this opportunity. She died in 1972. Her work on the shell model is now basic fact in nuclear textbooks; interestingly enough, her prenuclear physics papers are today cited as often as the work which earned her the 1963 Nobel Prize in Physics.

YVETTE CAUCHOIS

In Yvette Cauchois (1908–   ) we have an example of an eminent physicist whose value is chiefly observable through the citation index. Dr. Cauchois, currently a professor at the University of Paris and director of the Laboratoire de Chimie-Physique, has steadfastly refused to give any biographical details to books of the "Who's Who" type, and the author of this chapter was unable to obtain any such details from her. A study of the Science Citation Index shows, however, a thesis from Paris dated 1933 and numerous recent references to papers dated 1932 to the present.

Professor Cauchois's major work has been the application of X-ray spectroscopy to solids. She is an experimentalist who has measured and reported on a spectacular number of materials. The book of atomic properties which she and H. Hulubei collected, *Longeurs d'Onde des Emissions X et des Discontinuities d'Absorption X*,[40] is widely quoted; it was the natural outgrowth of many earlier studies by Dr. Cauchois on this subject. One paper frequently quoted recently,[41] which deals with X-ray emissions from lead, thallium, and gold, dates from several years earlier and contains many references to papers published before 1943.

Most properties of solids are determined by the energy levels of the electrons inside them, and these are related *both* to the energy levels of the electrons in isolated atoms of the substance *and* to the way in which the atoms combine to form the solid. Typically, the only way to learn about these energy levels is to "poke" the atom or solid under investigation with radiation of the right wavelength to provoke some transitions within the allowed levels. Dr. Cauchois concentrated for many years on measurements of this sort in solids and their implications for understanding the effects peculiar to solids.

In 1948 she published a monograph, *Les Spectres de Rayons X et La Structure Electronique de la Matiere*,[42] described in a preface by the eminent solid state physicist Neville Mott as being the first book devoted to the subject of the X-ray emission and absorption spectra of solids. The book, which is at a level that might have been used for an advanced graduate course of lectures (in fact it was originally given as two sets of lectures to the Société Francaise de Physique in 1946), gives us a good look into the solid-state physics of the time. From a historical point of view, we note that it is almost contemporaneous with Brillouin's *Wave Propagation in Periodic Structures*, a classic which explains basics needed to translate knowledge of the electron energies of atoms into properties of solids. A later review paper on the same field, "Electronic Band Structure of Solids by X-Ray Spectroscopy," by L. G. Paratt[43] quotes Cauchois's book as one of the earlier reviews of the subject which allowed Parratt to avoid the history and experimental techniques of the field and devote his paper to current topics.

This book was not, however, the culmination of Dr. Cauchois's interest in the topic. Rather, it marks the beginning of a long and fruitful series of investigations along these lines. She wrote many papers; again I will mention only those quoted most widely today. A 1949 paper by Cauchois and Mott, "The Interpretation of X-Ray Absorption Spectra of Solids,"[44] discusses properties of the fine structure of the X-ray absorption edges of solids in light of the exciton model. The collaboration between Cauchois and Mott (respectively primarily experimentalist and theoretician) continued for some time. A later paper by Cauchois in the same journal, "The L Spectra of Nickel and Copper,"[45] is followed by an interpretation from Mott, "Note on the Electronic Structure of Transition Metals."[46]

Meanwhile Cauchois was engaged in a comprehensive work which might be viewed as a continuation of the tabulation appearing in her book with Hulubei. Two major papers, "Energy Levels of Heavy Atoms"[47] and "Energy Levels of Atoms with Atomic Number Less than 70,"[48] describe the electron level structure of elements from atomic number 3 to atomic number 92.

Like most outstanding scientists, Dr. Cauchois has been eager to try new techniques. Among those of her papers referenced frequently is one coauthored with C. Bonnell and G. Missoni which describes a new study of the absorption spectrum of aluminum using X-rays from the high-energy synchroton at Frascati.[49] The next year, 1964, saw the publication of a very clear text by Cauchois and Heno, *Cheminement des Particules Chargées*,[50] as the first of three volumes in a series for physicists and physical chemists using radiation as a tool.

Professor Cauchois has had a very active career and has been quite important in the development of X-ray spectroscopy as a tool for studies of the band structure of solids. The Science Citation Index for 1980–84 contains over 125 references to her papers, covering the period 1932 to 1979. Nevertheless, she is relatively unknown in the United States. At one large university with an enormous solid-state physics effort, only professors themselves engaged in solid-state spectroscopy had heard of her; their students were unlikely to read either of her books because, although copies are in the university library, they have been banished to "cold storage" in a local warehouse due to lack of space on the shelves. The tables of Cauchois and Hulubei stand in the reserve section, in a spot where books regarded as "next candidates" for this banishment are kept. The reason for this treatment is quite simple: the books are written in French and hence their readership among American graduate students (whose native language may possibly be Fortran!) is limited.

For our purposes in this chapter, Yvette Cauchois illustrates a very interesting point. Even without biographical information, her work is recognizable and stands by itself. She has collaborated with many people, but with the exception of Mott these people are generally not so well known as she. The mere fact that she is a professor at the University of Paris and the director of a lab indicates substantial professional stature; citations of her papers show that this stature was developed by continuous work and publications over a period of more than forty years.

CHIEN-SHIUNG WU

Chien-Shiung Wu (1913–   ) has been one of the more outstanding women in modern experimental nuclear and elementary particle physics.[51] She recently retired from more than thirty years as professor of physics at Columbia, reducing the number of women physics professors there to 0. Her career from graduate student at Berkeley to professor at Columbia has been marked by a continuity and stability, and a position of power, achieved by few earlier women in physics. Let us examine her work to see how and why she managed where others failed.

By far her most important and famous work is the experiment demonstrating lack of parity conservation in the weak interaction. This study, reported in *Physical Review*,[52] was done by Wu along with four physicists from the National Bureau of Standards, E. Ambler, R. W. Hayward, D. D. Hoppes, and R. P. Hudson. Lee and Yang received the Nobel Prize in 1957 for their theory of the weak interaction, which predicted parity nonconservation and more or less inspired the experiment. Some persons in the physics community feel that Madame Wu and the experimental group should have been honored along with Lee and Yang, because her experiment was exceedingly timely, intricate, and clean. However, this seems to be one of the many cases where not all possible people were mentioned. It is clear, however, that it was her experiment which confirmed their theory.

Let me expand a bit on the physics involved here to show you how far the science has progressed since the work of Curie and Noether. Many of the radioactive substances investigated by Madame Curie emitted "beta rays"—i.e., electrons from the nucleus. By the 1950s, a great deal was known about the force which causes this radioactivity; it was called the "weak" interaction because (although the force is "turned on" all the time) radioactive substances take a fair amount of time to decay. (The simplest case, the decay of a free neutron outside a nucleus $n \rightarrow pe^- \nu$, takes about 15 minutes on the average.) By contrast, the force which holds protons and neutrons together inside the nucleus is called the "strong" interaction; its "strength" parameter is many orders of magnitude greater than that of the weak interaction.

A very successful quantum field theory of the electromagnetic interaction had been developed, and theoretical physicists believed (as they still do) that if they could write a correct Lagrangian containing all the interactions they would "understand" everything. As we mentioned with regard to Noether's work, the Lagrangian is constructed with an eye to the desired symmetries of the theory. For example, since all interactions are expected to be Lorentz invariant (i.e., to behave properly with respect to relativistic transformations), this property is included when the theory is written down. By Noether's theorem, it leads to conservation of the energy-momentum tensor.

For many years, it was also believed that the world on a submicroscopic level was unchanged under reflection or inversion of the axes. This is a "discrete" symmetry ($\vec{r} \rightarrow -\vec{r}$), rather than the continuous transformation of Noether's theorem, and discrete symmetries are not always related to a conserved quantum number. Nevertheless, all classical physics possesses this so-called "parity" invariance, electromagnetism (classical and quantum) has it, and we now believe the strong interaction has it.

Chien-Shung Wu adjusts her apparatus. Courtesy of the American Institute of Physics, Niels Bohr Library.

Once again, progress was made by a heretical assumption—the concept
that the weak interaction might *not* have this property. The Lagrangian of
the weak interactions does have a discrete symmetry (chiral invariance)
but this is incompatible with reflection invariance. In particular, certain
particles are always emitted with their spins pointing in the direction of
motion—never pointing in the opposite direction. To measure this sort of
effect requires very detailed measurements; unless you make enough momentum and spin direction measurements each time the reaction happens,
you cannot detect the difference between parity conservation and nonconservation. This is the basis of Wu's experiment—she and her co-workers
were the first to imagine the detailed measurements necessary and to carry
them out in a particular case.

All of Madame Wu's work has been marked by precise measurement
of correlations in various types of interactions. For example, one of her
early papers, on "The Angular Correlation of Scattered Annihilation Radiation," with I. Shaknov,[53] reports measurements on the spin polarization
of the photons emitted in the reaction $e^+ e^- \rightarrow \gamma\gamma$. This is a classic process

in the quantum field theory of electromagnetism, by far our most successful field theory. Careful checks here are crucial, not only to understand electromagnetism but also to test the whole underpinning of quantum Lagrangian construction.

Wu has done many further experiments on weak interactions in nuclei, and her review article "The Universal Fermi Interaction and the Conserved Vector Current in Beta Decay"[54] is frequently quoted. In addition, there was a period in which her laboratory made careful measurements on the X-rays and gamma rays emitted by muonic atoms (atoms in which one of the external electrons is replaced by a particle 200 times heavier called the muon). The measurement of these quantities allows one to determine various things about the properties of the nuclei. Her expertise in this area led to a monumental review article with Lawrence Wilets, "Muonic Atoms and Nuclear Structure."[55]

We thus have a picture of a long career of careful experimentation on such relatively delicate matters as spin correlations, and of a continuing effort to determine the properties of the nuclear weak interaction. As anyone would tell you, however, doing good work is not enough to achieve the status held by Madame Wu in the physics community today. She is one of those people who have a place in the cult of modern physics; that is, she has taken on a symbolic position which is not directly related to her personality or actions. For instance, she is almost universally referred to as "Madame Wu," despite the fact that Wu is her own name and her husband's surname is Yuan. She is reputed to be very smart, and very fierce—she has fought hard to control her turf and to defend her position.

MARY BETH STEARNS

For our last example, we take Mary Beth Stearns (1925–   ). She is representative of a generation of women physicists younger than Madame Wu and Yvette Cauchois; within this age bracket there are a large number of rather distinguished women. It is impossible to discuss the important contributions of all these people within the confines of this chapter. For this reason, I have chosen Dr. Stearns, not because she is more outstanding than many of the other women of her cohort (some of whom are listed below) but rather because she provides additional balance to our earlier sample. This balance comes in two areas. First, she is in solid-state physics now (her most widely quoted papers are in magnetism), thus balancing to some extent the large number of nuclear physicists mentioned previously; and second, most of her career has been in industry rather than academe.

Mary Beth Stearns received her Ph.D. from Cornell University in 1952. After doing research at the Carnegie Institute of Technology from 1952 to 1956, she worked for General Atomic from 1958 to 1960; in 1960 she went to the Ford Motor Company, where she was principal scientist on the scientific research staff for many years. Later Stearns left Ford to become a professor at the University of Arizona. While her early work included photonuclear reactions and meson spectroscopy, she moved from nuclear interactions into Mössbauer-effect studies of solids. This led to her studies of magnetism in solids.

The permanent magnets we all use get their "pull" from the spins (and hence the magnetic moments) of many ions in the substance pointing in the same direction. In most other substances this cannot be achieved. The solid-state physicist must struggle to understand why. Some of Dr. Stearns's most quoted works[56] use the Mössbauer effect, in which one sends in gamma rays of a known wavelength (emitted by one nucleus) and these are first absorbed and then emitted by iron nuclei. Information about the energy levels of the iron atom in its environment in the solid can be deduced from the behavior of the emitted γ ray.

She next went on to other types of nuclear magnetic resonance experiments, again designed to probe the energy level structure of the magnetic ions.[57] Her aim throughout was to determine the exact orientations of all the atoms in the substance. With this data in hand, in later papers[58] she has worked hard on models to fit the data and explain the detailed quantum structure of magnetic solids.

For those interested in collaborations, a large fraction of her papers have been published under her name alone. The papers are principally experimental, but they include some theoretical calculations.

A very good review of the history of thought on magnetism was presented by Stearns herself in *Physics Today* in 1978.[59] In this article, "Why Is Iron Magnetic?", Stearns explores the various possible ways in which those electrons in iron which are not in filled shells can contribute to the phenomenon of ferromagnetism (where spins at all the atomic sites line up to provide a very large magnetic field). She explains why various types of experiments (many of which are exemplified in the papers mentioned above) have led to the belief that this is accomplished by a few of the electrons from the 3d shells.

White and Geballe's book *Long Range Order in Solids*[60] gives a more technical overview of Stearns's contributions to this field, and her reasoning why iron, cobalt, and nickel are ferromagnetic, whereas manganese (which is next to iron in the periodic table) is antiferromagnetic.

SOME LATER WOMEN PHYSICISTS

It would be easy to go on for many more pages describing the contributions of women physicists. As we get to the group of women who were born later than 1915, there are so many very active people that their contributions form a very large body of work, and this small chapter is hardly the place to describe it. Also, these people have a number of years before them in which to expand their contributions. For this reason, rather than attempt to include all the contributions of currently publishing physicists, I have chosen only a few as samples. These include Yvette Cauchois, Chien-Shiung Wu, and Mary Beth Stearns, whose work I have discussed above. There are very many others, whose work I had to abbreviate or simply leave out for lack of space. Let me take a few paragraphs here just to mention a few of these people.

Mildred Dresselhaus is another very well known experimental solid-state physicist. Her research contributions have included microwave properties of superconductors, the electronic structure of semimetals, and the magnetic phases of magnetic semiconductors as probed by Raman scattering. She is best known for her work on graphite intercalation compounds. After a period on the research staff at MIT's Lincoln Laboratory, she became a professor at MIT and the director of its Materials Research Lab.

Gertrude Sharff-Goldhaber has had a long and influential career in nuclear physics. At Brookhaven National Lab, she has been involved in both theoretical and experimental work to determine the detailed properties of nuclear energy levels and magnetic moments, and from these to gain a better grasp of nuclear structure. We are still unable to calculate the nuclear properties with the accuracy with which we can determine atomic energy levels. Her sister-in-law, Sulamith Goldhaber, who died in 1965, was very important in the studies done by the Lawrence Berkeley Lab's Bubble Chamber group on A_1 production and on the production mechanisms for other elementary particle resonances. Fay Ajzenberg-Selove, a nuclear physicist, has become well known for her handling of the tables of the energy levels of nuclei; she has published these as a monumental series of papers in the journal *Nuclear Physics*.

Leona Marshal Libby did early work on neutron and proton scattering. Anneke Levelt Sengers, at the National Bureau of Standards, is well known for her work on critical phenomena. Esther Marley Conwell, a solid-state theorist, has published over a hundred papers on the properties of semiconductors, and how they can be affected by subjecting the substances to outside perturbations like high electric fields. From 1963 to 1972 she was manager of the physics department at GTE. After a year as the Abby Rockefeller Mauze Professor at MIT, she joined the staff at the Xerox Webster Research Center as principal scientist.

We must also not forget Melba Philipps, coauthor with W. Panofsky of the famous textbook *Classical Electricity and Magnetism*, and Edith Quimby, who did pioneering work in radiological physics. Kathryn McCarthy, professor at Tufts University, is well known for her work on the properties of optical materials.

In France, Henriette Faraggi, a nuclear physicist, is director of the Orme des Merisiers lab at Saclay; and Aniuta Winter, known for her studies on the stability of glass, became director of the French National Glass lab.

There also are a whole host of younger women in physics, at all stages of their careers. I will not even try to mention names in this group, because there are so many and because their contributions are so numerous.

CONCLUSION

Women in physics have suffered their share of discrimination, disappointment, and trouble. It has been less than a hundred years since women were even allowed in the laboratory in some countries. During that time we have seen more than one shift in the position of women in the physics community.

The earliest women were all singular cases who were able to "do their thing" only because their talent was eventually recognized by some influential man who insisted on their being allowed to do it. In the case of Noether, this role was probably played by Hilbert; for Meitner, it was played by Hahn; for Madame Curie, it was played by Monsieur Curie; and for Katharine Blodgett, it was played by Irving Langmuir. The only one of these people who attained a position of political power in the scientific community was Madame Curie, and this only because she stepped into her husband's shoes. The others were tolerated as "hangers-on," and means of support were found for them, but their influence was purely intellectual.

We thus have an early period in which a few outstanding women are tolerated within the intellectual group, but generally they were not considered real scientists by outsiders looking in. (The biographies of these early women all contain stories such as those of editors making requests for articles which were then withdrawn when they discovered that the person in question was a woman.)

During the 1950s, according to data of the National Research Council, the percentage of doctorates awarded to women in physics actually decreased.[61] At this time the role of scientists and engineers as "part of the war machine" was very much in the public eye, and large expenditures were being made with this in mind. Thus a climate in which this field was considered important enough to have high pay came along with a period of discrimination against women entering this field.

The percentage of women entering physics and related engineering disciplines increased steadily through the 1970s and the 1980s. Our courses at the University of Illinois (Urbana) are probably representative of the national situation. In 1974, my course in elementary mechanics (800 students) had about 10 percent women; in 1981 the same course (600 students) had about 20 percent women. A similar increase has been noted in the percentage of women obtaining Ph.D.'s in physics; this number reached 8 or 9 percent in the mid 1980s.

By the 1990s these people should have an influence on the permanent jobs in the field. There is certainly a long way to go—a survey in the early 1980s by the American Physical Society's Committee on the Status of Women found that only 1.9 percent of the university faculty were women, although more than 3 percent of the available pool of Ph.D. physicists were women.[62] (The survey covered all 171 departments which grant Ph.D.'s in physics; 125 of these departments had no women faculty at all.)

Discrimination against women in physics has become much less marked, as society has realized the importance of using everyone's talents. It is not particulary easy for anyone, male or female, to take up the discipline of physics as a career; but the very large additional barriers which confronted many women in the past have been considerably reduced. It is possible for competent women physicists to have a "real job" with "real money" and sometimes "real power."

The next step in attitude adjustment will come when some woman is recognized as a "real genius." By this I mean that she will be attributed with a discovery of importance which is not tainted by the words "yes, but"—"yes, but her father was a mathematician" (Noether); "yes, but her husband was the real genius" (Marie Curie); "yes, but she worked with the big names" (Meitner and Goeppert-Mayer); etc., etc. Since very few *male* physicists have achieved such a level, and since women make up a small fraction of the professional pool, it is statistically unlikely that we will reach this state of affairs in the next year or even the next decade. However, we will probably have to wait until then to cast off the last vestiges of that nineteenth-century thinking which makes it necessary to write a book on the contributions of women.

NOTES

1. Vera Kistiakowsky, "Women in Physics: unnecessary, injurious and out of place?" *Physics Today* 33, #2, p. 32 (1980).

2. Two figures are of relevance here: the fraction of practicing physicists who are women, and the fraction of students receiving new Ph.D.'s. In 1972, less than 3 percent of the Ph.D. physicists in the United States were women ("Women in Physics," Report of the Committee on Women in Physics, *Bull. Am. Phys. Soc.* 17 (1972):740). The trend is increasing, fortunately. In 1978, less than 5 percent of the new Ph.D. degrees awarded in physics went to women (Kistiakowsky, "Women in Physics . . . "); whereas data collected in 1985 and 1986 showed female new degree holders at 7.7 percent and 8.8 percent of the total (I thank Susanne D. Ellis of the American Institute of Physics Manpower Statistics Division for providing these figures prior to their publication).

3. Auguste Dick, "Emmy Noether 1882–1935," Beihefte #13 zur *Elemente der Mathematik*, 1970 (Basel: Birkhauser). Note also the memorial address in English by Hermann Weyl at the end of this monograph. A translation of the entire book, by H. I. Blocher, is also available (Boston: Birkhauser, 1981).

4. Ibid., p. 60.

5. Ibid., pp. 14 and 59.

6. Ève Curie, *Madame Curie*, translated by Vincent Sheean (Pocket Books, Inc.).

7. The series was produced by BBC Polytel, and broadcast on U.S. public television.

8. Deborah Crawford, *Lise Meitner, Atomic Pioneer* (New York: Crown Publishers, 1969).

9. *Nature* 143 (Feb. 11, 1939):239.

10. E. U. Condon and H. Odishaw, eds., *Handbook of Physics* (New York: McGraw-Hill, 1958), pp. 9–50.

11. L. Meitner, "Die γ-Strahlung der Actiniumreihe und der Nachweis, dass die γ-Strahlen erst nach erfolgtem Atomzerfall emittiert werden." *Zeitschrift für Physik* 34(1925):807.

12. O. Hahn and L. Meitner, "Die β-Strahlspektran von Radioactinium und seinen Zerfallsprodukten," *Zeitschrift für Physik* 34(1925):795.

13. Published in Stuttgart by Ferdinand Enke in 1926.

14. I. Noddack, "Über das Element 93," *Zeitschrift für Angewandte Chemie* 37(1934):653.

15. L. M. Libby, *The Uranium People* (New York: Crane Russak, 1979), p. 43.

16. O. Hahn and F. Straussman, *Naturwissenschaften* 27(1939):11.

17. L. Meitner, F. Strassmann, and O. Hahn, Künstliche Umwandlungsproresse bei Bestrahlung des Thoriums mit Neutronen; Auftreten isomerer Reihen durch Abspaltung von α-Strahlen," *Zeitschrift für Physik* 109(1938):538; also, Hahn and Straussmann, *Naturwissenschaften* 26 (1938):755.

18. E. Yost, *Women of Modern Science* (New York: Dodd Mead & Co., 1962), p. 27.

19. Curie Correspondence (Marie, Irène), Choix de Lettres (1905–1934), Presentation par Gillette Ziegler, Les Editeurs Francais Reunis, Paris, 1974.

20. Ibid., p. 83.

21. Maurice Goldsmith, *Frédéric Joliot-Curie* (London: Lawrence & Wishart, 1976).

22. Frédéric et Irène Joliot-Curie, *Oeuvres Scientifiques Complètes*, (Paris: Presses Universitaires de France, 1961).

23. "L'Avenir de la Physique Nucleaire en France," *Atomes* 100 (1954).

24. Curie Correspondence, p. 281.

25. E. Yost, *American Women of Science* (Stokes Co., 1943).

26. *J. Am. Chemical Society* 57(1935):1007.

27. *Phys. Rev.* 51(1937):964.

28. I am indebted to George Wise, R & D Historian of General Electric Corporation, for supplying me with information about Katharine Blodgett.

29. Margaret W. Rossiter, "Women Scientists in American before 1920," *American Scientist* 62(1974):312.

30. *Nobel Lectures in Physics 1963–1970* (Amsterdam: Elsevier, 1972), p. 38.

31. *Annalen der Physik* 9(1931):273.

32. For a much more complete biography of Maria Goeppert-Mayer, see *A Life of One's Own* by Joan Dash (New York: Harper & Row, 1973).

33. *J. Chem. Phys.* 6(1938):645.

34. M. Goeppert-Mayer and J. Mayer, *Statistical Mechanics* (New York: Wiley & Sons, 1948; second edition, 1976).

35. *Phys. Rev.* 60(1941):184.

36. *Phys. Rev.* 74(1948):235.

37. Letters to the Editor, *Phys. Rev.* 75(1949):1969.

38. Dash, *A Life of One's Own.*

39. Ibid.

40. Paris: Hermann et Cie, 1947.

41. "Nouvelles donées relatives aux atomes de numero atomique 82(Pb), 81(Tl) et 79(Au). Emissions L faibles et niveaux extérieures," *Comptes Rendus de l'Academie des Sciences* 216(1943):762.

42. Paris: Gauthier-Villars, 1948.

43. *Reviews of Modern Physics* 31(1959):616.

44. *Philosophical Magazine* 40(1949):1260.

45. *Philosophical Magazine* 44(1953):173.

46. *Philosophical Magazine* 44(1953):187.

47. Y. Cauchois, "Les Niveaux d'energie des atomes Lourdes," *J. Phys. Radium* 13(1952):113.

48. Y. Cauchois, "Les Niveaus d'Energie des Atomes de Numero Atomique Inférieure à 70," *J. Phys. Radium* 16(1955):233.

49. Y. Cauchois, C. Bonnelle, and G. Missoni, "Premiers spectres X du synchrotron de Frascati," *Comptes Rendus d'Academie Francais* 257(1963):409.

50. Paris: Gauthier-Villars, 1964.

51. A short biography of C-S. Wu is contained in *Women of Modern Science* by E. Yost (New York: Dodd, Mead & Co., 1962), pp. 80–93.

52. C-S. Wu, E. Ambler, R. W. Hayward, D. D. Hoppes, and R. P. Hudson, "Experimental Test of Parity Conservation in Beta Decay," in Letters to the Editor, *Phys. Rev.* 105(1957):1413.

53. In Letters to the Editor, *Phys. Rev.* 77(1950):136.

54. *Reviews of Modern Physics* 36(1964):618.

55. *Annual Reviews of Nuclear Science* 19(1969):527.

56. M. B. Stearns, "Internal Magnetic Fields, Isomer Shifts and Relative Abundance of the Various Fe Sites in Fe Si Alloys," *Phys. Rev.* 129(1963):1136; M. B. Stearns, "Spin-Density Oscillations in Ferromagnetic Alloys I. Localized Solute Atoms: Al, Si, Mn, V and Cr. in Fe," *Phys. Rev.* 147(1966):439.

57. M. B. Stearns, "Spin-Echo and Free-Induction-Decay Measurements in Pure Fe and Fe-Rich Ferromagnetic Alloys: Domain-Wall Dynamics," *Phys. Rev.* 162(1967):496.

58. M. B. Stearns, "On the Origin of Ferromagnetism and the Hyperfine Fields in Fe, Co and Ni," *Phys. Rev.* B8(1973):4383; M. B. Stearns, "Interant-3d-Electron Spin-Density Oscillations Surrounding Solute Atoms of Fe," *Phys. Rev.* B13(1976):1183; M. B. Stearns, "Hyperfine Fields at Nonmagnetic Elements in Ferromagnetic Metal Hosts," *Phys. Rev.* B13(1976):4180.

59. M. B. Stearns, "Why is Iron Magnetic?" *Physics Today* 31, #4 (1978):34.

60. R. M. White and T. H. Geballe, *Long Range Order in Solids* (New York: Academic Press, 1979), p. 61.

61. See Kistiakowsky, "Women in Physics."

62. 4.9 percent of Assistant Professors, 1.5 percent of Associate and Full Professors. *Physics Today*, Feb. 1982, p. 99.

G. KASS-SIMON

Biology Is Destiny

The amazing thing is that there are so many of them. In almost every journal in every month since the mid-nineteenth century, the names of women biologists abound. In almost every area within the discipline women have pursued and published their research. In every class of every year (since its inception in 1888), women have represented no less than 40 to 60 percent of the total number of students at the Marine Biological Laboratory (MBL) at Woods Hole. And yet, when we think of the history of biology, when we try to recount the concepts and techniques that have moved the science forward, we end up finally by shrugging our shoulders and asking, "But where were the women? What did they do?" Is it possible that for all their work and all their efforts most of these women, whose careers often spanned many decades, have in fact contributed so little to their science that for all intents and purposes they might never have existed? Or, is it more probable, as I would like to argue, that the work of these scientists has succumbed to the same fate as has the work of women artists and composers; that is, although their work has frequently been woven into the fabric of the discipline, it has only rarely been associated with the names of the women responsible for it.

To be sure, when knowledge becomes old, the names of those who produced it disappear from the body of the literature—and this is true for men's research as well as women's. Nonetheless I would like to propose that the work of women scientists tends to become dissociated from their persons earlier and more frequently than it does for men scientists.

I do not mean to imply that there has always been a deliberate failure to give due credit to a particular woman who may have performed this or that experiment (although there are instances of that too), but rather to suggest that in biology, as elsewhere in our society, we have been accustomed to perceive women and their work simply as adjuncts of men and their work. If women's work was a mere extension of men's work it is hardly surprising that the research done by the women in a discipline was considered inherently and fundamentally to be the property of its men; nor was it anything but "natural" to persistently ignore the names of the women whose work we have incorporated into our literature.

It is therefore the intent of this chapter to take note of some of the earlier women biologists; to look at their contributions—some of which were indeed germinal—and to insist on their proprietorship of their own research.

Because biology is so diverse, and its history so old, it was necessary to limit this discussion to those areas and times which seemed easily accessible to me; for this reason, of the many disciplines which could have been included there are four which I have chosen to consider. These are developmental biology, genetics, physiology (in particular, nerve physiology), and some natural history. I have also found it necessary to focus my discussion on American women scientists and have included only those women of other nationalities whose work happened to form part of my own background and training.

WOMEN'S CONTRIBUTIONS TO DEVELOPMENTAL BIOLOGY: GRADIENTS, ORGANIZERS, AND DIFFERENTIATION

Developmental biology, as the term is understood today, is that area of experimental biology aimed at explaining how the cells, tissues, and organs of an individual, which begin life as part of a mass of more or less identical cells, acquire the functional and anatomical characteristics which distinguish them from all other types of cells, tissues, and organs.

This was not always the objective of developmentalists. At first embryologists were concerned, as were all other life-scientists, with the precise depiction of nature's diversity. Differences and similarities were compared and often used as evidence with which to extol a god's wisdom. Later, with the appearance of Darwin's and Wallace's theory of evolution in the early 1860s, biologists, especially embryologists and comparative anatomists, became intent upon gathering data to use as arguments for or against evolutionary theory. Haeckel's Biogenetic Law, that ontogeny (the progressive changes occurring during embryonic development) recapitulates phylogeny (the progressive changes occurring during the evolutionary development of a species), epitomizes the philosophical *raison d'etre* for much if not all of the developmental biology of the second half of the nineteenth century.[1]

With the appearance of the experimental work of Wilhelm Roux (1850–1924)[2] and Hans Driesch (1864–1941),[3] the emphasis shifted; the essentially descriptive gave way to the experimental and the analytical. The problem was no longer how do species arise, but rather by what mechanisms do cells acquire their ultimate identities. One may set the birthdate of contemporary experimental embryology roughly at 1894, which marks the year that the first volume of *Roux' Archiv für Entwicklungsmechanik* appeared.[4]

At that time (roughly the twenty years before and the twenty years after the turn of the century), women were beginning to be allowed to pursue their intellectual activities in academic institutions, and this is reflected in the growing numbers of students of embryology who were women. But before looking to see what these women embryologists were doing, it is necessary to set the philosophical framework of the new experimentalism a little more precisely.

In trying to grasp the nature of differentiation, three principles emerged which appeared to explain the observations and experimental findings. The first was that embryonic cells seemed to possess an inherent developmental potential which permitted but did not necessarily determine the ultimate fate of the developing cells. The cellular location of this potential became the focus of much of the research in embryology. Besides Roux and Driesch, the names of C. O. Whitman (1842–1910), E. G. Conklin (1863–1952), E. B. Wilson (1856–1939), T. H. Morgan (1866–1945), and Jacques Loeb (1859–1924) come to mind in association with this idea.

The second and somewhat later concept that arose was that there was a hierarchy of differentiation (particularly a chemical or metabolic hierarchy); differentiated structures became dominant and controlled the fate of still undifferentiated ones in a sort of chemical gradient of determination. Although the concept of a gradient seems to have first been published by Theodore Boveri (1862–1915), it is irrevocably associated with the name of C. M. Child (1869–1954), who made it the cornerstone of his scientific work.

Finally, the third and most compelling idea was that there were embryonic cells which seemed to control differentiation; they were thought to produce chemical substances by which they induced and directed the development of other cells. For this concept of embryonic induction by organizers, Hans Spemann (1869–1941) was awarded the Nobel Prize in 1935. The crucial experiments which demonstrated that a second embryo could be induced in an already developing amphibian egg by grafts of specified areas of tissue from another egg were performed in Spemann's laboratory by his student Hilde Mangold and published in 1924.[5]

But, most ideas in science, unlike Minerva, do not spring fully grown from Zeus's head; what often appears to be the brainchild of a particular scientist is, in reality, the final though perhaps new and perceptive synthesis of what has gone before. This is especially true for the ideas in experimental embryology.

ETHEL BROWNE HARVEY: INDUCTION AND MEROGONY

In 1909 Ethel Nicholson Browne (1885–1965) while a predoctoral fellow at Columbia University (working with T. H. Morgan, whom she thanks for suggesting the work) published a paper entitled "The Production of New Hydranths in *Hydra* by the Insertion of Small Grafts."[6] In this paper she describes a series of experiments by which she induced the formation of a new hydra (a freshwater polyp related to jelly-fish and anemones) in the body column of another hydra by implanting a tentacle cut near its

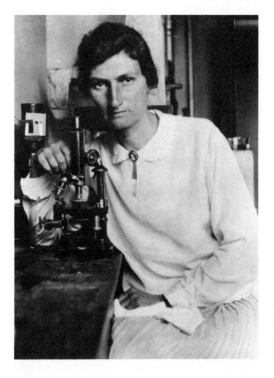

Ethel Brown Harvey at the Marine Biological Laboratory. Courtesy of the Library, MBL, Woods Hole.

base (at the oral cone or peristome) into the column of the second animal (fig. 1).

As far as I can tell, this is the first time that the induction of a differentiated animal by a specific tissue was demonstrated and correctly analyzed. In the conclusion of her paper Browne writes: "A new hydranth can be formed by a hydra in any part of its body region, except in the tentacle region, when the necessary stimulation is given by a grafted piece."[7] These grafts may be of "peristome tissue at the base of the tentacle . . . , material of a regenerating hydranth," or "material of a bud. Neither wound nor the graft of any other kind of tissue will stimulate the stock to send out a new hydranth."[8]

These experiments preceded those of Hans Spemann and Hilde Mangold by some twelve years, but are never cited in reviews that recount the development of the concept of induction. Spemann himself traces the evoion of some of his ideas to Boveri (1901) and Roux (1888)[9] and to his own earlier experiment on optic lens induction (1900).[10]

Although the hydra experiments should have been provocative, they seem to have gone largely unnoticed by the embryological community; Ethel Browne herself did nothing further with them. The reasons for this can only be guessed at. Morgan had not yet turned to genetics and was

PRODUCTION OF NEW HYDRANTHS IN HYDRA
Ethel Nicholson Browne
PLATE III

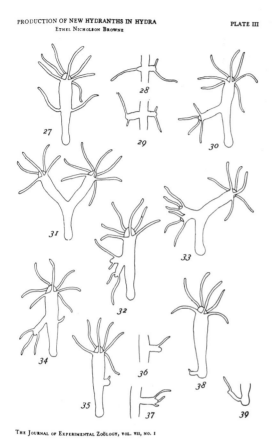

Figure 1. Ethel Browne Harvey's induction experiments (1909). 27–31: Two implanted tentacles induce the formation of two new hydranths, in the midsection of the hydra. One hydranth subsequently absorbed, the other develops into a full-sized head. 32–33: Two heads are induced by two tentacles. 34: Only one hydranth is induced by the tentacle implant, the other tentacle is resorbed. 35–37 and 38–39: Two other experiments. Courtesy of the Library, MBL, Woods Hole.

still avidly working on embryo development and hydroid and planarian regeneration; he seems to have assigned regeneration problems to a number of his Bryn Mawr students, including Helen D. King, Florence Peebles, and later Nettie Stevens. Nonetheless, neither he nor the many other embryologists of the time seem to have been aware of the significance of Browne's experiments.

Be that as it may, Ethel Browne, like the many brilliant male developmental biologists at the time, turned her attention to the problems of sea-urchin embryology. Among her many contributions to these problems was the application of centrifugal force to the study of differentiation in the developing egg.[11] Although the use of the centrifuge to study the physical properties of the egg's membrane and protoplasm appears to have been pioneered by her husband, E. N. Harvey,[12] it was Ethel's studies on the effects of redistributing the various parts of an egg's contents by centrifugal force that were more important to embryology. With this method, she was able to demonstrate that even eggs without nuclei could, if placed in the proper medium, be made to behave as though they had been fertilized. The term "parthenogenic merogony," by which is meant the cleavage and development of an unfertilized enucleated egg, is hers.[13]

Ethel Browne Harvey is also responsible for developing the method by which sex is ascertained in sea urchins[14]—an important contribution for those who need to obtain eggs and sperm with which to study the process of differentiation and development. But perhaps her most useful contribution is the book she published in 1956, *The American Arbacia and Other Sea Urchins*.[15] This volume, which is still a standard reference work for sea-urchin embryologists, is a fascinating compilation of natural, cultural, and experimental sea-urchin history and useful experimental recipes.

During her long career, Ethel Browne Harvey became variously a school teacher, a research assistant, and an instructor. In 1916 she married E. Newton Harvey, himself a rising physiologist. At first they appear to have worked collaboratively on the questions of sea-urchin embryology, but ultimately E. N. Harvey's work became more and more centered on questions of the physics of biological phenomena, while Ethel Harvey's work remained focused on differentiation and development. Although her work appears to have been recognized and acknowledged during her lifetime (she became a member of the Board of Trustees of the MBL in 1950, was a member of numerous national and international embryological societies, and was accorded the singular honor of being invited to deliver a Friday Evening Lecture at the MBL in 1944, one of only three women to have been so invited between the years 1888, when the MBL was founded, and 1946), she nonetheless was never appointed to a full faculty position at Princeton, where she remained an investigator for more than 25 years, and where her husband had held a tenured professorship since 1920 and before that an assistant professorship from 1916 to 1920.[16]

FLORENCE PEEBLES:
REGENERATION, DIFFERENTIATION, AND DETERMINATION

Of the number of other well-known biologists who owe their training to the famous genetisist T. H. Morgan, there were several women who were his students while he was a teacher at Bryn Mawr. Among them was Florence Peebles (1874–?). Although Peebles's name is known today only among the real aficionados of hydra regeneration, her experiments and ideas deserve to be reexamined, because the views she expressed over fifty years ago are just now finding currency among developmental biologists.

Beginning with her predoctoral publications,[17] Peebles worked for the next thirty years on problems of tissue determination during regeneration and development. She focused her attention on how external influences, especially those of the immediate environment, affected differentiation.

Florence Peebles, 1921. Courtesy
of the Library, MBL, Woods Hole.

Among her most interesting observations were those she published on
regeneration in *Hydra*. In these experiments she found that tissue which
had originally been differentiated into tentacles could be reorganized to
look and behave as though it were a part of the lower body column.[18] This
is a truly interesting finding, for it implies that the parts of this simple
animal—whose nerve cells barely comprise a nervous system—somehow
take into account their relationship to each other and alter their behavior
accordingly.

Although Peebles did not pursue these experiments, she continued to
publish papers on regeneration and cell and tissue regulation.[19] In her 1931
paper, "Growth Regulation in *Tubularia*,"[20] she addresses the by-then
familiar question of what factors determine how a regenerating structure
will be differentiated, or, to put it in terms of a regenerating hydroid stalk,
how did the cut pieces know which end was up. To Spemann's organizer
theory and Child's gradient hypothesis,[21] Peebles adds something which
had not really been recognized by other experimental embryologists at that
time—namely, the idea that an animal's or a cell's characteristics can be
modified by external factors;[22] she suggests that the nature of the response
that a tissue makes to a regeneration stimulus is directly determined by its
prior exposure to a set of distinct conditions during its development. In

1981, the Nobel Prize in Medicine was awarded to Hubel and Wiesel in part for their demonstration that a kitten's eye deprived of light stimulation during a critical period in its development becomes permanently blind. Peebles's early insistence on the role of external stimuli in fixing the nature of the cell is therefore important and worthy of remembrance.

Although Peebles must certainly have come into contact with the most prominent embryologists of her time, both in Europe where she studied and at Woods Hole where she did research, her work seems rarely to have been cited by her peers. This may have been because her early work did not in fact have the necessary philosophical basis which would have allowed it to become part of the emerging theoretical framework, while her later work, which was far from extensive, does not, even now, appear on the surface to be significantly different from that of her more prolific and better-known male contemporaries.

If her work now goes virtually unacknowledged, in her lifetime Peebles attained some measure of professional success. She held professorships at Bryn Mawr, Tulane, and Lewis and Clark universities. In 1954 Goucher awarded her an honorary LL.D. She continued to be listed in *American Men of Science* [23] until 1954 and in *Who Was Who in America* until 1973.[24]

Libbie H. Hyman:
Hierarchies—Physiological and Taxonomic

Before leaving the problems of developmental biology, it is necessary to mention the work of Libbie H. Hyman in this context. Hyman (1888–1969) is famous for her work in another area of biology—that of invertebrate zoology. She is internationally acclaimed for her five-volume treatise, *The Invertebrates*.[25] This work, which is a survey of invertebrate morphology, physiology, embryology, and taxonomy, is the only one of this scope in the English language; it serves the same function for English-speaking biologists as the great German handbooks serve for European scholars. It took Hyman some thirty-three years to write this five-volume survey, and although advancing age finally prevented her from completing the work as originally planned (she published her last volume when she was seventy-eight), it still remains a monumental achievement. It is especially amazing when one considers that there are about a million known invertebrates and, except in the final volume, when she had the help of an illustrator, Hyman executed all the illustrations herself from living or prepared specimens.

It is to be noted that before Hyman went to the American Museum of Natural History in the 1930s as an unsalaried staff member to work on her book,[26] she had spent the first sixteen years of her professional life as a developmental physiologist in the laboratory of C. M. Child at the University of Chicago.

While at Chicago she published a number of papers on vertebrate embryonic development,[27] on regeneration in hydroids[28] and flatworms,[29] as well as several works in taxonomy.[30] Almost all of these papers are under her sole authorship. In themselves, the developmental papers cannot be regarded as much more than attempts at experimental confirmation of Child's gradient theory. But, notice should be taken of them for several reasons. First, the work is of interest in its own right, and secondly, when viewed in the context of the history of embryology and in relation to Hyman's own later success, it says a great deal about the way the scientific community accords ownership to scientific research.

Although Hyman's experiments are now all but forgotten, at the time they represented the best experimental evidence for Child's theory. They were performed in his laboratory and were explicitly aimed at confirming his gradient hypothesis. But all that is remembered about them today is that together with Child's own experiments, hers validated *his* theory. What is not recognized is that the experiments must have been Hyman's to a very large extent (else why the single authorship?) and that, even if they were directed at proving Child's gradient hypothesis they contain elements of experimental design which could only have been fundamentally and essentially hers. Hyman has said that she brought to Child's laboratory a degree of chemical expertise which he himself did not initially have.[31] Therefore, although Child may at first have formulated his ideas of metabolic gradients on the basis of what now seem rather crude observations that various axial regions were differentially susceptible to metabolic poisons, it was Hyman's more sophisticated knowledge of chemistry that allowed experiments to be designed in which precise and accurate measurements of differential oxygen-use could be made.[32] Her experiments are clearly the more convincing.

It is evident that had Hyman not left the field of developmental biology, her scientific contributions would always have remained in possession of C. M. Child. In part this would have been because Hyman herself viewed her research as an extension of his work, and in part because Child's name was so tightly associated with the ideas of the gradient theory that even if the experiments were hers she would have received little credit for them. Moreover, Child's gradient theory was then so compelling and occupied such a central position in embryological thought that it is unlikely that Hyman would have been able to divorce herself from it even if she had continued on in her own laboratory.

But fortunately for Hyman and for biology, Hyman's interests lay in an area of biology which was not dominated by a unifying hypothesis attached to the name of a particular individual. Her treatise in invertebrate zoology has stood alone and has been duly recognized as hers.

During her life, Hyman was accorded many honors. In 1939, after the publication of only the first of her five volumes, she was given an honorary doctorate by the University of Chicago. In 1954 she was elected to the

National Academy of Science, and in 1960 she received the gold medal of the Linnean Society.[33] Her book is still the primary resource for students of invertebrate zoology.

WOMEN IN GENETICS:
SEX, THE DOUBLE X, AND OTHER FACTORS

Beginning with the twenty-five years before the turn of the century, at the same time that biologists were trying to understand how the characteristics of individual cells were determined, they were also trying to come to grips with the question of how individuals of one generation transmitted their characteristics to the individuals of the next generation. Indeed, the question of how a hand became a hand was perceived, if often only vaguely stated, as not very different from the question of how a species became a species—similar cellular mechanisms could well be operating to achieve both feats—and it is not really a coincidence that many of the great early geneticists began their careers as developmental biologists, e.g., T. H. Morgan, E. B. Wilson, Nettie M. Stevens.

The answer to the questions of how cells and species were determined obviously lay in some fundamental property of individual cells, and the problem for both questions became a matter of finding out what substance was doing the determining, where it was located, and how it worked.

That the cellular determinants of specific traits existed in the cell nucleus was first simultaneously perceived during the years 1884–1888 by four people: August Weismann (1834–1914), Oscar Hertwig (1849–1922), Albert von Kölliker (1817–1905), and Eduard Strasburger (1844–1912).[34]

That there were two factors which determined a particular trait in an individual, and that for every trait one factor was transmitted from the mother and one from the father, was first recognized by Gregor Mendel (1822–1884). Mendel also showed that each factor was transmitted independently of the others and that only one such factor (either the mother's or the father's) from an individual could be given to each future offspring.[35] Although Mendel published his findings in 1866 and 1869, it was not until some thirty years later that the scientific community was made aware of his work by the independent publications of three botanists, Carl Correns (1864–1933),[36] Hugo DeVries (1848–1935),[37] and Erich von Tschermak (1871–1962).[38] Each of these men interpreted their own results in relation to Mendel's work.

Thus, by the turn of the century Mendel's work had provided the rules by which traits were inherited, and Weisman, Hertwig, Kölliker, and Strasburger had localized the trait carriers in the cell nucleus. It now remained to be found out on what structures these factors were located. Wilhelm Roux and August Weismann had already suggested in the late 1880s that the hereditary material was probably carried in "the nuclear thread," which at certain periods appears in the form of "loops or rods";[39] that is, the material was carried on the chromosomes. But this immediately led to a problem: There were far more traits than there were chromosomes so that it was not at all certain whether or not the chromosomes were in fact the bearers of the hereditary material. The evidence that they were came in the years 1901 to 1905.[40]

A number of biologists had observed that female flies and bugs had one extra or accessory chromosome in their cells. This led to the strong suspicion, first voiced by McClung (1902), that the accessory chromosome was responsible for determining the sex in insects.[41] If that were so, then evidence was at hand that the chromosomes were indeed the anatomical structures carrying Mendel's hereditary factors.

NETTIE STEVENS: AND THE SEX CHROMOSOME

Definitive proof that the chromosome was the basis for sex determination came in 1905 with independent work of Nettie M. Stevens (1861–1912) at Bryn Mawr and her more famous contemporary E. B. Wilson (1856–1939) at Columbia.

Nettie Stevens was born in Cavendish, Vermont—the daughter of a carpenter. At the age of thirty-five, after having been graduated from the Normal School in Westfield, Vermont, she enrolled as a physiology major at Stanford University. She earned her B.A. in 1899 and her M.A. in 1900. In the fall of 1900, she went to Bryn Mawr as a graduate student of T. H. Morgan. Stevens remained at Bryn Mawr for most of the rest of her life, first as Research Fellow in Biology, then as Reader (1905), and finally as an associate (1905–1912); this last post was a faculty research position created especially for her by Bryn Mawr.[42]

Her predoctoral work was, as expected, concerned with questions of morphology and embryology. (Peebles cites her as the first person to demonstrate that flatworms were capable of regeneration.)[43] But in 1903, Stevens and Morgan began a joint study on sex determination in aphids.[44] At that time, most biologists thought that external influences such as food and temperature invariably determined the sex of offspring. In fact, Morgan and Stevens had specifically set out to study the effects of varying external conditions on the sex of aphid-offspring. According to Morgan, he was to do the experimental work, Stevens the histological, i.e., she was to analyze the sperm and egg cells under the microscope.[45] As it turned out, "the experimental work gave only negative results" while the histological work "yielded the results that Miss Stevens published in 1905."[46]

In her papers, which are two monographs of 75 pages, Stevens showed that in a beetle, *Tenebrio*, the male produced two kinds of sperm, one carrying a large (X) chromosome and the other a smaller (Y) chromosome.[47] The unfertilized egg, however, contained only two X chromosomes. Stevens correctly concluded that eggs fertilized by X-carrying sperms produced females. Further, she recognized that her results satisfied Mendel's rules of inheritance and his two-factor requirement for the determination of a particular trait. In her discussion, Stevens writes that her work provides "a mass of evidence in favor of the belief in both morphological and physiological individuality of the chromosomes. . . ."[48]

Similar results for other insects[49] were also published in 1905 by E. B. Wilson in a two-page paper in *Science*. In his paper, Wilson says that his findings "are in agreement with the observations of Stevens on the beetle *Tenebrio*."[50]

It is interesting to note that apparently Wilson was aware of Stevens's work and explicitly acknowledges it. But it is even more interesting to see what happens to the relative importance of Stevens's work and Wilson's work in the hands of later writers.

During her life, Stevens's findings were given immediate recognition. Wilson[51] and Sturtevant and Morgan[52] cite her discovery as being simultaneous with Wilson's. But later, and this is especially so in books on general genetics and the history of biology, Wilson's work is usually cited first and explained in some detail, while Stevens is simply given credit for also having made the same discovery. Stevens and Wilson are usually cited in the same order as are Wallace and Darwin when both are given "equal" credit for discovering the mechanisms of evolution.

Furthermore, while the discussion in Wilson's paper is frequently quoted at length and often imbued with a theoretical significance[53] that it may or may not deserve, Stevens's discussion is rarely quoted, even though both Wilson and Stevens make almost an identical analysis of their respective data and offer essentially the same sorts of theoretical speculations. It is curious to note in this regard that in an ostensibly laudatory memorial article on Stevens's work Morgan damns her with the following praise: "Her contributions are models of brevity—a brevity amounting at times almost to meagerness. Empirically productive, philosophically she was careful to a degree that makes her work appear at times wanting in that sort of inspiration that utilizes the plain fact of discovery for wider vision."[54] Although it may have been possible to thus diminish Stevens's abilities, it has not been possible to completely obliterate her name from her work.

In her brief nine-year professional career (she received her Ph.D. in 1903 and died in 1912) she published some 40 papers. She was the first person to describe the chromosomes of the European vinegar fly, *Drosophila melanogaster*.[55] (This has been *the* model animal for the study of chromosomal inheritance in complex animals since Morgan and his followers first used it to analyze Mendel's laws.) She was also the first person to establish the fact that chromosomes exist as paired structures in body cells (instead of long loops and threads) and the first to ascertain that certain insects have supernumerary chromosomes.[56]

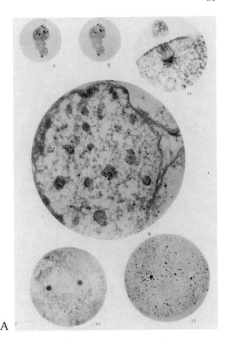
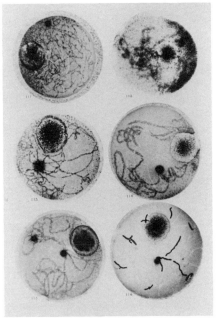

A B

Figure 2. Foot and Strobell's photomicrographs of developing oocytes. (A) From their 1898 paper, "Further notes on the egg of Allolobophora foetida." (B) From their 1905 paper in the *American Journal of Anatomy* (IV), on the same subject. Note the definition of the individual chromatids in 114, 115, and 116. Photographs reproduced from Foot and Strobell's Cytological Studies, 1894–1917. Courtesy of the Library, MBL, Woods Hole.

KATHERINE FOOT AND ELLA STROBELL: "WORTH A THOUSAND WORDS"

Although the evidence produced by Stevens and Wilson was strong that chromosomes existed as individual structures which bore Mendel's hereditary factors, there were those (including Child himself) who resisted this idea and who mounted persuasive arguments against it. Among those who fought most adamantly on the side of the opposition were Katherine Foot (b. 1852) and her associate Ella C. Strobell (d. 1920).[57] They argued that the chromosomes were too variable in size and shape to be considered the specific individual structures that Stevens and others were saying they were. And, furthermore, they could support the contentions with photographs (Figs. 2A and 2B).

Although their theoretical views were wrong, Foot and Strobell's work gave biology the sort of impetus that fundamentally changes the nature of a science by changing its technology. At a time when everyone, almost without exception, was drawing pictures of what they saw through the microscope, Katherine Foot and Ella Strobell began to take photographs of their preparations.[58] These photomicrographs, of which they published hundreds, are remarkable for their clarity and detail. But more than that, they represent the first systematic attempts to remove the subjectivity of the observer from the material observed.

In her preface to *Cytological Studies, 1894–1917*, which is a collection of Foot and Strobell's reprints, Foot says, "If these results are of any scientific value . . . , it is due to the use of photography as a method of demonstration; to our persistent determination to make no statement we could not support by photographs of the material under investigation, and thus present the facts free from distortion which the personal element makes inevitable in a drawing made by even the most conscientious observer."[59]

Foot and Strobell's advocacy of photomicrography borders on the militant. In one of their earlier papers they plead their cause in this way: "A dozen photographs of a variety of features can be taken in the time required to reproduce any one of them by a careful drawing. The printed photographs can be kept in a form serviceable for frequent reference and the impression first made by a preparation not allowed to fade." After comparing two hundred sketches with as many photographs, they conclude: "Of the relative values of these two methods there can be no question, in every case the photographs proving to be the more valuable aid in recalling the preparations."[60]

Such proselytizing suggests that their work must at first have been met with considerable reservation, if not downright skepticism. (Some of this may have been because their technology was far less straightforward than they imply.) Nevertheless, the new technique prevailed. Today no one draws diagrams instead of taking photographs.

Besides working out the many technical problems of photomicrography, Foot and Strobell also devised a method for making extremely thin sections of material at very low temperatures.[61] Although modern methods of preparing thin sections for the microscope supercede their techniques, they were among the first to develop such procedures.

As far as I can tell, neither Foot nor Strobell seem to have been affiliated with any academic institution. Foot's name first appears on the rolls of the Marine Biological Laboratory in 1892, when at the age of forty she is listed as "receiving instruction." Her address at that time was listed as Denver, Colorado, and the next year, Evanston, Illinois. She returned to the MBL in following years, until 1897, as an investigator. In *American Men of Science*, her education is given as "private schools."[62] In 1896, Ella Church Strobell joined her and they continued to collaborate until Strobell's death (sometime between 1918 and 1920).

If it is true that Foot and Strobell were not officially affiliated with an academic or research institution, then it is possible that they funded at least

A class at the Marine Biological Laboratory in the late 1890s. Katherine Foot is seated left of C. O. Whitman, director of the Laboratory (center, third row), and next to Cornelia Clapp, long-time member of the MBL Board of Trustees and professor at Mount Holyoke. About one-third of the class are women. Courtesy of the Library, MBL, Woods Hole.

part of their research themselves. This is given some credibility by Foot's 1920 paper, which is presented as a "contribution from the Foot and Strobell laboratory because of a legacy left her by her friend Miss E. C. Strobell."[63]

In their lifetime, their work was judged to be significant by their peers. (Foot's name is starred in the 1944 edition of *American Men of Science*.) And although Morgan argued against their interpretations, he nonetheless reproduced one of their photomicrographs in his 1914 book, *Heredity and Sex*.[64] It is the only photomicrograph in the book.

Today, no one has ever heard of Foot and Strobell. When I showed their photographs to some of my colleagues in genetics and cytology, they were amazed, not only because the photographs were so good and bore such an early date, but because their existence had been completely unknown.

By 1915, with the publication of Morgan, Sturtevant, and Bridges's historic book, *Mechanisms of Heredity*,[65] not only was the chromosome theory of inheritance completely accepted, but Morgan and his associates had described the characteristics of the theoretical gene so vividly that it had become a familiar object to biologists. Genes were discrete physical entities—beads lined up on chromosomes; their positions could be mapped and manipulated; they could be duplicated, deleted, and exchanged for homologous genes on other chromosomes. In short, biologists had imparted a physical reality to Mendel's inheritance factors in the same way that physicists do to their inferred particles.

BARBARA McCLINTOCK: JUMPING GENES

Among those scientists who wrote the fundamental laws of gene and chromosome behavior is the now-famous scientist Barbara McClintock (b. 1902).[66] Her work, which was begun fifty years ago, is responsible not only for much that is considered classical genetic theory, but also for what has just now come to be recognized as a fundamental and revolutionary concept of gene functioning. In recognition of this, McClintock was awarded the 1983 Nobel Prize in Medicine and Physiology. The work for which she was cited was published over thirty-five years ago, but because it demanded that biologists unlearn the picture of the gene they had just so painstakingly finished painting, this aspect of her research was simply not understood, and, like Mendel's work, was ignored until the rest of the world could catch up.

In 1951, McClintock gave a paper at the Cold Spring Harbor Symposium for Quantitative Biology entitled "Chromosome Organization and Genic Expression."[67] In this paper she explained that some genes appeared to be shifting their positions on the chromosomes in a random way from one generation to the next. She referred to these factors as transposable elements. These elements seemed to control other genes; they had the ability to turn adjacent genes on or off and so could permanently modify the way a particular gene expressed itself in the developing organism. But their most interesting and unorthodox behavior lay in the fact that they refused to stay put and seemed to move about in response to external stimuli received from the developing organism. In this, McClintock felt, lay a possible mechanism by which cells could differentially differentiate even though they might all have the same complement of genetic material. "The numerous different phenotypic expressions attributable to a change at one locus [on the chromosome] need not be related in each case to changes in the genic components at the locus, but rather to changes in the mechanisms of association and interaction of a number of individual chromosome components with which the factor or factors at the locus are associated."[68]

McClintock arrived at this formulation by observing the changes in color patterns in kernels of Indian corn and correlating these changes with changes in chromosome structure. In the 1960s, after biologists had learned that the genetic material was DNA (deoxyribonucleic acid) and had figured out some of the rules by which it determined inheritance, McClintock's proposed transposable elements were verified again and again in simpler bacterial systems with more direct biochemical techniques.

It might be recalled that the concept of a fixed complement of genes whose expression could be altered by the external influences to which a developing organism was exposed was similar to the sort of thing Peebles (1931) seemed to have had in mind when she sought to explain differences in regeneration ability of various parts of the hydroid *Tubularia*.

McClintock's paper was published at a time when geneticists were not yet prepared to accept the concept of shifting genetic associations as a mechanism for determining how genes could produce differences in particular traits. How is it, then, that McClintock's revolutionary work was remembered some twenty or thirty years after it had been published, while the important work of other women often seems not to have been? Why is it that McClintock's work escaped becoming anonymous?

It is difficult to answer this question. McClintock had already achieved a large measure of distinction for her classical work on chromosome behavior—she was elected to the National Academy of Science in 1944. Her paper "A Correlation of Cytological and Genetical Crossing-Over in *Zea Mays*"[69] became an immediate classic. This paper provided the first methodology by which changes in the appearances of the offspring could be correlated to an interchromosomal exchange of material. (Curt Stern, at about the same time, independently published a similar demonstration with *Drosophila* chromosomes.)[70]

In subsequent work, McClintock continued to address the problem of correlating observable characteristics with changes in chromosome behavior. Her research was widely presented and published in numerous journals.

Because her work was almost if not quite expected (in the sense that it could fit with the emerging picture of how genes and chromosomes worked) and unique (in the sense that it was clever and original), it was regarded with interest, if not complete acceptance. Because she tended to work independently, on an experimental object that was somewhat out of the mainstream (she used corn, when many of the well-known laboratories seemed to have preferred *Drosophila*), her reputation became hers alone. When it was time, therefore, to accept the idea of transposable controlling elements, it was also possible to acknowledge McClintock's ownership of it.

Whether or not the recognition of McClintock's "new" science is also to be attributed to the feminist movement which began in the mid 1960s is not really possible to say. It is true that many of McClintock's most prestigious awards came after 1965: the Kimber Genetics Award, 1967; the

National Medal of Science, 1970; the Rosenstiel Award, 1978; and the Lasker Award, 1981. But it is also true that the discovery of movable elements in bacterial systems occurred at just this time, and that she had been elected to the National Academy in 1944. Perhaps because the social revolution coincided with the scientific one, the members of the scientific hierarchy were now genuinely pleased to applaud the work of a member of the newly enfranchised social class.

RUTH SAGER: "CYTOPLASMIC INHERITANCE"

In the meantime, while almost everyone was focusing their attention on the mechanisms of Mendelian inheritance, another revolution was quietly taking place. There were some biologists—a very few—who began to suggest that perhaps not all of the cell's inheritance factors were located on chromosomes inside the nucleus. Indeed, they claimed there was strong evidence that some inheritance factors resided in the cytoplasm and that these did not obey the Mendelian rules of inheritance.

Ironically, the existence of these non-Mendelian genes was first proposed by Carl Correns (1909), one of the three geneticists who had rediscovered Mendel's laws. Correns demonstrated that mutation in the color pattern of leaves in a certain plant was always inherited according to the pattern of the mother's leaves, never according to that of the father's. This maternal effect persisted in subsequent matings of the offspring, so that the trait could not be made to vary according to the ratios predicted by Mendel's two-factor analysis.[71]

For the first fifty years of its discovery, non-Mendelian inheritance had a dubious respectability. Reputable geneticists accorded it but scornful recognition, and it was treated very much like the heresy of the French biologist Lamarck (1744–1774), who asserted that acquired characteristics could be inherited. Nevertheless, despite the bias against it, reports of cytoplasmic inheritance kept coming in; finally, when the evidence could no longer be discounted, the idea of cytoplasmic, non-Mendelian inheritance was embraced by geneticists as though it had been a long-lost friend.

The person who is almost entirely responsible for this acceptance and who presented the first best experimental evidence for cytoplasmic heredity is the well-known geneticist Ruth Sager (b. 1918).[72] In the 1950s, Sager, using the single-celled algae *Chlamydomonas*, demonstrated not only that cytoplasmic genes existed, but that they alone could be made to mutate when treated with streptomycin. Moreover, they could be mapped on a "cytoplasmic chromosome" and the maternal effect they produced could be altered by irradiation with ultraviolet light. Sager's evidence, based on classical genetic techniques, was convincing and was supported by results from other organisms such as yeast (Ephrussi, 1950–1955)[73] and the mold *Neurospora* (Mitchell, 1952).[74] But the final acceptance of cytoplasmic inheritance came only with the repeated demonstration in the last twenty years that DNA also existed in cytoplasmic structures. The first demonstration that nonnuclear DNA existed in chloroplasts was made by Sager herself in 1963,[75] and independently by Chun, Vaughn, and Rich,[76] and by Nass and Nass in mitochondria.[77]

The realization that genes existed outside of the nucleus placed an entirely new perspective both on the mechanisms of heredity and on the whole question of the nature of evolution. Was it possible that in some sense Lamarck was right after all? Sager believes that cytoplasmic genes exist because they provide the organism with a degree of flexibility: chloroplasts and mitochondria replicate *in response to environmental stress.*[78]

Evolutionists like Lynn Margulis (b. 1938) use the finding that cytoplasmic organelles have their own hereditary machinery to argue that these organelles had an independent existence billions of years ago. At some point in the course of evolution, they say, these organelles were joined with other independent cells to form a symbiotic relationship that finally developed into a permanent interdependence. These joined cells then became the ancestors of what biologists up to now have called "single-celled" organisms and these, in turn, became the ancestors of more complex, multicelled organisms.[79]

That Sager's work is directly responsible for changing the way biologists think about cell heredity[80] and their perception of the processes of evolution is widely recognized today. Sager's work is well-acknowledged by her peers. In 1977 she was elected to the National Academy of Science. At present she is Chief of the Genetics Division of the Sidney Farber Cancer Research Institute and Professor of Cellular Genetics at Harvard, a post she has held only since 1975. Before that time, Sager was Professor of Biology at Hunter College in New York (1966–1975).

Interestingly, in the eighteen years before she became a full professor at Hunter, Sager's only other positions were those of research fellows. After receiving her Ph.D. in 1948 in genetics from Columbia University, Sager became a Merck research fellow (1949–1951); she then served as an assistant biochemist at the Rockefeller Institute from 1951 to 1955. In 1955 she returned to Columbia as Research Associate and was promoted to Senior Research Associate in 1961. She held this post until her appointment to Hunter.[81] During this time, Sager had published more than fifty articles— many of which appeared in the most prestigious scientific journals—and had co-authored her first book, *Cell Heredity.* (Her second book, *Cytoplasmic Genes and Organelles,*[82] was published while she was still at Hunter.)

Again, the question must be asked whether the long delay in recognizing Sager's achievements was due to the initial unpopularity of cytoplasmic inheritance or to a similar unpopularity of women in science. As we have already noted for Barbara McClintock, many of Sager's accolades came coincidentally with the rise of the new feminism. Her election to the National Academy in 1977, her appointment first to Hunter and then to

Harvard, and, not least, her invitation to deliver one of the prestigious Friday Evening Lectures at the MBL (summer 1981) all occurred after the beginning of the current feminist movement. One cannot help but think, therefore, that the changing social and political circumstances were instrumental in teaching the scientific community how better to assess the work of its women members.

That scientific society has not always been able to properly regard the work of its women is nowhere given more poignant testimony than in the now-infamous story of the discovery of the molecular mechanism of heredity. The discovery of how DNA works to transmit heredity information from cell to cell and from generation to generation is perhaps the most significant scientific finding of the twentieth century. For this discovery, Francis Crick, James Watson, and Maurice Wilkins received the Nobel Prize in 1962.

ROSALIND FRANKLIN: "ONE PICTURE . . . "

The story of Rosalind Franklin (1920–1958), Maurice Wilkins, DNA, and Watson and Crick has been recounted many times. The best and, for the most part, most penetrating account is Horace Freeland Judson's *The Eighth Day of Creation*.[83] The admittedly subjective summary that follows is drawn primarily from that book, from Anne Sayre's *Rosalind Franklin and DNA*,[84] and from James Watson's *The Double Helix*.[85]

By the late 1940s the fundamentals of Mendelian inheritance had become commonplace. Much of genetics had assumed the characteristics of a chess game: the questions that were raised could penetrate no deeper into the nature of heredity than can an analysis of chess moves reveal the nature of the brain that produces them. What was known was that chromosomes carried hereditary information; that within the nucleus of a cell there were two substances, proteins and nucleic acids, and that somehow either or both could be responsible for telling cells and organisms what they should look like. The question that had not been asked and the question that most scientists did not know how to ask was, what was it about the chemistry or physics of either proteins or nucleic acids that gave these their special ability to transmit information about an organism's traits? The first step was to determine which substance was the important one.

In 1944, Oswald Avery and his colleagues published the ground-breaking experiment of modern molecular genetics[86] which showed that nucleic acids and not proteins were responsible for determining the traits in pneumococcal bacteria. The question of how chromosomes worked could now be rewritten to read, how do nucleic acids work?

In 1950, Erwin Chargaff demonstrated that different DNA's from different species were biochemically dissimilar.[87] But in all the DNA's from all the animals tested, there was a peculiar consistency in that the ratio of purine to pyrimidine bases was always close to one. (DNA is composed of the sugar ribose, with attached phosphates, and four bases [alkaline substances]: two purines and two pyrimidines.) This important finding was later used by Watson and Crick in constructing their final model of DNA.

At about this time, as part of a group of biophysicists engaged in applying physics to the problems of biology, Maurice Wilkins, at Kings College, London, also began to work on the structure of DNA. When Wilkins realized that he did not have the expertise to continue the X-ray crystallography he had started, the head of the group, Sir John Randall, hired Rosalind Franklin, to work out its structure.[88] Wilkins was thus apparently under the impression that Franklin had been hired to help him with his work, a belief that later came to be shared by Watson[89] and probably also by Crick.

But Franklin was under the impression that the DNA project had been assigned to her, and that she had been hired not as Wilkins's assistant, nor even as his collaborator, but rather simply as his colleague. The letter in which Randall asks Franklin to work on DNA is sufficiently ambiguous to allow either interpretation.[90] That two interpretations were made is certain and points to a problem that plagues many a junior scientist: Does the work belong to the scientist doing it, or to the laboratory that assigned it?[91]

In the meantime, Crick and Watson in Cambridge had now also started to work on DNA and had begun to stake a claim on it.

The question of whose DNA was it, anyway, is central to understanding the sequence of events that finally led to the solution of DNA's structure: Not only did Wilkins believe that the DNA problem was his and Franklin believe it to be hers, but, in a deep and fundamental sense, it appears that Watson and Crick felt that the problem was in reality theirs; not because they had been working on it for so long (for they had not), but because, as they seem to have thought, they alone understood its significance. They had, as it were, a moral right to it, or to use Watson's terminology, they had come to "deserve it."[92] In recalling the race to solve DNA, Watson says, "Then DNA was still a mystery, up for grabs, and no one was sure who would get it and whether he would deserve it if it proved as exciting as we semi-secretly believed."[93] Now, one might well ask, how was this moral right conferred on Crick and Watson?

Watson, after finishing his Ph.D., had set out to solve the gene. In the spring of 1951, he heard a talk by Wilkins which made him see that in studying DNA with X-ray crystallography it was possible to get at its structure and that "when the structure of DNA was known, we might be in a better position to understand how genes worked."[94]

By the time Watson met Crick in Cambridge in October 1951, Crick had already come to the realization that the structure of a cellular protein was specifically defined by the structure of the genetic material. Thus both Watson and Crick had arrived at the understanding that in the structure, not the biochemistry of DNA, lay its method of functioning.[95] So, because, at the time, they alone seemed really to know this, it was, I think, an easy matter for them to consider DNA theirs.

In the meantime, Franklin was continuing to work on DNA in London. She had discovered that there were two forms, A and B, and had begun to detail the properties of the A form. Among its most important characteristics was the fact that the bases faced inward toward each other from their phosphate backbones. She presented her findings before a colloquium which was attended by Watson and also wrote a progress report which included a series of exact and significant measurements that, although it had not been formally published, was openly distributed outside Kings College. Watson and Crick obtained a copy in early 1952.

At this time, Franklin, deeply involved in trying to figure out DNA's structure, had apparently decided to do this step by step. While working on the A form, she had put in her desk drawer an especially clear X-ray diffraction photograph of the B form—the one which indicated unequivocally that the structure was a helix which made a full turn every 34 angstroms.[96]

Although Watson and Crick had now been working on the DNA puzzle for a number of months, and may even have decided to construct a model of it, they were not quite in a position to do so. Models, like castles, cannot be built of thin air; concrete—in this case conceptual and numerical—building blocks are required. And these belonged to Franklin. Some of them were hers exclusively—or so they should have been. But then there was Maurice Wilkins, who somehow must still have felt that DNA and Franklin's expertise along with it were in reality his property. For how else is one to explain the curious events that took place at Kings College in 1952–1953?

About a year after Franklin had given her colloquium, and about six months after Watson had taken his only X-ray crystallograph of RNA, Watson came to London to visit Wilkins (January 1953), who, "on this occasion, suddenly became very confiding." He told Watson that he and his assistant "had been duplicating some of Rosalind's X-ray work on the quiet," and that "Franklin had for some time had evidence of a new three dimensional form of DNA. . . . When Watson asked what the X-ray diffraction pattern of this new form was like, Wilkins showed him the picture."[97] This was the beautiful picture of the B form in Franklin's desk drawer, on which the helix and its precise conformational relationships could be seen.

The rest of the story is anticlimactical: Watson and Crick built their DNA model; they paired the bases to fit Chargaff's ratios and spaced them according to Franklin's measurements and Crick's theoretical calculations. Franklin's externally placed phosphate backbones now twined about each other in her helix, the double strands making a full revolution every 34 angstroms. Watson and Crick were able to suggest how, when the two strands separated, each could be capable of producing its complementary thread. And so they showed the world how genes could reproduce themselves and how they could act as templates for generations of cellular proteins. In 1962 Crick and Watson shared the Nobel Prize with Wilkins. Franklin had been dead for four years.

Had Franklin not had her work secretly taken from her and had she thus been allowed enough time to use her data to solve her puzzle, there is hardly any doubt that she would have unraveled the helix—perhaps even before Crick and Watson. For, after all, Watson and Crick would then have had to have made their own unequivocal photographs of the DNA helix. This they had not succeeded in doing.

The argument has been made that neither Franklin nor Wilkins really perceived the significance of the specific relationship between DNA's structure and its genetic functioning[98] and thus did not deserve to solve the puzzle. Certainly it is true that they had not yet fully understood that relationship at the time Watson and Crick presented the solved problem. But it is also true that as soon as Franklin (or for that matter anyone) was shown the DNA structure, its method of functioning immediately became obvious. What the race would finally have boiled down to (had Crick and Watson been obliged to gather their own data) is this: Who would have been faster? Franklin, in understanding the biological significance of her solved double helix, or Crick and Watson, in making the X-ray crystallographs from which to obtain the relationships needed to construct their model?

Whatever the answer to this, the fact is that Franklin's work *was* taken from her and she did *not* solve DNA. But fortunately, the measure of a scientist does not usually rest on a single piece of work (Watson may be the exception here) and Franklin's other research, before and after this episode, is today considered to have been outstanding; much of it continues to be fundamentally and essentially significant. (See Julian's chapter for more about this.)

CONTRIBUTIONS OF WOMEN TO PHYSIOLOGY: THE NERVE OF SOME WOMEN

If it was not until the latter half of the nineteenth century that biologists learned to ask how cells and species were determined, the question of how animals worked and organs functioned had come into existence long before then. Speculations as to what nerves and veins, hearts and livers were doing had been the subject of natural philosophy since well before Aristotle. Unlike experimental embryology and genetics, which really only came into being in the second half of the nineteenth century, physiology had existed as an integral part of medicine and natural science for several centuries.

By the middle of the eighteenth century, the repertoire of physiological analysis included observation, dissection, vivisection, and experimentation. The nineteenth century added the methods and language of chemistry and physics to the growing perception of the mechanistic nature of living organisms.

If the eighteenth century had taught scientists that even animals could produce electricity (Galvani, 1737–1798; Volta, 1745–1827), and that muscles shortened when they contracted not because they became swollen with fluid but because their fibers became shorter (Swammerdam, 1637–1680), then it was the nineteenth century that taught them that nerves and muscles functioned because of the electrical currents they themselves produced and that these currents were caused by the selective movement of ions, unequally distributed across their membranes (Hermann, 1839–1914; Dubois-Reymond, 1818–1896; Bernstein, 1839–1917).

Superimposed on the physical-chemical characterization of organ functioning was the all-pervasive realization, brought on by the nineteenth century *Naturphilosophen* and given historical validation by Darwin's theory, that among species each organ system was to be understood as a series of variations of an underlying biological theme. Thus the nineteenth century taught students of neurophysiology not only to look to structural and functional comparisons in related species for clues about how "the" nervous system worked, but to use such comparisons to construct a scenario for the evolution of the nervous system.[99]

Until the latter half of the nineteenth century, physiology was virtually the sole province of male biologists. In part this was because physiology, especially human physiology, was taught in medical schools from which, at that time, women were barred (at least in America and Northern Europe).[100] But, as the nineteenth century progressed and more and more women undertook formal studies in mathematics and the natural sciences, physiology took its place beside embryology and natural history as an appropriate domain for women scientists.

IDA HYDE: AND THE MICROELECTRODE

The first woman to be elected to the American Physiological Society was Ida H. Hyde (1854–1945). She was also the first woman to be awarded a doctorate in physiology from a German university (the first woman to be graduated from the University of Heidelberg) and the first woman to do research at the Harvard Medical School. Her major contribution to physiology—the development of the microelectrode—has never been acknowledged as hers. (As recently as 1983,[101] accounts of the invention of the microelectrode do not include Hyde's name.) In addition to her scientific legacy, Hyde has left us with a first-hand account of the obstacles placed before women who wished to enter the scientific community.

Hyde was born in Ohio in 1857 of German parents. She first attended

Ida Hyde. Reprinted from *The Physiologist* Vol. 24 (1981) with permission.

college at the age of twenty-four and, after some years of teaching, entered Cornell University. In 1891, at thirty-four she obtained a B.S. degree. After Cornell, Hyde went to Bryn Mawr as a student of the renowned embryologist Jacques Loeb, who had begun his career as a nerve physiologist and anatomist. In 1893 she left Bryn Mawr to study physiology in Germany.[102] The account that follows is taken from her article "Before Women Were Human Beings . . . Adventures of an American Fellow in German Universities in the '90's." It was published in 1938 in the *Journal of the American Association of University Women.*[103]

While Hyde was at Bryn Mawr some of her experiments caught the attention of Professor Goette at the University of Strasbourg, who had worked on the same problem. She was invited to work with him in Strasbourg and was awarded a fellowship in 1893 by the Association of Collegiate Alumnae for Study in Foreign Universities. Unfortunately, at that time few women had matriculated at a German university, and indeed, as a rule, women were not permitted to do so. Nevertheless, Hyde persisted. She petitioned the Ministry of Education for admission, but before they could act, she withdrew her petition because she had learned that the chairman of the Examining Committee in Natural Science at Strasbourg would refuse to allow a woman to take the examination. Hyde then tried to gain admission to the University of Heidelberg, and in 1894 succeeded. In doing so she opened the University door to all women, despite the University's declaration that Hyde's acceptance was not to be construed as a precedent to permit other women to demand admittance to the University.

When Hyde arrived at Heidelberg, a new obstacle presented itself in the form of Professor Willy Kühne. Kühne, one of the most influential and renowned physiologists of the day, refused to allow Hyde to attend his lectures. But at the same time, with what appears to us an almost inexplicable perversity, Kühne invited Hyde to attend his special Saturday morning laboratory sessions and arranged to have his two chief assistants take lecture notes for her.

Hyde's final examinations took place on the sixth of February, 1896, and although she passed with highest honors, Kühne objected to granting this distinction to a woman. As a compromise, she was awarded instead her Ph.D. *Multa cum Lauda Superavit*, "with greater than high honors," a new title created for the purpose. With the same seeming perversity—but which was probably in reality an expression of Kühne's sense of scientific integrity and perhaps an underlying sense of fair play, which, from time to time, appears to have won out over an equally strong nineteenth-century sense of correct social order and propriety—Kühne now granted Hyde another privilege and was responsible for her receiving still another honor. Upon obtaining her degree in March, Hyde was officially informed by the Physiology Department (i.e., by its director, Professor Kühne), that the use of the Physiology Department was cordially placed at her disposal and that Professor Kühne and his associates would be pleased to welcome her there. Later, Professor Kühne was instrumental in having the University Research Table at the Naples Marine Station, given each year to an honor student, awarded her. Upon Hyde's return from Naples, Kühne obtained an invitation for her to spend a year at the University of Bern.

In her research, for the most part, Ida Hyde was a child of her time. Her early studies on the anatomy and physiology of nervous systems were often addressed to the problems of determining phylogenetic relationships. Her approach, like that of her contemporaries, was to analyze an organ system with respect to its evolutionary development. In her first publication (1892), under the direction of Jacques Loeb, she determined that breathing in the horseshoe crab was controlled by the abdominal nerve cord and, unlike the situation among vertebrates, did not require the involvement of a brain.[104] Later she found that grasshoppers could be cut into several independently breathing sections, and in 1904 she showed that while the respiratory center of skates (a lower fish related to sharks) was located in the brain, it had, during the course of its embryological development, arisen from a segmented stucture. She concluded, therefore, that the skate represented an intermediate form between the segmented centers of higher invertebrates (horseshoe crabs and grasshoppers) and the nonsegmented centers of higher vertebrates (higher fishes, amphibians, and mammals).[105]

Although Hyde made a number of valuable contributions to neurobiology,[106] perhaps her most interesting work is that involving her contribution to the development of the microelectrode. Hyde invented the first microelectrode for intracellular work. The microelectrode, which is today routinely used to deliver discrete electrical or chemical stimuli to a cell and to record the electrical activity from within individual nerve and muscle cells, has been the single most useful and powerful tool in electrophysiology. Its invention revolutionized neurophysiology; without it very little would have been learned about how nerves and muscles function. The electrode consists of an ultrafine salt-filled capillary tube connected by a metallic salt-metal or other nonpolarizable interface to a metal wire leading to an electrical amplifier or stimulator.

In 1918 and 1919, while Hyde was working on single-celled organisms and the eggs of sea urchins, she found it necessary to have an apparatus which could not only inject or withdraw small amounts of fluid from these cells, but which could also be used to stimulate them electrically. Although an electrode made from two concentrically mounted capillary tubes had been used to stimulate individual muscle fibers by Pratt in 1917, the diameter of the concentric tubes was large and could not be inserted into a single cell.[107] Earlier, in 1912, Barber had devised an ingenious method by which to dissect individual cells,[108] modified versions of which are still in routine use. His apparatus consisted of a drawn-out and bent capillary tube held in a pipette holder which was mounted on a movable microscope stage adapted for the holder. With this apparatus it was possible to maneuver the bent capillaries into and out of single cells while at the same time viewing them under a microscope. But neither Barber's nor Pratt's instruments could be used to simultaneously stimulate the inside of a cell while injecting or withdrawing substances. Using Barber's apparatus, Hyde constructed a salt-solution-filled electrode of very small diameter (inner diameter 3 μm) which was connected to a small column of mercury whose level could be altered by the introduction of varying amounts of positive or negative electrical current (see fig. 3). The mercury in turn could force the salt solution toward or away from the tip of the capillary (thus providing a means of injecting or removing substances from the cell) and could also transmit electrical stimuli to the cell through the salt solution.[109]

With this method, Hyde gave the first evidence that the recently discovered principle of all-or-nothing contraction[110] was not universally true for all contractile cells. In her 1921 paper she reported that *Vorticella*, a single-celled organism, did not contract in an all-or-nothing fashion. (The all-or-nothing principle states that if an electrical stimulus is given to a

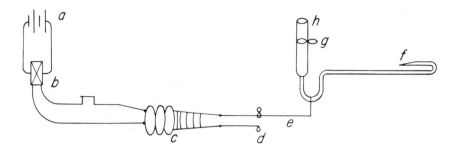

a, battery; *b*, commutator; *c*, induction coil; *d*, clamp; *e*, platinum wire; *f*, tip of pipette; *g*, clamp; *h*, rubber tubing

Figure 3. Ida Hyde's Microelectrode (1921). The outer diameter of the tip was 8μm. The electrode is bent because the cells were suspended in a hanging drop. Redrawn from Hyde's paper in *Biological Bulletin* 40 (1921) page 132.

muscle, it will cause the muscle to contract completely, if the stimulus is strong enough to make it contract at all. Today, this principle has been found to hold true for most but *not all* contractile cells and also to hold true for the size and duration of some but *not all* electrical impulses conducted in nerve and muscle cells.)

In her lifetime, Ida Hyde gained a marked degree of recognition. In 1898, after spending two years working part-time at the Harvard Medical School with William Townsend Porter, and at the same time teaching zoology in three preparatory schools, Hyde joined the faculty of the University of Kansas as an assistant professor of zoology. The following year she was made associate professor of physiology and four years later, when the college created a separate department of physiology, Hyde was appointed full professor and made head of the department. In 1902 she was elected to the American Physiological Society as its first woman member and was the only one until 1914.[111]

Throughout her life Hyde remained committed to securing equal scientific opportunities for women. In 1896, upon her return from Europe she led a group of academic women in creating an organization to subsidize a Woman's Table at the Naples Zoological Station. Among the members of this organization was Ellen Swallow Richards,[112] and among its first users was Florence Peebles. Hyde also endowed scholarships for women students of science at the University of Kansas and at Cornell, and before her death in 1945 she gave $25,000 to establish the Ida H. Hyde Woman's International Fellowship of the American Association of University Women.

In 1926, Hyde endowed a scholarship for the women of the University of Kansas to work at the Marine Biological Laboratory at Woods Hole. The director, F. R. Lillie, acknowledged her gift as follows: "The older members of the Marine Biological Laboratory do not need to be reminded that Miss Hyde was a regular investigator at the Marine Biological Laboratory in the nineties of the last century and after that frequently up to the time of the War. Miss Hyde's investigations were then well known."[113] Although Lillie then briefly notes her devotion to the progress of science, and her loyalty to the University of Kansas and the MBL, he makes no further mention of the extent or importance of her work.

During the first decades of the twentieth century, as physiologists tried to understand how nerves and muscles functioned, and as they became increasingly concerned with defining the chemical and physical properties of these cells, it became obvious that a method had to be found which would not only allow individual cells to be stimulated, but which would permit their individual electrical response to be monitored. Although it was already possible to record locally from the outside of groups of cells in a nerve or muscle bundle, information from such recording could at best be inferential. To analyze exactly what was happening when a nerve or muscle conducted an electrical impulse, it was necessary to observe the electrical changes directly in the cell itself.

Thus, in the ten odd years immediately before and after the Second World War (from the 1930s through the 1950s) physiologists came to the realization that if they wanted to study individual nerve or muscle fibers, they would either have to build a smaller electrode (with its concomitant electrical recording devices) or find a larger nerve cell. As it turned out, they did both.

By 1921, Hyde had developed a solution-filled stimulating electrode that was small enough to be inserted into a cell. In 1925, apparently unaware of Hyde's electrode, Ettisch and Peterfi manufactured a recording electrode filled with a salt-agar gel.[114] This electrode was essentially the same as Hyde's except that it was wider at the tip (10 μm) and was filled with a gel rather than a solution. With this electrode, Ettisch and Peterfi recorded the resting potential—i.e., the voltage difference between the inside and the outside—of the single-celled organism, the *Amoeba*.[115] In 1927 Taylor and Whitaker reported measuring the pH of *Nitella* cells with a glass-sheathed platinum microelectrode (inner diameter, 1 μm) inserted into the cells and a potassium-chloride-saturated agar-filled micropipette as the external reference electrode.[116] Eight years later, in 1933, Kamada modified the electrode again by replacing the gel with a salt solution and thus essentially reconstructed Hyde's electrode. Kamada used this electrode to record the resting potential of another single-celled organism, the *Paramecium*.[117]

For some reason—perhaps because they were done on lower organisms, perhaps because the experimenters were not well known—these experiments did not become part of the repository of available neurophysiological techniques, so that in the years during the Second World War, the microelectrode had once more to be reinvented.

JUDITH GRAHAM POOL: THE MICROELECTRODE AGAIN

The first report that a microelectrode had been used to record resting potentials from the membranes of a frog muscle was published in 1942 by Judith Graham (1919–1975), G. R. Carlson, and R. W. Gerard.[118] The electrodes they constructed were also bent capillary tubes drawn out to a tip of about 2–3 μm in diameter. The capillaries were filled with an isotonic salt solution (potassium chloride) which made contact with a mercury chloride or a silver–silver chloride interface connected by a metal wire to the electronic instruments. A switch permitted the electrode to be connected either to a recording or to a stimulating apparatus. With these electrodes, Graham and Gerard were able to record resting potentials and action potentials (impulses conducted along nerve and muscle fibers) which they evoked with their electrode.[119] Later, in 1947, Gilbert Ling and Gerard produced an even smaller electrode (less than 0.25 μm in diameter), which they filled by immersing the electrode in boiling salt solution, thereby evacuating the air as the capillary tube filled.[120]

In the 1950s Ralph Gerard was nominated for a Nobel Prize for developing the microelectrode. According to Gerard he had been looking for a simplified preparation with which to study spontaneous rhythms and activities. He chose the isolated frog's brain, but in the course of his studies it became apparent that he needed a really small microelectrode. Gerard says: "In the brain studies, I kept searching for a true microelectrode, again finding the way from my earlier adventure with striated muscle as a histology student. With Judith Graham, I developed a salt-filled capillary with a tip small enough (up to five microns) that a muscle fiber could be impaled without excessive damage. Gilbert Ling soon picked up these studies, and the electrode was pushed down to a few tenths of a micron. . . . "[121]

Although she appears to have played a major role in the rediscovery of the microelectrode, Judith Graham (Pool) is rarely acknowledged for this. Until recently, the microelectrode was frequently referred to in the literature as the Ling-Gerard electrode[122] and for many years it was only in the World's Who's Who in Science that Graham was cited as having "invented the microelectrode" and here the source is, as is typical of all such listings, herself.[123]

Judith Graham Pool (1919–1975) began her career as a student in physiology at the University of Chicago. She received her B.S. degree in 1939 and her Ph.D. in 1946 working in the laboratory of R. W. Gerard. It was during this time that the microelectrode work was done. In 1939 she published an abstract with F. J. M. Sichel on the physiology of isolated frog muscle fibers. In this paper they demonstrated that isolated muscle fibers contracted in response to the injection or local application of calcium chloride through a micropipette.[124] Earlier, Elsa Keil (Sichel) and F. J. M. Sichel had used a micropipette to similarly apply acetylcholine onto a muscle fiber.[125] In 1942 Graham, Carlson, and Gerard published their first microelectrode paper. According to Bretag, Gerard had learned how to manufacture Peterfi's glass electrodes from Umrath in Czechoslovakia and had brought the technique back to Chicago.[126]

After leaving the University of Chicago, Graham held various positions as researcher and instructor; at one point she taught physics at Hobart and William Smith Colleges (1943–1945). In 1949 her husband accepted a post at Stanford University and the family moved to California. In 1950 she became a research associate at Stanford, a position she held until 1970. She was promoted to senior research associate in 1960 and after twenty years of outstanding independent research, was made a senior scientist in 1970 and a full professor in 1972—three years before her death.[127]

At Stanford her work was perforce redirected from muscle physiology to blood physiology. As before, it resulted in significant contributions to the field. She published a number of germinal papers on hemophilia and blood coagulation and was responsible for developing the method of isolating the antihemophilic factor in blood. Today this method is used for transfusions to correct bleeding in hemophiliac patients. Her research helped to revolutionize the treatment of hemophilia, and her procedure, published in 1964, of cold precipitation of the antihemophilic factor[128] has since become standard.

As with McClintock and Sager, the 1970s began to bring Pool adequate recognition of her later achievements. In 1974 she was invited to give the Paul M. Aggeler Memorial Lecture. She received the Murray Thelin Award from the National Hemophilia Foundation in 1968. In 1975 she received the Professional Achievement Award from the University of Chicago; after her death, the National Hemophilia Foundation named its research fellowship the Judith Graham Pool Research Fellowship.[129]

But to return to the subject of the microelectrode. If, as it appears, Judith Graham Pool was so instrumental in redeveloping it, the question must be asked, why was this fact not more widely acknowledged? Indeed, the fundamental question that should really be asked is why was it necessary in the first place to reinvent Ida Hyde's ingenious microelectrode so many times?

Certainly, one must acknowledge that neurophysiologists did not actually seem to want a microelectrode until almost twenty years after Hyde had published her paper. For one thing, the required electronics had yet to be developed,[130] and for another, as we have noted, Hyde, Ettisch and Peterfi, and Kamada all worked on anthropomorphically less acceptable single-celled organisms. Certainly, too, one must take into account Hyde's own failure to publish no more than a single paper about her electrode.

But, just as these factors were undoubtedly instrumental in burying Hyde's invention, one cannot help but perceive Lillie's dismissal of Hyde's work as having been "then well known" as diagnostic of the way women's contributions were expected to be overlooked by the scientific community.

ANGELIQUE ARVANITAKI: SNAILS AND SQUIDS

In any event, by the 1940s neurobiologists had learned to penetrate electrically active cells with microelectrodes. At about the same time, they were also beginning to uncover another approach to the inside of the nerve. They realized, for the first time, that if they wanted to study the properties of nerves from within a single cell, they had better find a very large nerve into which to insert their electrodes. And furthermore, they realized that if they wanted to study groups of interacting nerves, they had better find not only very large nerves but ones that were easily accessible and consistently recognizable from preparation to preparation. Now, the problem with vertebrate nerves is that they are exceedingly small and thin, and they are housed in a brain and nervous system, which, until the last decade, was virtually inaccessible to systematic experimental analysis. This, however, was not true of the large nerve cells of certain invertebrates.

The first neurobiologists to exploit large invertebrate nerves were, on the one hand, K. C. Cole and Curtis,[131] and Hodgkin and Huxley,[132] who used the giant axon of the squid to analyze the properties of the nerve impulse, and, on the other hand, Angelique Arvanitaki, who developed the ganglion preparation of large identifiable nerves in the snails *Aplysia* (the sea hare) and *Helix* (the land snail).[133]

Between them, the squid axon and the *Aplysia* ganglion have been responsible for revealing an uncountably large number of fundamental concepts about nerve functioning. (The 1963 Nobel Prize in Physiology and Medicine was awarded to Hodgkin and Huxley for solving the problem of how nerves worked by analyzing the ionic basis of the action potential in the squid axon.) They are, in every sense, classical and model preparations, for they have long been among the best experimental objects for studying nerve functioning.

Arvanitaki developed the snail preparation as a result of her interest in cellular rhythmicity. As a student at the marine biological station in Tamaris, France, Arvanitaki had been studying the problem of repetitive firing in isolated nerves. She discovered that in low-calcium solutions, isolated nerve fibers of the cuttlefish *Sepia* (a relative of octopuses) produced regular electrical oscillations which periodically became larger and larger, until from time to time the nerve fired a series of action potentials.[134] This was the first demonstration that spontaneous, rhythmically recurring activity could be an inherent property of a single nerve and did not require an entire neuronal circuit to generate it. (Cells that can fire spontaneously at regular intervals are called pacemakers; they are found all through the animal kingdom and control many functions, including heartbeat and breathing.) According to Kandel, in order to find out whether neurons in the central nervous system could also fire spontaneously under normal circumstances, Arvanitaki began to search for an animal that had "comfortably large nerve cells."[135] She found that the nerve cells of *Helix* and *Aplysia* were ideally suited for her purpose and in this way the snail ganglion preparation came to be introduced into the neurophysiological repertoire.[136] Arvanitaki found that the abdominal nerve cells of *Aplysia* were especially interesting because many of the cells had characteristic firing patterns that could be analyzed. The realization that the cells in each ganglion could be recognized from preparation to preparation[137] gave neurobiologists another of their most useful tools—the till-then unprecedented chance to study reliably and exhaustively the many properties of known cells and cell networks.

In addition to these ground-breaking discoveries, Arvanitaki also found that when two or more nerves run close together, the activity in one nerve can entrain the activity in its neighbor. She designated the close anatomical apposition and electrical interaction of two nerves, an ephapse, a term that until very recently was used to describe such phenomena. Her paper, "Effects evoked in an axon by the activity of a contiguous one," was completed in 1940 but was not published until 1942, and then without her own corrections because "correspondence concerning this manuscript were prevented by war developments."[138] Today we know that the activity in many nerves is influenced indirectly by the electrical milieu surrounding them. The effect is significant and in some cases is the primary mechanism for switching nerves on and off.

While some neurobiologists were beginning to analyze the nature of the electrical activity of nerve cells, and others were learning how nerve networks functioned, there were still others who approached the questions of the nervous system from the viewpoint of the developmental biologist; rather than ask how nerves worked, they wanted to know instead how nerve fibers developed—where did they come from and how did they know where to go?

RITA LEVI-MONTALCINI: NERVE GROWTH FACTOR

In 1907, Ross G. Harrison (1870–1959) performed an experiment on developing nerve cells that was to give biology one of its most powerful and useful experimental tools. He removed immature nerve cells (those which did not yet have axons) from a frog embryo, suspended them in a drop of frog lymphatic fluid, and watched as the immature nerve cells made axons which grew out in long thin fibers. Thus he demonstrated that axons were produced as a single outgrowth of a developing nerve cell and not, as had been previously proposed, by a joining together of several cells.

More importantly, with this experiment he introduced the technique of tissue culture to biology.[139]

Among those who sought to understand the mechanisms by which growing nerve fibers arrived at their destinations was Viktor Hamburger (b. 1900), a former student of Hans Spemann who had settled in the United States just before the Second World War. In 1934 he had published a paper, "The Effects of Wing-bud Extirpation on the Development of the Central Nervous System."[140] In this article he demonstrated that nerves appeared to be guided to their target organs by the organs themselves. The paper was to have, in a not altogether anticipated way, a far-reaching effect on the future course of developmental neurobiology. For it became the "bible and inspiration" of a recently graduated physician, Rita Levi-Montalcini (b. 1909). Because she was Jewish, Levi-Montalcini had in 1938, as had all "non-Aryans," been barred from academic pursuits in Mussolini's Italy, and had, in consequence, decided to pursue a clandestine career as an experimental neuroembryologist. With Hamburger's article to guide her, Rita Levi-Montalcini set up a laboratory in her "bedroom with a few indispensable pieces of equipment, such as an incubator, a light, a stereo-microscope, and a microtome. The object of choice was the chick embryo, and the instruments consisted of sewing needles transformed with the help of a sharpening stone into microinstruments."[141]

In the early years of the Second World War, joined by her former professor, Guiseppe Levi, Levi-Montalcini studied the effects of destroying the peripheral target organs on the development of the nerves of the central nervous system. Their paper, published in 1942,[142] later attracted the attention of Viktor Hamburger himself, and in 1947 he invited Levi-Montalcini to come to the United States to work with him.

One of her first assignments was to repeat an experiment done in 1948. One of Hamburger's students, Bueker, had shown that when a rapidly growing mouse tumor had been implanted into the side of a developing chick embryo, the developing nervous system responded by sending out a profusion of fibers. Bueker, and later Levi-Montalcini and Hamburger (1951),[143] interpreted this finding to mean that the nerve fibers had adjusted their growth rate to correspond to that of the rapidly growing tumor. But then, on looking at the results more carefully, Levi-Montalcini noticed that instead of innervating the implanted mouse tumor, the nerve fibers had simply grown in great profusion and had spread out all over from their origin. With this, she realized that the nerves were not merely keeping pace with the growing tumor, but that the tumor was in fact releasing some substance which caused the nerves to proliferate.[144]

Levi-Montalcini then decided to test her hypothesis by resorting to the still somewhat esoteric technique of tissue culture as a means of observing the effect directly. She had been introduced to the method by Levi. With her friend Herta Meyer, Levi-Montalcini demonstrated that the substance (now called Nerve Growth Factor) which was produced by the tumor cells could induce certain nerve cells in culture to produce a dense felt-work of nerve fibers.[145]

Since then, Nerve Growth Factor has been shown by Levi-Montalcini and others to be the substance produced by many target tissues. It is absolutely essential for guiding certain nerves (i.e., sympathetic nerves which produce adrenaline) to their destinations and for keeping those nerves alive; without it, the nerves simply degenerate. It has therefore also become an indispensable tool for growing sympathetic nerves in tissue culture. Because many of its actions have now been analyzed, Nerve Growth Factor has played a vital role in our understanding of how nerves produce their axons and what the function of certain structures within those axons are.[146] And it has opened the door to an understanding of the role of chemical factors in developmental control.

In 1987, the Nobel Prize in Physiology and Medicine was awarded to Levi-Montalcini and Stanley Cohen for their work on growth factors.

BERTA AND ERNST SCHARRER: NEUROSECRETION

Complex animals are comprised of cells and organs which, for the most part, are geographically isolated from one another. Nonetheless, it is an absolute requirement that the activities of these separated entities be co-ordinated and integrated. Animals have evolved two systems of internal communication for this purpose. The one is fast, electrical, and hooked up in a system of cable-like nerve fibers. The other is slower, chemical, and connected together by the veins and arteries of the circulatory system. For many years (at least until the 1920s) it was thought that the nervous system and the endocrine (glandular) system were categorically separate and distinct anatomical and functional entities.

Gradually this view began to change. In 1922, a Polish biologist, Kopec, showed that the brain of butterflies released a chemical substance which caused them to pupate.[147] Three years earlier, Speidel had described gland-like cells in the spinal cord of skates and had correctly interpreted these as nerve cells.[148] In 1928, as a result of his doctoral research, Ernst Scharrer (1906–1965) showed that certain nerves in the fish brain seemed to produce chemical substances which were directly secreted into the blood.[149] With these findings, the long-established dichotomy between nerves and glands began to disappear and the concept of neurosecretion was established.

In 1934, Berta Vogel (b. 1906), a fellow student at the University of Munich, married Ernst Scharrer. Thus the team Scharrer and Scharrer was formed. Together, for the next thirty years, the Scharrers worked on the problem of neurosecretion, and, as a result of their research, helped to create a new discipline in physiology, that of neuroendocrinology.

In her recollections, Berta Scharrer says of their collaboration: "After our marriage in 1934 . . . , the road seemed open for a joint effort to probe the role of neurosecretion. We decided to divide the animal kingdom; Ernst would continue his studies on vertebrates, and I would set out to search for comparable phenomena among invertebrates."[150] The papers that were to come out of their research reflect this self-imposed division of labor, and are only from time to time co-authored by both of them.

Because biologists had been so used to viewing the nervous system and the endocrine system as anatomically and physiologically distinct, the idea that nerves could secrete hormones into the blood was greeted with considerable resistance. First, the evidence for it was still meager, and second, the need for such a mechanism was not at all self-evident. The Scharrers met the first of these criticisms by demonstrating the existence of neurosecretory cells in a wide range of species. The second criticism was answered by Ernst, who, in 1952, proposed that nerve cells in the brain secreted hormonal substances in order to translate fast electrical nerve signals into slow chemical hormonal signals. Long-lasting regulatory effects, typical of chemical but not of electrical systems, could thus be triggered by fast incoming nerve signals.[151]

Among the most important of the Scharrers' findings was the discovery that in both mammals and insects there were two completely analogous neuroendocrine organ systems, each of which controlled a variety of non-nervous processes.[152] In both mammals and insects, hormones were found to be produced by a part of the brain and stored in a glandlike structure from which they could later be released into the circulatory system. Other neurons in both mammals and insects were found to produce chemical hormones that were circulated without storage. These were respectively responsible for controlling a large number of day-to-day metabolic processes and the growth and differentiation occurring during the molting cycle.

Although Hanstrom had earlier (1941) indicated an analogy between these two systems,[153] it was Ernst and Berta Scharrer who gave them their functional significance as nerve-controlled mediators of chemical regulation. It was also as a result of Berta's work that the *corpus cardiaca* was identified as a hormonal storage organ in insects, and indeed that the brain itself was shown to be a neurosecretory organ.[154]

In the 1950s and 60s, the concept of neurosecretion finally began to take hold. Both Berta and Ernst continued to make important contributions to the new field, each describing and defining various aspects of their respective domains. Later, in the 1970s, after Ernst's death (1965), when it began to become apparent that secreting neurons were far more ubiquitous and important than had at first been suspected, Berta's work continued to refine and reshape the concepts of neurosecretion and neurosecretory communication.

Although from their writings it is perfectly clear that the concept of neurosecretion and the understanding of its physiological significance could only have been arrived at by pooling the information from each of their individual researches, it is also very evident that Berta's and Ernst's scientific partnership was something less than equal. Indeed, it is Berta herself who goes to considerable lengths to make this inequality apparent.[155]

To be sure, there were good scientific and intellectual reasons for this small imbalance. Ernst's discovery of endocrine-like nerve cells was scientifically exciting—one not to be ignored. Furthermore, it was Ernst who had first recognized the neuroendocrine function of the vertebrate hypothalamus, and it was Ernst who later realized the importance of the neurosecretory system as a means of translating electrical signals into chemical ones. Nonetheless, by choosing to devote their careers to the joint study of neurosecretion, the Scharrers made a de facto decision to pursue the scientific questions that belonged, at least initially, to Ernst. In going over their work and their careers, one cannot shake the impression that the small but real inequality that existed in their scientific collaboration reflects nothing so much as the social hierarchy which was imposed, often self-imposed, on every female-male partnership in our culture and which especially dominated the marital one.

Berta recalls that "in the early thirties, prospects for an academic career in Germany were bleak, and for a woman, virtually non-existent,"[156] so that Ernst decided to obtain an additional degree in medicine, while Berta studied to become a secondary school teacher. Later, Ernst received several research appointments, which, though they included Berta, were not hers in her own right.[157] In 1940, Ernst obtained his first teaching appointment at Western Reserve and Berta followed him there to work as a fellow and instructor in the department of anatomy. In 1946, Berta again went with Ernst, this time to the University of Colorado Medical School, where she held a series of fellowships and research appointments. In 1955, they both moved to Albert Einstein, where Ernst had now been appointed chairman of the Department of Anatomy, a position he held until his death. Here, "no longer subject to the rule of nepotism," she received her "first regular academic appointment"[158] as professor of anatomy. Since then, her career has been clearly seen as her own and she has been accorded many honors and accolades. She is a member of the American National Academy of Science and of several foreign national science academies.

Reflecting upon the effect of her social position on her science, Berta wrote in 1975: "Yet conditions that might appear to have been restrictive for me during the two preceding decades had their positive side. It was a privileged existence; free from official duties and other pressures, and much encouraged by my husband, I was allowed to develop and pursue my research program. The fact that much of it was carried out on a lowly laboratory animal, the cockroach, a choice originally dictated by the limited facilities available to a 'guest investigator,' likewise turned into an asset."[159]

Be that as it may, in our anthropomorphically determined order of bio-logical importance, the mammalian brain takes precedence over the cock-roach brain (indeed, the initial goal that the Scharrers had set for themselves was the elucidation of mammalian hypothalamic function). And this was, from the first, Ernst's and not Berta's domain.

THE "REAL" BIOLOGISTS:
WOMEN IN NATURAL HISTORY AND
RELATED DISCIPLINES

All of biology began life as natural history. Developmental biology, ge-netics, physiology, animal behavior, ecology, and systematics share with physics, astronomy, and geology a common ancestor whose purpose and method were the description and ordering of the physical universe. As these became more quantitative and experimental, they branched off from their ancestral stock to develop questions and techniques of their own. In the meantime, the central core itself developed specialized areas of inquiry. Natural history gave way to zoology and botany, and from zoology sprang mammalogy, ornithology, herpetology, ichthyology, entomology, parasi-tology, and invertebrate zoology. Today these exist as separate but recog-nizably related disciplines whose subject matters differ but whose shared methodology remains deeply rooted in observation, description, and clas-sification and whose theoretical framework rests on the concept of the fundamental similarity and inherent relationship of all living organisms whose differences evolved as inherited adaptations to external conditions.

The pursuit of natural history can be as complex as arranging plants and animals into a "natural" system,[160] i.e., one which reflects their in-herent relationships (Linnaeus, 1707–1778), but it can also be as simple as holding a magnifying glass up to a beetle. It can, therefore, readily be pursued by anyone who will take the time to be trained to the required level of expertise. Today, as in the past, it is possible to become a student of classification and of the many subdisciplines of natural history without acquiring an extraordinary degree of formal or technical training. Because this is so, bird watching and insect collecting have always been the domain of brainy young girls and keen old ladies and anyone else who either by personal choice or by social design pursues the science of biology in social or intellectual isolation.

Of the very many women naturalists who could have been included here, I have chosen to focus on only a few whose contributions were either significant because they addressed the general theories of natural history (e.g., Hildegard), or because their works exercised an influence beyond their own spheres of knowledge (e.g., Anna Comstock and Rachel Carson).

HILDEGARD VON BINGEN: THE FIRST NATURAL HISTORIAN OF THE MIDDLE AGES

From the time of Aristotle's first catalogues of nature until Linnaeus's natural system some 2000 years elapsed. During this period natural historians described and listed the numerous natural objects of their universe. Although premedieval practioners of natural history often took their lessons from life, by the time of the Middle Ages many of the descriptions and drawings made by natural historians were copies—and sometimes very poor ones—ancient of works, and not infrequently they were fabrications made entirely of whole cloth; physics and metaphysics were interwoven into a conceptual continuum which did not really begin to unravel until the seventeenth century.

The repositories of intellectual activity in the Middle Ages were the convents and monasteries. Science, medicine, art, and music were nurtured behind the walls of these communes. Besides Héloïse (1101–1164) and Abelard (1078–1142), and later Albertus Magnus (1206?–1280), one of the greatest and best known medieval practioners of medicine and philosophy was Hildegard of Bingen (1098?–1179), who became abbess of the Benedictine abbey St. Ruperts on the Rhine in 1147.[161]

Hildegard is considered by many to have been the first and foremost natural historian of her age. She was also among its most successful and renowned practioners of medicine, and wielded a political influence that extended to kings and popes.

Hildegard spent most of her life within the convent. At the age of eight, she had been sent to study with her aunt Yutta, abbess at Disibodenburg; she remained there until she was forty-nine. When she was thirty she succeeded her aunt as abbess of the convent. In 1147 Hildegard and some of her followers moved to Bingen, where she had been permitted to establish a new convent at Rupertsberg. She remained abbess at Bingen until her death at the age of eighty-one. Hildegard is said to have written fourteen books in prose and poetry, many in several volumes. Her works include discourses on natural science, medicine, and religious philosophy.

Throughout her life, Hildegard was beset by dreams and visions, which she considered to have been of divine origin. They were, she says, not only the source of her prodigious healing powers, but also the inspiration of all of her physical and metaphysical perceptions. But despite her typical medieval mysticism, Hildegard, unlike her contemporaries, drew a large measure of her observations neither from her imagination nor from the literature, but rather from nature itself.[162] In this she stood well ahead of her fellow scholastics, and, because of this, her work becomes the first reintroduction of Aristotle's scientific methodology into Western thought.[163] (Cf. Wasmann and Singer on this point.) Hildegard's natural history is contained in a nine-volume treatise entitled *Subtilitatum diversarum naturarum creaturarium libri IX*. In this there are a great number of direct observations of local plants and animals, and also a recounting of their folk history. There are references to nearly five hundred plants, metals, stones, and animals, including birds, fish, reptiles, and mammals, and discussions of their use in medicine.[164]

In these volumes, Hildegard begins by describing animals which are not native to her country. Her accounts include such beasts as dragons, griffins, and unicorns, as well as unknown (but real) animals such as lions and tigers. Here she relies entirely on the literature and tells fables and stories about them in typical medieval fashion. But then, in each book, she turns her attention to the accurate and first-hand account of the birds, beasts, trees, and flowers around her. In her Botany, she describes the several varieties of grain, medicinal plants, and trees in her region, and refers to the ability of the Dictamnus plant to burst into flame if its flowers are ignited (this is due to an oil of ether which envelops the flower on dry, windless days). The discovery is usually attributed to Linneaus's daughter.[165]

In Hildegard's book on fishes more than thirty local fishes are identified and described, while in her 72-chapter book on birds she categorizes and names a great number of the birds common to her area, including doves, fowl, sparrows, birds of prey, crows, and gulls. In her book of birds she includes the flying insects, and lists bees, bumble bees, wasps, flies, mosquitoes, and locusts in this group—a logical pre-Linnean classification on her part. Her book on mammals begins with fables of elephants, camels, unicorns, and bears and then proceeds to accurate descriptions of endemic species such as ermines, weasels, badgers, otters, and so forth. As an addendum to the book she adds a section on ants. Wasmann points out that when her observations conflict with the current mythology about a particular animal, her observations win out.[166]

Although Hildegard's eminence as a naturalist rests mainly on her empiricism, like all great naturalists she places her observations within a framework of theory and natural philosophy. Much of her *Weltanshauung* is simply that of the Middle Ages, but some of it appears to arise from her own peculiar vision of the universe.[167] For example, although the theory of the universe as a series of concentric spheres was known before Hildegard, she adds details to the philosophy which are her own. She conceives of the universe as having five powers of fire, four of air, fifteen of water, and seven of earth, and believes that winds, for example, emanate from each of the four concentric heavens thought to surround the earth. These rings consist of a skinlike shadow of fire, followed by ether, then water, and then air.

Although Hildegard does not appear to add any essentially new principle to the theory of the universe, Singer says her natural philosophy is important because it represents an "early attempt at something like a coherent philosophy intended to cover the appearances of the material universe. As such her work is in fact science and we have left the dark ages and the dawn has begun."[168] In her lifetime, Hildegard won great acclaim, and although the Catholic Church never officially canonized her, she is often referred to as St. Hildegard.

Some Eighteenth- and Nineteenth-Century Naturalists

In the centuries that followed, naturalists became ever more empirical. Noteworthy among them was the Bavarian naturalist Maria Sibylla Merian, who in the seventeenth century produced several treatises on insects and flowers.[169] Her book on insects, published in 1699, is the first on the subject. She became noted for her accurate observations; her zoological drawings were considered among the best. In America, Jane Colden (1724–1766) became the first woman to classify plants according to the Linnaean system. She catalogued over three hundred local plants, described and named, among others, the *Gardenia*, and became known in both America and Europe for her botanical classifications.[170] Another American, Maria Martin (1796–1863), became an associate of John James Audubon. Many of the backgrounds and habitats of his best-known works are hers. Working from nature, she painted numerous flower and plant backgrounds for Audubon's *Birds of America* (The Elephan Folio). She also worked on *Viviparous Quadrupeds of North America* (1846–1854) and contributed to John Edward Holbrook's *North American Herpetology* (1836–1842).[171]

Among naturalists whose work significantly affected the society in which they lived was American naturalist Margaretta Hare Morris (1797–1867). Her work was instrumental in controlling two devasting agricultural pests—the seventeen-year locust and the Hessian fly. Morris was the first woman to become a member of the Philadelphia Academy of Science.[172]

She showed that the larvae of the seventeen-year locust entered the roots of fruit trees, so that with time the trees became weaker and weaker until ultimately, after several seasons, they failed to bear fruit.[173] Earlier Morris had shown that the larvae of the Hessian fly, which had previously been classed as a single species, was in reality two species. The larvae of one of the flies could become entrapped in the developing wheat and inadvertently transported from one country to another in the straw.[174]

When Darwin's *On the Origin of Species* was published in 1859, it raised a hue and cry throughout the world. In Britain, the battle was carried forward by Darwin's champion, Thomas Henry Huxley (1825–1895), who on numerous occasions defended Darwin's thesis against the onslaughts of conservative scientists and churchmen. In France, Darwin's banner was carried forward by Clemence Augustine Royer (1830–1902).[175] Royer made the first French translation of *On the Origin of Species* in 1862. This was expanded in 1870 and published as *The Origin of Man and Society*.[176] She was often compared to such famous natural historians and theoreticians as Vogt, Buchner, and Haeckel. Her writing created much controversy in scientific and religious circles. Two theoretical works, *Two Hypotheses of Heredity* (1877)[177] and *New Principles of Natural Philsophy* (1900), are noteworthy among her publications. During her lifetime her works were well read and influential. Mozans says that she was once described as "almost a man of genius."[178]

ANNA COMSTOCK: AND THE AGRICULTURE MOVEMENT

As the nineteenth century progressed and as the Industrial Revolution began to force the replacement of a rural-agricultural society with an urban one, there were those who sought to stem the ever-increasing flight from farm to city. In an effort to reverse the trend, New York State instituted a program designed to revitalize interest in agriculture among rural children. Among those who became a leading force in the movement was Anna Botsford Comstock (1854–1930). Anna Botsford, who had begun her career as a student of language and literature at Cornell, left school in her sophomore year to marry one of her former instructors, John Henry Comstock (1849–1930). Comstock was an entomologist and a new member of the zoology faculty.[179]

Anna became John's assistant, helping him with his clerical work and with various diagrams for his lectures. In 1880 she produced the drawings of scale insects for *Report of the Entomologist*,[180] and to better help her husband she reentered Cornell, where in 1885 she received a Bachelor of Science degree in natural history. In 1888, John Comstock published *An Introduction to Entomology*.[181] The wood engravings for the volume were entirely Anna's. In that year she became one of the first four women to be initiated into Sigma Xi, the national scientific honor society.

In 1895, Anna Comstock was named to the privately sponsored Committee for the Promotion of Agriculture in New York. Under the committee's aegis, she undertook the teaching of nature study in the Westchester school system. As a result of her efforts, the Cornell College of Agriculture was given funds in 1896 by New York State to expand the project. Comstock was appointed Assistant Professor for the 1899 summer session, the first woman to achieve this rank at Cornell. In 1900 her position was downgraded to Lecturer. But in 1913, she was again made an assistant professor. Two years before her retirement from full-time teaching, she was made a full professor.

Comstock's many books and leaflets on nature study for students and teachers were widely used. Her best-known work, *The Handbook of Nature Study*, first published in 1911, saw twenty-four editions (the last in 1939).[182] It was translated into eight languages and is still in frequent use today.

The book is a detailed study guide to American plants and animals. The descriptions permit the identification of numerous species and the small-scale experiments and exercises give students a knowledge of natural history which, as children of the twentieth century, they could no longer expect to automatically acquire.

Her influence as a naturalist was so great that in 1923 the League of Women Voters named Comstock as one of the twelve greatest living American women.

RACHEL CARSON: THE IDEA OF ENVIRONMENT

If it is true that the effect and indeed the function of science is to alter our perception of the universe, then it is also true that not all great scientists produce the data from which they create the new perception. Old data is often only differently organized into what then becomes our new vision of the "real" world. Further, one might argue that the importance of a scientific endeavor rests precisely on its ability to create such a new perception.

If, now, we were to compile a list of the great biologists based on this argument, we must immediately include the name of Rachel Carson (1907–1964)—whom, on first consideration, we might have been inclined to overlook, for Carson's achievements as a writer tend to obscure her real contribution as a scientific thinker.

Although a trained biologist—she received her master's degree in zoology from Johns Hopkins in 1932 and worked as a professional biologist both as a teacher at the University of Maryland (1931–1936) and as an aquatic biologist at the United States Department of Fish and Wildlife (1936–1952)—her work was essentially that of instructor and reporter. Indeed, she seems to have seen herself primarily as a writer and an intrepreter of nature rather than an original researcher: She once wrote to a friend, "If I could choose what seems to me the ideal existence, it would be just to live by writing," and "biology," she had earlier said, "has given me something to write about."[183]

While with the Fish and Wildlife Department, she produced two large monographs, "Food from the Sea: Fish and Shellfish of New England" (1943), and "Food from the Sea: Fish and Shellfish of the South Atlantic and Gulf Coasts" (1944).[184] Both monographs, written during World War II, were aimed at finding new food sources and preventing the overconsumption of certain species of food fish. They are detailed, accurate descriptions of the anatomy, behavior, and habitat of twenty-six varieties of fish and shellfish as well as an evaluation of their usefulness as food.

In 1941 Carson published her first book, *Under the Sea Wind—A Naturalist's Picture of Ocean Life*.[185] In contrast to the unembellished factual monographs, her first book was a highly romanticized, anthropomorphic portrayal of the lives and habits of a number of aquatic species. Nevertheless, it was well received and, ten years later, it became a best-seller. Two books followed: *The Sea Around Us* (1951), and *The Edge of the Sea*, (1955).[186] Like neither the monographs nor *Under the Sea Wind*, these books retain the best characteristics of both; they give a vivid and penetrating account of the physics, chemistry, and biology of the ocean and its shores. Their language is at the same time lyrical and precise and carries none of the melodramatic, sometimes saccharine overtones of her first book. But instructive as they are, these works do not force upon readers a reconstruction of their world. That accomplishment was left to Rachel Carson's last book, *Silent Spring* (1962).[187]

The view that all of nature is interdependent, in the strictest sense, and that human industrialized activity almost invariably causes permanent damage to the earth is a view which today is so commonplace that to reiterate it is almost boring. It is hard to believe, therefore, that when *Silent Spring* first appeared, the idea that modern technology could annihilate us by irretrievably destroying our habitat was an unheard of revolutionary thought—one that provoked debate and discussion throughout the society and was greeted with alarm and controversy in government and industry.[188]

It is further hard to believe that until the publication of *Silent Spring*, not only was it hardly known that small perturbations in the food chain could have severe, far-reaching, and unpredicted ecological consequences, but also that the deleterious genetic and physiological effects of unintended chemical consumption could have been so little suspected. It required some 350 pages (including 55 pages of references) to build an argument that showed how the indiscriminate spraying of pesticides like DDT caused widespread permanent biological destruction. Not only did Carson have to cite reports of such destruction, she also had to present and explain many of the biological and ecological principles which accounted for the destruction. Today we know these principles so well that, as with our own language, we are no longer aware that we had to be taught them.

The immediate effect of Carson's book was the banning of DDT as an insecticide; the long-term effect, as one editor put it, was to change the world: "A few thousand words from her, and the world took a new direction."[189]

In 1964, two years after the publication of *Silent Spring*, Rachel Carson died of cancer at the age of 56.

CONCLUSION

The women whose work we have discussed in the preceding pages by no means represent all of the major women contributors to their fields; nor do they represent the many women working in those areas of biology we have not investigated. Nonetheless, it must be apparent from even these few examples that women were not in fact absent from the practice of biology. Nor were their contributions at all second-rate. Often, as with Foot and Strobell, (Graham-)Pool, Franklin, McClintock, Levi-Montalcini, Sager, and Carson, their work changed the direction of their fields or propelled them forward. Even so, not infrequently, as with Hyde or Browne-Harvey or Stevens, their germinal discoveries only became an impelling force after these had again been made by more influential male contemporaries. Nor was their science, as often imagined and frequently implied by their inferior academic status, to be subsumed under their husbands' accomplishments; rather, as with Scharrer and Comstock, their work is to be seen as an original and personal achievement, even though it may initially have sprung from their husbands' scientific interests.

From all this, it must be obvious that the glaring absence of women from the recounting of biology and, indeed, from much of its bibliography rests not so much on their actual physical absence from its science, but on the persistent, almost routine expurgation of their existence from its thought and literature.

NOTES

1. For an excellent account of the development of nineteenth-century biology, see Garland Allen's *Life Science in the Twentieth Century*, chapters 1 and 2. Cambridge History of Science Series, Cambridge University Press, Cambridge, England. 1978.

2. Wilhelm Roux, Die Entwicklungsmechanik der Organism, eine anatomische Wissenschaft der Zukunft. In *Gesammelte Abhandlungen Über Entwickelungmechanik der Organismen*, Wilhelm Engelmann, Leipzig, Germany (1895), Vol. II, pp. 24–54.

3. Hans Driesch, 1892. Entwicklungsmechanische Studien, I. Der Werth der beiden ersten Fürchungszellen in der Echinodermentwicklung. Experimentelle Erzeugen von Theil-und Doppelbildung. *Zeitschrift für Wissenschaftliche Zoologie.* 53:160–178, 183–184.

4. Allen 1978, op. cit. (note 1), chap. 2.

5. Hans Spemann and Hilde Mangold, 1924. Über Induktion von Embryonalanlagen durch Implantation artfremder Organisatoren. *Wilhelm Roux Arch Entwicklungsmech. Organ.* 100:599–638.

6. *Exp. Zool.* (1909) 8:1–23.

7. Ibid., p. 13.

8. Ibid., p. 21.

9. Hans Spemann, 1938. *Embryonic Development and Induction*. Yale University Press, New Haven, Conn., pp. 142–143.

10. Hans Spemann, 1901. Über Korrelationen in der Entwicklung des Auges. *Verh. Anat. Ges.* 15. Bonn, pp. 61–79. H. M. Lenhoff has recently suggested that Spemann was familiar with Browne's work prior to the publication of his and Mangold's 1924 paper. 1988 *American Zoologist* 28:92a.

11. Ethel Browne Harvey, 1935. Cleavage without nuclei. *Science* 82:277.

12. E. Newton Harvey, 1931. Observations on living cells made with the microscope-centrifuge. *J. Exp. Biol.* 8:267–274.

13. Harvey 1935, op. cit. (note 11), p. 277. Parthenogenesis, first described by Leeuwenhoek (1632–1723), refers to the birth of offspring from unfertilized eggs. It is a naturally occurring phenomenon among some species such as rotifers. A merogone refers to an organism that can be made to develop without participation of the maternal nucleus; the paternal nucleus is injected into the egg and the maternal nucleus withdrawn.

14. Ethel Browne Harvey, 1941. Cross fertilization of echinoderms. *Science* 94:90–91.

15. Ethel Browne Harvey, 1956. *The American Arbacia and Other Sea Urchins.* Princeton University Press, Princeton, New Jersey.

16. *American Men of Science*, 1955. R. R. Bowker Company, New York, NY.

17. Florence Peebles, 1897. Experimental studies on *Hydra. Roux Arch.* 5:794–819.

Florence Peebles, 1898. Some experiments on the primitive streak of the chick. *Roux Arch.* 7:405–429.

18. Peebles 1897, op. cit.

19. Florence Peebles, 1912. Regeneration and regulation in *Paramecium caudatum. Biol. Bull.* 23:154–170.

Florence Peebles, 1910. On the exchange of the limbs of the chick by transplantation. *Biol. Bull.* 20:14–18.

Florence Peebles, 1931. Growth regulation in *Tubularia. Physiol. Zool.* 4:1–36.

20. Peebles 1931, op. cit.

21. C. M. Child, 1924. *Physiological Foundations of Behavior.* Henry Holt and Co.

22. Erasmus Darwin (1731–1802) and Jean Baptiste Lamarck (1744–1829) had both maintained that animal forms were directly determined by their environment—i.e., characteristics could be acquired and inherited. Peebles's version simply says that characteristics could be differentially called forth by different environments and is directly related to the then-emerging ideas of the expression of genetic dominance and penetrance.

23. *American Men of Science*, 1954. R. R. Bowker Company, New York, NY.

24. *Who Was Who in America*, 1973. Vol. V. N. Marquis.

25. L. H. Hyman, 1940–1967. *The Invertebrates*, Vol. I-V. McGraw Hill Co., New York.

26. She was able to support herself with the royalties of her laboratory manual, *A Laboratory Manual for Comparative Vertebrate Anatomy*, published in 1922 (University of Chicago Press), which remained a best-seller for premedical courses in comparative anatomy for some thirty years.

27. L. H. Hyman, 1926. The metabolic gradients of vertebrate embryos. *J. Morph. Physiol.* 42:11–114.

28. L. H. Hyman, 1919. The axial gradients in hyrozoa. I. *Hydra. Biol. Bull.* 36:182–223.

29. L. H. Hyman, 1930. The effect of oxygen tension on oxygen consumption in planaria and echinoderms. *Physiol. Zool.* 2:505–534.

30. L. H. Hyman, 1929, 1930, 1931. Taxonomic studies in the hydras of North America. I, II, III. *Trans. Am. Microsc. Soc.* 48:406–415; 49:322–333; 50:20–29. For a complete bibliography of Hyman's work see *Biology of the Turbellaria* (1974), Nathan Riser and M. Patricia Morse, eds., McGraw Hill Book Company, New York.

31. Edna Yost, 1943. *American Women of Science.* Frederick Stokes Co., Philadelphia.

32. Hyman, 1930, op. cit. (note 29).

33. *American Men of Science.* 11th ed. 1966. R. R. Bowker Co., New York.

34. Mordecai Gabriel and Seymour Fogel, eds., 1955. *Great Experiments in Biology*. Prentice Hall Inc., Englewood Cliffs, N.J., p. 226.

35. Gregor Mendel, 1865. Versuch über Pflanzenhybriden. *Verh. des Naturf. Vereins Brünn* 4:3–47.

36. Carl Correns, 1900. G. Mendels Regel über das Verhalten der Nachkommenschaft der Rassenbastarde. *Berichte d. Deutsch, Botan. Gesellsch. Bd.* 18:158–168.

37. Hugo DeVries, 1900. Sur la loi de disjonction des hybrides. *Compt. rend. de. l'Acad. d. Sc. Paris*, 20 Mars.

Hugo DeVries, 1900. Das Spaltungsgesetz der Bastarde. *Berichte d. Deutsch. Botan. Gesellsch. Bd.* 18:83–90.

38. E. Tschermak, 1900. Über Künstliche Kreutzunger bei *Pisum Satirum*. *Zeitschr. landw. Versuchswesen in österreich.* 5.

39. Bruce R. Voeller, 1968. *The Chromosome Theory of Inheritance*. Appleton-Century-Crofts, New York, p. 61.

40. Ibid., p. 76.

41. C. E. McClung, 1902. The accessory chromosome—sex determinant? *Biol. Bull.* 3:43–84.

42. T. H. Morgan, 1912. The scientific work of Miss N. M. Stevens, *Science* 6:468–470.

43. Florence Peebles, 1913. Regeneration Acoeler Plattwürmer. I. Aphanostoma diversicolor. *Bull. L'Inst. Oceanograph. Monaco* 1–5.

44. T. H. Morgan, 1909. A biological and cytological study of sex determination in phyllozerans and aphids. *J. Exp. Zool.* 6:240–345.

45. N. M. Stevens, 1905. Studies in spermatogenesis with especial reference to the "accessory chromosome." *Carnegie Inst. Publ.* 36.

46. Morgan 1909, op. cit. (note 44), p. 240.

47. Stevens 1905, op. cit. (note 45).

48. N. M. Stevens, 1906. Studies in spermatogenesis. Part II. Sex determination. *Carnegie Inst. Publ.* 36. p. 56.

49. Later Stevens extended her own work to cover some varieties of insects.

50. E. B. Wilson, 1905. The chromosomes in relation to the determination of sex in insects. *Science* 22:500–502.

51. Ibid.

52. T. H. Morgan, A. H. Sturtevant, H. J. Muller, and C. B. Bridges, 1915. *The Mechanism of Mendelian Heredity*, Henry Holt & Co., New York. 1923 Revised Ed.

53. Voeller 1968, op. cit. (note 39), pp. 254–255, p. 254n.

54. Morgan 1912, op. cit. (note 42), p. 470.

55. Ursula Mittwoch, 1967. *Sex Chromosomes*, Academic Press, New York. pp. 53, 66.

56. Edward James, 1971. *Notable American Women 1607–1950*. Vol. II, Belknap Press of Harvard University Press, Cambridge, Mass. p. 372.

57. We were unable to establish a death date for Foot; 1943 was the last year she was listed as a member of the Marine Biological Laboratory Corporation at Woods Hole. We were unable to establish a birth date for Strobell.

58. In the introduction to "Further notes on the egg of *Allolophora foetida*" (*Zool. Bull.* 2:130–151) Foot notes that the very first photomicrographs of developing eggs were taken of her sections by her friend Dr. Charles G. Fuller of Chicago.

59. K. Foot and E. Strobell, *Cytological Studies 1894–1917*. Preface. A collection of reprints, privately published. A copy is in the MBL Library. Stack #7394.

60. K. Foot and E. Strobell, 1898. Further notes on the egg of *Allolobophora foetida*. *Biol. Bull.* 2:143.

61. K. A. Foot and E. C. Strobell, 1905. Sectioning paraffine at a temperature of 25°F. *Biol. Bull.* 9:281–286.

62. *American Men of Science*, 1944. R. R. Bowker Company, New York. Apart from the listing in *American Men of Science* (last listing 1944), we were unable to obtain further biographical information for Foot or Strobell.

63. K. A. Foot, 1920. Notes on *pediculus vestimenti*. *Biol. Bull.* 39:261.

64. T. H. Morgan, 1914. *Heredity and Sex*. Columbia University Press, New York.

65. T. H. Morgan, A. H. Sturtevant, and C. B. Bridges, 1915. *Mechanisms of Heredity*, Henry Holt, & Co., New York.

66. Evelyn Fox Keller, October 1981, McClintock's Maize, *Science 81* 2:55–58.
E. Fox Keller, 1983. *A Feeling for the Organism. The Life and Work of Barbara McClintock*. W. H. Freeman and Co., New York.

67. Barbara McClintock, 1951. *Cold Spring Harbor Symp. Quant. Biol.* 16:13–47.

68. Ibid., p. 34.

69. H. Creighton and B. McClintock, 1931. *Proc. Natl. Acad. Sci.* 17:492–497.

70. C. Stern, 1931. Zytologish-Genetische Untersuchungen als. Beweise für die Morganische Theorie des Faktorenaustauschs. *Bio. Zentrl. Blatt.* 51:547–587.

71. C. Correns, 1909. Zur Kenntnis der Rolle von Kern und Plasma bei der Verebung. *Z. Verebungslehre* 2:331–340.

72. For a review of Sager's papers and complete references see R. Sager, 1972. *Cytoplasmic Genes and Organelles*, chap. 3. Academic Press, New York.

73. B. Ephrussi, 1958. The cytoplasm and somatic cell variation. *Jour. Cell Physiol. Supplement*, 52:35–53.

74. M. B. Mitchell and H. K. Mitchell, 1952. Observations on the behavior of suppressors in Neurospora. *Proc. Natl. Acad. Sci.* 38:205–214.

75. R. Sager and M. Ishida, 1963. Chloroplast DNA in *Chlamydomonas. Proc. Natl. Acad. Sci. US.* 50:725.

76. E. Chun, M. Vaughn, and A. Rich, 1963. The isolation and characterization of DNA associated with chloroplast preparations. *J. Molec. Biol.* 7:130–141.

77. M. M. Nass and S. Nass, 1963. Intramitochondrial fibres with DNA characteristics. I. Fixation and electron staining reaction. *J. Cell Biol.* 19:593–611.
S. Nass and M. M. Nass, 1963. Intramitochondrial fibers with DNA characteristics. II. Enzymatic and other hydrolytic treatments. *J. Cell Biol.* 19:613–629.

78. R. Sager, 1965. Genes outside the chromosome. *Sci. Amer.* 212:70–81.

79. L. Margulis, 1971. Symbiosis and evolution. *Sci. Amer.* 225:48–57.

80. R. Sager and Francis Ryan, 1961. *Cell Heredity*. John Wiley & Sons, New York.

81. *Current Biography, 1967*. Chas. Molitz, ed. H. W. Wilson Company, New York. pp. 367–370.

82. R. Sager, 1972. *Cytoplasmic Genes and Organelles*. Academic Press, New York.

83. H. F. Judson, 1979. *The Eighth Day of Creation*. Simon and Schuster, New York.

84. S. Sayre, 1975. *Rosalind Franklin and DNA*. W. W. Norton & Co., Inc., New York.

85. J. Watson, 1968. *The Double Helix*. Atheneum, New York.
The three books are listed in reverse chronological order (the order in which I read them); but for a truly eye-opening experience, it is recommended that they be read in chronological order, i.e., Watson, Sayre, Judson.

86. O. T. Avery, C. M. MacLeod, and M. McCarty, 1944. Induction of transformation deoxyribonucleic acid fraction isolated from pneumococcus type III. *J. Exp. Med.* 79:137–158.

87. E. Chargaff, 1950. Chemical specificity of nucleic acids and mechanism of their enzymatic degradation. *Experientia* 6:201–209.

88. Judson 1979, 99–101.

89. Watson 1968, p. 16.

90. Judson 1979, p. 103.

91. This is a problem which often leads to deep acrimony and recrimination and, although it confronts both men and women, it has been particularly a problem faced by women scientists, for it is they, more often than not, who are the junior scientists and it is they who were relegated to the social underclass of research associate for many if not all the years of their professional lives. (Cf. the careers of Ethel Browne Harvey, Ruth Sager, and Nettie Stevens.) Many laboratories alleviate the problem by allowing the scientist who does the work to publish as first author, while the scientist who assigned the problem, or the director of the laboratory, is listed as last author.

92. Watson 1968, p. 4.

93. Ibid., p. 32.

94. Judson 1979, pp. 110–116.

95. Ibid., p. 136.

96. Ibid.

97. Sayre 1975, p. 150.

98. That is, they did not ask, as Watson and Crick appear to have done, what a molecule must look like in order to replicate itself and in order to direct the synthesis of a protein.

99. Students of biology from the time of Aristotle had, more or less correctly, tacitly assumed that many anatomical systems were basically the same and had used animal models to explore human anatomy and physiology.

100. H. M. J. Mozans, 1913. *Women in Science*. D. Appleton & Co., New York. p. 295.

101. A. Bretag, 1983. Who did invent the microelectrode? *Trends in Neurosci.* 6:365.

102. *Notable American Women, 1607–1950*. Vol. II, pp. 247–249.

103. I. H. Hyde, 1938. Before women were human beings . . . *J. Amer. Assoc. Univ. Women*, 31:226–236.

104. I. H. Hyde, 1894. The nervous mechanism of the respiratory movements in *Limulus polyphemus*. *J. Morph.* 9:431–448.

105. I. H. Hyde, 1904. Localization of the respiratory center in the skate. *Am. Jour. Physiol.* 5:236–258.

106. I. H. Hyde, 1902. The nervous system of *Gonionemus Murbachii*. *Biol. Bull.* 4:40–45.

L. H. Hyman, 1940. *The Invertebrates: Protozoa through Ctenophora*. McGraw Hill Co., Inc., New York & London. p. 422.

107. F. H. Pratt, 1917. The excitation of microscopic areas: a non-polarizable capillary electrode. *Am. Jour. Physiol.* 43:159–164.

108. M.A. Barber, 1904. A new method of isolating microorganisms. *Jour. Kans. Med. Soc.* 4:487.

109. I. H. Hyde, 1921. A micro-electrode and unicellular stimulation. *Biol. Bull.* 40:130–133. A modification of this approach is used today as a standard technique to inject various charged substances into or on top of cells. Thus in 1954 Katz and del Castillo demonstrated that synthesized acetylcholine produces exactly the same effects on muscle cells as the chemical transmitter naturally released by nerves. They used an electrode which released the positively charged acetylcholine onto the muscle cells when a positive current was conducted through the electrode. See J. del Castillo and B. Katz, 1954. Electrophoretic application of acetylcholine to the two sides of the end-plate membrane. *J. Physiol.* 125:16–17.

110. Keith Lucas, 1909. The "all-or-none" contraction of amphibian skeletal muscle. *J. Physiol.* 38:113.

111. *Notable American Women*, op. cit. (note 56).

112. See Trescott's chapter on Women in Engineering.

113. F. R. Lillie, 1927. 29th Annual Report. *Biol. Bull.* 53:21–22. By 1926, Hyde, who was still an active scientist, had published or directed the publication of more than twenty papers and had written a textbook and laboratory manual in experimental physiology. She had also been a staff member of the Physiology Department of the MBL in the summers of 1903, 1906, and 1907.

114. G. Ettisch and T. Peterfi, 1925. Zur Methodik der Elektrometrie der Zelle. *Archiv. F. Physiologie.* 208:454–466.

115. Structurally, recording and stimulating electrodes are the same except that in the former case impulses are led from the cell into an electrical recording instrument, while in the latter, the impulses are transmitted from the electrical source to the cell. In both cases, suitable electronic instrumentation must be available and the interface between the solution in the pipette and the metal wire leading to the instruments must be such that a charge cannot be accumulated on it with repeated use.

116. C. V. Taylor and D. M. Whitaker, 1927. Potentiometric determinations in the protoplasm and cell-sap of *Nitella*. *Protoplasma* 3:1–6.

117. T. Kamada, 1934. Some observations on potential differences across the ectoplasm membrane of *Paramecium*. *J. Exp. Biol.* 11:94–102.

118. J. Graham, G. R. Carlson, and R. W. Gerard, 1942. Membrane and injury potentials of single muscle fibers. *Federation Proceedings* 1:31.

119. R. W. Gerard and J. Graham, 1942. Excitation and membrane potentials of single muscle fibers. *Fed. Proc.* 1:29.

J. Graham and R. W. Gerard, 1946. Membrane potentials and excitation of impaled single muscle fibers. *J. Cell Comp. Physiol.* 23:99–117.

120. G. Ling and R. W. Gerard, 1949. The normal membrane potential of frog sartorius fibers. *J. Cell Comp. Physiol.* 34:383–396. The microelectrode in use today is essentially the same as this, except that it usually contains a concentrated salt solution, and a fine capillary thread within the tube allows it to be filled without boiling.

121. R. W. Gerard, 1975. The minute experiment and the large picture in the neurosciences. In *Path of Discovery.* F. G. Worden, J. P. Swazey, and G. Adelman, eds. M.I.T. Press, Boston.

122. C. Edwards, 1983. Who invented the intracellular microelectrode? *Trends in Neurosci.* 6:44.

123. *World's Who's Who in Science*, A. N. Marquis, Allan Debus, ed., 1968. p. 1362.

124. Judith E. Graham and F. J. M. Sichel, 1939. Response of frog striated muscle to CaCl$_2$. *Biol. Bull.* 77:332.

125. Elsa M. Keil and F. J. M. Sichel, 1936. The injections of aqueous solutions, including acetylcholine, into the isolated muscle fiber. *Biol. Bull.* 71:402.

126. Bretag, 1983, op. cit. (note 101).

127. *Notable American Women. The Modern Period.* Barbara Sicherman and Carol Green, eds. The Belknap Press, Cambridge, Mass., pp. 553–554.

128. Judith G. Pool and Edward J. Hershgold, 1964. High-potency antihaemophilic factor concentrate prepared from cryoglobulin precipitate. *Nature* 203:312.

129. *Notable American Women*, op. cit. (note 127).

130. I am indebted to L. M. Passano for this observation.

131. K. S. Cole and H. J. Curtis, 1939. Electric impedence of the squid giant axon during activity. *J. Gen. Physiol.* 22:649–670.

132. A. L. Hodgkin and A. F. Huxley, 1952. Currents carried by sodium and potassium ions through the membrane of the giant axon of *Loligo*. *J. Physiol.* 116:449–472.

A. L. Hodgkin and A. F. Huxley, 1952. The components of membrane conductance of the giant axon of *Loligo. J. Physiol.* 116:473–496.

A. L. Hodgkin and A. F. Huxley, 1952. The dual effect of membrane potential on sodium conductance in the giant axon of *Loligo. J. Physiol.* 116:497–506.

133. A. Arvanitaki and H. Cardot, 1941a. Contribution à la morphologie du systeme nerveaux des Gasteropodes. Isolement, à l'état vivant, de crops neuroniques. *C. R. Seances Soc. Bio. Fil.* 135:965–968.

A. Arvanitaki and H. Cardot, 1941b. Les characteristique de l'activite rythmique ganglionnaire "spontanee" chez l'Aplysie. *C. R. Seances Soc. Biol. Fil.* 135:1207–1211.

134. A. Arvanitake, 1939. Recherches sur la réponse oscillatoire locale de l'axone géant isolé de 'Sepia'. *Arch. Int. Physiol.* 49:209–256.

135. E. R. Kandel, 1976. *Cellular Basis of Behavior.* W. H. Freeman and Co., San Francisco, p. 56.

136. Arvanitaki and Cardot, 1941a, 1941b, op. cit. (note 133).

137. A. Arvanitaki and S. H. Tchou, 1942. Les lois de la croissance relative individuelle des cellules nerveuses chez l'Aplysie. *Bull. Histol. Appl. Physiol. Pathol.* 19:224–256.

138. A. Arvanitaki, 1942. *Neurophysiol.* 5:89–108, p.89n. Somewhat earlier (1939) Berhnard Katz and O. H. Schmitt had completed a similar study on crab nerves and had published their findings in 1940. See B. Katz and O. H. Schmitt, 1940. Electric interactions between two adjacent nerve fibres. *J. Physiol.* 97:471–488.

139. Ross Harrison, 1907. Observations on the living developing nerve fiber. *Anat. Rec.* 1:116–118.

140. V. Hamburger, 1934. The effects of wing-bud extirpation on the development of the central nervous system. *J. Exp. Zool.* 68:449–494.

141. R. Levi-Montalcini, 1975. NGF: An uncharted route. In *Paths of Discovery*, op. cit. (note 121), p. 247.

142. R. Levi-Montalcini and G. Levi, 1942. Les conséquences de la destruction d'un territoire d'innervation peripherique sur le développement des centre nerveuz correspondants dans l'embryon de poulet. *Arch. Biol.* (Liege) 53:537–545.

143. R. Levi-Montalcini and V. Hamburger, 1951. Selective growth stimulating effects of mouse sarcoma on the sensory and sympathetic nervous system of the chick embryo. *J. Exp. Zool.* 116:321–361.

144. R. Levi-Montalcini and V. Hamburger, 1953. A diffusable agent of mouse sarcoma, producing hyperplasia of sympathetic ganglia and hyperneurotization of viscera in the chick embryo. *J. Exp. Zool.* 123:233–287.

145. R. Levi-Montalcini, H. Meyer, and V. Hamburger, 1954. *In vitro* experiments on the effects of mouse sarcomas 180 and 37 on the spinal and sympathetic ganglia of the chick embryo. *Cancer Res.* 14:49–57.

146. R. Levi-Montalcini and P. Calissano, 1979. The nerve-growth factor. *Sci. Amer.* 240:68–77.

147. S. Kopèc, 1922. Studies on the necessity of the brain for the inception of insect metamorphosis. *Biol. Bull.* 42:323–342.

148. C. C. Speidel, 1919. Gland cells of internal secretion in the spinal cord of the skate. *Papers from the Department of Marine Biology of the Carnegie Institution of Washington.* 13 (publ. 281):1–31.

149. E. Scharrer, 1928. Die Lichtempfindlichkeit blinder Elritzen (Untersuchungen über das Zwischenhirn der Fische. I.) *Z. Vergl. Physiol.* 7:1–38.

150. B. Scharrer, 1975. The concept of neurosecretion and its place in neurobiology. In *Paths of Discovery*, op. cit. (note 121), p. 232.

151. Ibid., p. 233.

152. B. Scharrer and E. Scharrer, 1944. Neurosecretion VI. A comparison be-

tween the intercerebalis-cardiacum-allatum system of the insects and the hypothalamo-hypophyseal system of the vertebrates. *Biol. Bull.* 87:242–251.

153. B. Hanström, 1941. Einige Parallelen im Bau und in der Herkunft der inkretorischen Organe der Arthropoden und der Vertebraten. *Lunds. Univ. Arsskr.* (N. F. Avd.) 37(4):1–19.

154. B. Scharrer, 1952, Über neuroendokrine Vorgänge bein insekten. *Pflügers, Archiv.* 255:154–163.

155. Scharrer 1975. op. cit. (note 150), pp. 231–243.

156. Ibid., p. 231.

157. Ibid., p. 233.

158. Ibid.

159. Ibid.

160. This concept was first introduced by Kaspar Bauhin (1560–1624), Joachim Jung (1584–1657), John Ray (1627–1705), and Joseph Pitton de Tournefort (1656–1708), who created a system of classification which was to reflect the natural relationships of living things (a *Systema Naturae* [1735–58] in which all plants and animals were classed according to their morphological similarities and differences). See Charles Singer, 1931. *A History of Biology.* Henry Schuman Inc., New York.

161. The following summary of Hildegard's life and natural history is taken from K. C. Hurd-Mead, 1938, *A History of Women in Medicine* (Milford House, Inc., Boston, 1973); Mozans, 1913, *Women in Science,* op. cit. (note 100); Charles Singer, 1913, essay on Hildegard in *From Magic to Science* (Dover Publications, New York, 1958); and E. Wasmann, 1913, Hildegard von Bingen als älteste deutsche Naturforscherin, *Biologisches Central Blatt.* 33:278–288.

162. Singer 1913, op. cit. (note 161); Wasmann 1913, op. cit. (note 161).

163. Roger Bacon (1214–1294) and Albertus Magnus (1206?-1280) did not come upon the scene for another 100 years and even then only Albertus rivals Hildegard's studies in natural history.

164. Hurd-Mead 1938, op. cit. (note 161), p. 187.

165. Wasmann 1913, op. cit. (note 161).

166. Ibid.

167. Lynn Thorndike, 1923. *Magic and Experiemental Science.* Columbia University Press, New York.

168. Singer 1913, op. cit. (note 161), p. 236.

169. Mozans 1913, op. cit. (note 100).

170. *Notable American Women,* 1607–1950. pp. 357–358.

171. Ibid., 505–506.

172. Clark Elliot, 1979, *Biographical Dictionary of American Science 17th–19th Century,* Greenwood Press, Westport, CT.

173. M. H. Morris, 1846. Letter in *Proc. Acad. Nat. Sci. Phil.* 3:132–134.

174. M. H. Morris, 1841. Observations on the development of the Hessian Fly. *Proc. Acad. Nat. Sci. Phil.* ??:66–88.

175. Mozans 1913, op. cit. (note 100), p. 251.

176. *Who's Who in Science* 1968.

177. C. Royer, 1877. Deux hypotheses sur L'heredite. *Revue D'Anthropol.* 6:443–484, 660–685.

178. Mozans 1913, op. cit. (note 100), p. 246.

179. Anna Botsford Comstock, 1953. *The Comstocks of Cornell: John Henry and Anna Botsford Comstock.* Comstock Publ. Assoc. Ithaca, N. Y. See also *Notable American Women,* 1971, p. 367–368.

180. John Comstock, 1880. *Report of the Entomologist.* U.S. Dept. of Agriculture.

181. John Comstock, 1888. *An Introduction to Entomology.* Comstock Publ. Co., Ithaca, N. Y.

182. Anna Comstock, 1939. *The Handbook of Nature Study.* Comstock Publ. Co., Ithaca, N. Y.

183. *Notable American Women, The Modern Period,* p. 139.

184. R. Carson, 1943. *From from the Sea: Fish and Shellfish of New England.* U.S. Dept. Int. Fish and Wildlife Service No. 33.

R. Carson, 1944. *Food from the Sea: Fish and Shellfish of the South Atlantic and Gulf Coasts.* U.S. Dept. Int. Fish and Wildlife Service No. 37.

185. R. Carson, 1941. *Under the Sea Wind—A Naturalist's Picture of Ocean Life.* Oxford University Press, New York.

186. R. Carson, 1951. *The Sea Around Us.* Oxford University Press, New York.

R. Carson, 1955. *The Edge of the Sea.* Houghton Mifflin Co., Boston. Both books were widely acclaimed. *The Sea Around Us* won the John Burroughs Medal and the National Book Award. Interestingly, one reader wrote: "I assume from the author's knowledge that he must be a man." See *Notable American Women: The Modern Period,* p. 139.

187. R. Carson, 1962. *Silent Spring.* Houghton Mifflin Co., Boston.

188. Although the argument between environmentalists and government/industrialists continues today, it is no longer a question of whether industrialization and technology cause damage to the environment, but rather of how much damage is acceptable.

189. *Notable American Women, The Modern Period,* p. 140.

I would like to thank the Marine Biological Library at Woods Hole, where most of the research for this chapter was done. D. A. Nash contributed essential portions of the research for this chapter which was done at the University of Rhode Island Library and I thank her for many fruitful conversations.

PATRICIA FARNES

Women in Medical Science

In this chapter I will consider some contributions of women in the evolution of the discipline of medical science, and, where applicable, the roles of women as recipients of the developments of this discipline. It will be useful to pose some questions in this introduction to outline the scope and emphasis of that framework.

Is medicine a discipline whose conformation might have evolved differently if women and women's perspectives had been integrated? It is especially important to pose this question in relation to medical science, because the answers may turn out to be different and more complex than they are in relation to some of the other disciplines analyzed in this book.

It is tempting to respond, "of course!" There are numerous fields within medicine in which a female perspective might theoretically have altered directions and priorities of research; where scientific outcomes and their applications to applied clinical practice might have been different. Obvious examples include research about contraceptives, mental health, childbirth technology, and identification of missing information about female sexual physiology.[1] In fact, would not the whole attitude of women's sexuality and its disorders have been influenced by the participation of women at policymaking levels or in research positions of stature? The threads of some possible answers to these questions will appear in the subsequent discussions, which address three main areas:

(1) I will define some of the major advances in medicine, and determine whether women were there. If they were not, I will analyze reasons for their absence. If they *were* there, I will attempt to define whether their presence was important, and whether their contributions bore any relationship to their gender.

(2) Who are the women whose contributions were rediscovered, attributed to others, or essentially ignored? Rather than simply listing various women who did anything at all,[2] I will try to evaluate how important their discoveries were, what prevented their recognition, and what were the modes of their burial. This approach remains relevant because burial persists in our time and has consequences for the individual scientist far transcending the question of who gets credit.

(3) In the course of these discussions, I consider ways in which women's

perspectives might have changed the directions of the science, and analyze what happened when women were present or absent. These reviews will be more or less chronologically based.

In looking at the roles women have played in the past, we have to recognize that we will never identify the woman who looked up from a laboratory bench one day, muttered "Hmmm . . . paradigm shift" to her boss, and went back to work, unsung. In any event, as we shall see, it is far more likely these days that the muttering occurs over the dinner table between husband and wife—not necessarily in that order.

HISTORICAL PERSPECTIVES IN THE EVOLUTION OF MEDICAL SCIENCE

It would be difficult to define any particular moment in time when medicine as a trade, or an art, or a lay societal activity, became medical *science*. In early times, the vast majority of medical activity was purely and primitively empirical. A rather impressive number of physicians during this period were women.[3] This was true in France, in Italy, and in Greece, but the kinds of contributions these women made were almost entirely without lasting influence. Indeed, this was also true of most medical accomplishments, whether by men or by women, during the same period. A probable exception was Trotula, or Dame Trot, an eleventh-century gynecologist to whom many writings were attributed.[4]

Probably we can define medical science as part of our present continuum from about the beginning of the Renaissance, when Vesalius clarified some basic features of anatomy, Harvey described circulation of the blood, van Leeuwenhoek's discovery of the microscope was applied to cells and tissues, and diagnostic instruments (such as the thermometer) were designed. At these stages, women were not there. At least they were not there at the professional levels. Sometimes they were there as lay healers, who contributed important and exciting insights (some of which were developed at the professional level by men).

EARLY CONTRIBUTIONS OF UNKNOWN OR LITTLE-KNOWN WOMEN

I have identified a number of prototypic historical examples in this early period of the development of medical science in which women unquestionably played roles of sorts but were not in positions to develop their contributions. We will look at three such contributions: (1) the efficacy of onions in inflammatory reactions, (2) the discovery of digitalis, and (3) the development of smallpox vaccination.

The first example is a model of how a folk remedy was identified and then used extensively in Renaissance times. The explanation for its usefulness became clear four hundred years later. In 1552, Ambroise Pare was a surgeon in service in the French army.[5] He was a nonacademic, pragmatic practitioner who was willing to try any or all new treatments to facilitate wound healing. When an old woman advised him to apply raw chopped onions to burns, he performed a controlled experiment on a badly burned scullion. His results confirmed that the treatment was successful, and he continued to use it, even though it was contrary to Galenic doctrine. In 1952, it was discovered that onions contain an antimicrobial principle, which may have been responsible for their healing effect.[6] Complications of burns are now known to be in a large measure the result of bacterial superinfection.

In 1766, William Withering heard that an old woman from Shropshire had a secret remedy for the cure of dropsy, and he approached her to obtain the recipe.[7] The old woman's name has been variously given as Mother Hutton or Mrs. Hutton; she enjoyed considerable success in treating patients with fluid accumulation (edema) who had not responded to other treatments. Mother Hutton's recipe contained a number of herbs, and Withering suspected from the beginning that foxglove (*Digitalis purpura*) was responsible for the curative power of the recipe. Withering was ultimately responsible for standardizing digitalis preparations, working out dosages, and studying side effects. He published a definitive study on the usefulness of foxglove for heart failure.[8] Withering was not a simple opportunist. He was an accomplished botanist and, in 1776, published "A Botannical Arrangement of All the Vegetables Naturally Growing in Great Britain with Descriptions of the Genus and Species According to Linnaeus."

During the 1700s, milkmaids in England were well aware of the protective power of cowpox infection in conferring immunity to smallpox.[9] Edward Jenner's attention was drawn to this when, during his apprenticeship, one milkmaid was queried about smallpox and she said, "I cannot take it, for I have had cowpox." Later, in 1796, a dairy maid (Sarah Nelmes) appeared with cowpox, and Jenner made a test which would have violated every code of patient's rights as we endorse them today. He used the material from her wound to inject a boy, James Phipps, and found that later exposure to smallpox failed to result in infection. It was, of course, the beginning of vaccination. However, even Jenner's work represented a sort of rediscovery, since the concept of vaccination was intimately related to the variolation carried on earlier in Asia, the Middle East, and Europe. In this procedure smallpox debris which had been attenuated was used. The crucial difference between Jenner's work and the early variolation lay in the safety of the former as compared with the latter, which sometimes produced smallpox, rather than immunity to it. Jenner's method, in contrast to the earlier variolation techniques, provided security that the disease of smallpox itself would not occur after vaccination.

We can ask, were there any women physicians in these years who made contributions? There were probably many, but only a few survived in citation. In France, Louise Bourgeois was noted (in 1600) to be a prominent and respected physician who was "the first to describe the treatment of chlorosis with iron."[10] Chlorosis was the term for a severe iron deficiency anemia particularly prevalent among adolescent girls. Reference to this disorder continued into the nineteenth century medical literature.

It is important to appreciate, in terms of lost women, that in an overview of the development of medical history from the sixteenth century to the twentieth, literally an entire perspective of women's medicine became lost. Women's medicine was herbal medicine, which flourished in early Renaissance times, and was gradually displaced by an evolving system of bleeding and purging, which persisted as the dominant therapeutic mode into the nineteenth century.[11] In seventeenth-century England, "from 8s to 10s a lb Herbs could be bought from the physical herb-women in Newgate Market or Covent Garden."[12] Movements toward the new heroic medicine were pondered by some authorities, who wondered if herbal medicine didn't result in unexplained vicissitudes. In the late seventeenth century, Robert Boyle suggested that valuable clues to the treatment of disease might be obtained from "Midwives, Barbers, Old Women, Empiricles, and other illiterate Persons."[13] Two of the dissenters from the new medicine founded splinter sects which retained reliance on herbals. Samuel Thompson, who discovered the medical uses of lobelia and developed a whole herbal school, learned his basics from the widow Benton. John Shelton, another herbalist of the early 1800s, learned plant medicine as a child, from his grandmother.[14]

WOMEN VERSUS MEDICAL SCIENCE IN THE 1800S

By the nineteenth century in the United States, women were effectively almost completely excluded from medical science with respect to caregiving or potential academic scholarship. The domination of male physicians in the obstetrical-gynecological area had led to the displacement of midwives.[15] The persecution and extermination of women healers as witches, and the establishment by Florence Nightingale of nursing as a suitable role for women in medicine led to the relegation of women to subsidiary roles.

FLORENCE NIGHTINGALE

It is necessary to take a byroad to consider the impact of Florence Nightingale (1820–1910) on the evolution of the medical profession during the nineteenth century. Far from being *only* the dramatic "lady with the lamp" during a devastating year in the Crimea, or *only* the founder of modern nursing as an outgrowth of those experiences, she was also a visionary who pioneered the use of statistical approaches to the health problems of populations.

Florence Nightingale was born to a wealthy English family and educated at home. In her early twenties, and after refusing two marriage proposals, she relentlessly (to the horror of her family) pursued training and experience in nursing. Nursing was not a profession in the early 1800s, and nurses were drawn from the lowest social classes. They were not educated and not respectable. Nightingale nevertheless felt that she had been informed by God that her destiny was to develop new dimensions in nursing.

In 1854, England and France entered the Crimean War against Russia; the British Army was decimated and nearly destroyed by epidemics of cholera and typhus. The hospitals were filthy and wholly unequipped. Nightingale, invited to head a team of nurses, revolutionized the woefully inadequate system, and thus became confirmed in her destiny. On her return to England, she provided the data on which sanitary reform was later to be based.

For the next twenty years, Nightingale's activities were largely conducted from bed, or from as much seclusion as she could manage. In the course of her invalidism (interpreted by Pickering as psychoneurosis),[16] she did the following: founded the Nightingale School of Nursing at St. Thomas's Hospital (1860); consulted about and interpreted statistics relating to health in India; effected Poor Law Reform with respect to workhouse conditions; dealt with the problems of puerperal sepsis in lying-in institutions. In all of these instances, Nightingale introduced the application of statistics to the solutions of problems of populations.[17]

When she was about sixty, Nightingale reemerged into the world and worked until her death.

One of the interfaces of Nightingale with the medical world has only recently been analyzed by Monteiro.[18] As Monteiro points out, Nightingales's interaction with Elizabeth Blackwell (1821–1910; America's first female physician) may have "critically influenced the role of women in medicine." These women met and corresponded during the 1850s. There was a brief period in which they explored the possibility of establishing a program to train women in medicine. But the goals of the two women, as Monteiro points out, were wholly different. They could not agree on the nature or venue of their proposed institution and finally fell away from each other.

There are fleeting references again and again in accounts of Nightingale's life to her passion for and dedication to statistics as an approach to her life's mission. I. B. Cohen has recently emphasized Nightingale's contributions to population statistical analysis in terms of health care.[19] Hers was a step-by-step descriptive approach. At Scutari, she "systematized the chaotic record-keeping procedures." Later, with the help of William Farr (a professional statistician), she developed a more global approach to the analysis of the data, which would have a significant impact on public health. According to Cohen, "she invented polar-area charts, in which the statistic being represented is proportional to the area of a wedge in a circular diagram." Nightingale made people see disease, mortality, and insufficiency in health care by numbers. She made them, by graphs and charts, envision it. She made them look at who was dead, from all who had been alive, and at the reasons for the mortality.

NINETEENTH-CENTURY PROBLEMS OF WOMEN IN MEDICINE

The societal values in operation to exclude women from professional and academic pursuits may have had an even more profound effect on women in medicine than in certain other professional fields. This is because women's presumed physiology precluded their participation in such an indelicate and demanding field. The popular nineteenth-century concept that the ovaries were master organs of women governing all other body systems (especially the brain) constituted one reason why women failed to gain access to education. However, even more compelling arguments were put forth in support of excluding women from the practice of medicine. Ehrenreich and English cite the so-called romanticist argument that women were unfit for the profession.[20] They quote Dr. Augustus Gardiner, an American gynecologist, who stated in 1892, "More especially is medicine disgusting to women accustomed to softnesses and the downey side of life . . . fightings and tumults, the blood and mire, bad smells and bad words, and foul men, and more intolerable women she but rarely encounters. . . . "[21]

The crowning argument against women entering medicine was that women, whose hormones dictated all of their reactions to life's events, could not be responsible in matters involving life and death.

Some strong voices were raised in protest to this position. For example, Dr. Mary Putman Jacobi (1842–1906), buttressed by her prestigious husband, Abraham, won a prize for her essay on "The Effects of Rest on Menstruation."[22] Her conclusions, based on a large survey of women's experiences, did not support the prevailing views that menstruating women need be indisposed. Incidentally, the essays submitted in the contest were not identified by gender until after the judgment of the prize award.

During the 1800s, when women were fighting to gain admission to medical schools both here and abroad, the major advances in medicine concerned pathology, physiology, and identification of specific diseases, especially those of infectious nature. The women who practiced medicine during this period, and who tried to pave the way for their sisters who followed them, rarely achieved positions where scholarly activities were possible. Medical society memberships, university faculty appointments, and research opportunities were almost wholly out of reach. It is true that in 1899, Louise Taylor-Jones, a graduate of Wellesley College and Johns Hopkins Medical School, taught science at Columbia University; however, she was never paid, and when her name did not appear in the catalogue she dropped out as an instructor.[23]

A rare reference to a woman in medicine in this period was to Augusta Klumpke (1859–1927),[24] who described a new disorder, called Klumpke palsy, which resulted from injury to the lower brachial plexus nerve. The result is partial paralysis of the arm. Klumpke was born in the United States, but was educated in Switzerland and Paris after the usual difficulties in gaining admission to American medical schools in the 1870s. She and another woman, Blanche Edwards, ultimately qualified and, after almost incredible struggles, obtained appointments for postgraduate training. Later married to Jules Déjerine, Augusta collaborated with him on clinical neurologic research, and textbooks. For her 1885 paper on brachial plexus injury, she received the prize of the Academy of Medicine.

QUANTUM JUMPS AND WHO DID THE JUMPING

The turn of the twentieth century also represented the beginning of a sort of golden age of opportunity for women in academic roles in medical science. This is true even though women made up only 5 percent of full professorships in medical schools, and 15 percent of total medical school faculty members up to 1980.[25] During the early 1900s about a dozen women made substantial contributions which are still cited (sometimes even celebrated) today.

FLORENCE SABIN

Florence Rena Sabin (1871–1953) has been hailed as the outstanding woman scientist in the medical field in the first half of this century.[26] Dr. Sabin first found science intriguing at Smith College, where she graduated in 1893. She learned of the new Johns Hopkins medical program from the Smith College physician (a woman). In 1896, after three years of teaching at a boarding school to accumulate the necessary funds, Sabin entered the fourth class at Johns Hopkins. There were fifteen women among the forty-two class members. Sabin was attracted from the beginning toward research and published her first paper, relating to nuclei of cochlear and vestibular nerves, as a second-year student. In her fourth year she constructed a model of the brainstem in newborns, and, subsequently, an atlas of the medulla and midbrain. Her experiences that year crystallized her commitment to the laboratory, rather than to clinical practice.

Her first research efforts were in a controversial field: the origin of lymphatic vessels. By using the ingenious new approach of injecting lymphatic channels with india ink, she was able to demonstrate the derivation of lymphatic vessels from the venous system. During this time she became an academician and teacher and, in 1907, was appointed associate professor of anatomy. Her lymphatic work provoked considerable controversy, but was ultimately acclaimed as a highly significant contribution. Other important contributions by Sabin included the development of supravital staining techniques (for living cells), and the identification of the monocyte as a definitive type of white blood cell.[27] Sabin's interpretations about monocytes were only partly correct. She proposed a precursor cell, the monoblast, and this remains essentially correct. She proposed further that these cells might be derived from cells of blood vessels; this latter proved not to be the case (they are derived, instead, from bone marrow stem cells).[28]

In 1917, Franklin Mall died; he had been Sabin's mentor and friend, and head of the anatomy department at the Johns Hopkins Medical School. Sabin would have been the logical candidate for promotion to the position, but a man was appointed instead. However, shortly thereafter, she became the first woman to receive a full professorship with the title of Professor of Histology. In 1925 she was invited to the Rockefeller Institute, where she worked over the next decade on the pathogenesis of tuberculosis.

A curious career dichotomy took place in her later years, following retirement from the Rockefeller Institute.[29] Florence Sabin returned to her home state of Colorado to make her home with her sister. By the mid 1940s, she was deeply involved in the public health concerns of that state. Appointed to a state subcommittee on health, she mounted a successful attack on antiquated public health laws, and ultimately mobilized the public and the legislature to correct deficiencies in infectious disease control, milk control, and sewage disposal. The Sabin Health Bills dealing with these problems were enacted in 1947. Florence Sabin was the first woman to be elected to the National Academy of Science. In the course of her scientific career she received numerous honors, awards, and prizes, of which the most prestigious was the Lasker Award in 1951.

Florence Sabin was not the only outstanding woman in medical science who devoted a substantial portion of her career (in this case, her second career) to social and public health concerns. Many problems in these areas did not attract the attention and interest of men but were somehow considered natural for women. Indeed, the women who achieved the most ongoing recognition in the field of medical science were those who devoted some, or all, of their careers to areas where social concerns formed an integral part of medicine.

DOROTHY REED

A classmate of Sabin's, Dorothy Reed (1874–1964), became more formally memoralized when the characteristic cell of Hodgkin's disease, which

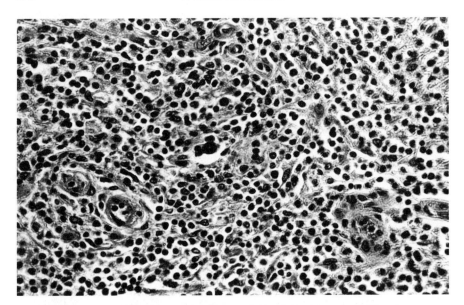

Lymph node of Hodgkin's disease, with Reed-Sternberg cell. The characteristic cell is the large dark one surrounded by the white space. Photo by P. Farnes.

she described in 1902, was named for her; that is, it was co-named the Reed-Sternberg cell (less often called the Sternberg-Reed cell).[30] How did Dorothy Reed manage to come out on the left-hand side of the hyphen most of the time when Carl Sternberg had described the cell four years earlier than she did? The answer probably lies in the fact that although he provided the first authentic and detailed description, he interpreted the cell as part of the expression of tuberculosis (which it is not), while Reed correctly interpreted her findings as an entity distinct from the disease. It is not enough to see something in a microscope. In discovery the conceptual integration is what counts. The presence of Reed-Sternberg cells in lymphoid tissue remains pathognomonic of Hodgkin's disease; they are now believed to be related to mononuclear phagocytes.[31] Dorothy Reed had made a brilliant beginning as a pathologist, even before she began her medical career. So it is especially interesting that Dorothy Reed (later Mendenhall) turned her career to problems of social concern, especially in feminist directions. In 1926 she authored a comprehensive and authoritative government publication on the importance of milk as child's food.[32] In the same year she traveled to Denmark and returned to advocate midwifery in this country. In Reed's case, a direct relationship between her experiences as a woman and her career direction seems inescapable since she "lost her first and only daughter to poor obstetrics."[33]

ALICE HAMILTON

Some fields in medical science were pioneered by women. The earliest example is industrial medicine, which did not exist as a distinct entity until about 1910. At the forefront of its development was Alice Hamilton (1869–1970), who has been termed the First Lady of industrial medicine.[34] Indeed, she could have been cited as the First *Person* of industrial medicine. Alice Hamilton received her medical education at the University of Michigan, and graduated in 1893 as one of thirteen women in a class of forty-seven. Her ongoing association with Jane Addams and Hull House gave her interests an increasingly social focus; she established medical education classes there, as well as a well-baby clinic. When a typhoid fever epidemic occurred in Chicago in 1902, Hamilton identified the relationship between ineffective sewage disposal and the role of flies in the transmission of disease. Her findings led to investigation and restructuring of the Chicago Health Department. Soon she became interested in problems relating to industrial hygiene, especially lead poisoning. Her efforts led to an improvement of working conditions in factories where lead dust was a serious hazard.

Jane Addams, writing in *Twenty Years at Hull House*,[35] gives an account of some of Dr. Hamilton's activities relating to the epidemic, and to the connection between housing conditions and tuberculosis in the area. As regards the fly transmission of typhoid fever, "her researches were so convincing that they have been incorporated into the body of scientific data supporting that theory. . . . " Hull House was a veritable hotbed of instigation of public health reforms at different levels while Dr. Hamilton was there. These public health issues included filthy conditions of food preparation, druggists who sold cocaine to minors, working conditions of factory girls in relation to tuberculosis infection, and child-labor laws. It was within this framework that Alice Hamilton developed her own background for her future work.

In 1911 she joined what would become the Bureau of Labor. Her activities addressed the hazards of lead, arsenic, mercury, organic solvents, and the radium used in watch dial manufacture, among others. By 1918, Harvard University concluded that Dr. Hamilton was the best qualified candidate for a new faculty position in industrial medicine in their School of Public Health. She was offered the position, with three stipulations. She might not use the Faculty Club, she would have no access to football tickets, and she was not to march in the commencement procession. Ultimately, she co-authored *Industrial Toxicology*, a comprehensive treatise of industrial hazards which is currently in its third edition.[36] At her death in 1970, she had only achieved the faculty rank of Assistant Professor Emeritus of Industrial Medicine. She attributed some of her success and opportunity to the fact that the field was new and of limited interest to men.

ALICE EVANS

At the outset of her career, Dr. Alice Evans (1881–1975) began investigating brucellosis. Brucellosis is a disease of protean manifestations, caused by any of three types of bacterial organisms designated *Brucella*. The first type was isolated in 1887, the second in 1897, and the third in 1914. Transmitted by milk from affected animals, the disorder is characterized by fever, arthritis, effects on the reticuloendothelial system, and acute and chronic illness. As cited in Davis et al., "only when Alice Evans of the U.S. Department of Agriculture recognized in the 1920s that all three of these organisms were closely related morphologically, antigenically, and metabolically, were they placed in a separate genus. . . . "[37] Pasteurization of milk has dispelled the risks of brucellosis in Western countries (Dr. Evans had insisted that it would), but the disease still exists in Third World nations.

WINIFRED ASHBY

Among the lasting contributions made by women to medical science before 1930 is the Ashby technique, an especially ingenious technique for the determination of red blood cell survival time in the human body. In 1919 Winifred Ashby (1879–1975) described her method, which depended upon mixing innocuous but different types of red blood cells in a patient and following the rates of disappearance of the respective cell types over time.[38] This technique is still described in current hematology texts. Ashby received her Ph.D. from the University of Minnesota. By 1921, she had studied more than one hundred patients. Decay states, "She was ahead of her time; her papers remained on library shelves largely unread and her technique was relatively unused until the late 1930s."[39] She died at age 95. She is remembered now, but much of her career was spent in obscurity.

MAUDE SLYE

Maude Slye (1879–1954) was a well-known cancer researcher in the early 1900s. The field was truly pioneer territory, and her contributions were extraordinarily important. However, the dogma she derived from her work proved to be erroneous and simplistic. In some sense Maude Slye was a victim of her time. She graduated from Brown University in 1899 and took her doctorate at the University of Chicago, where she became director of a cancer research unit. She wrote dozens of important papers about the spontaneous occurrences of mouse tumors, in collaboration with two colleagues; these served as the basis for further mouse tumor genetic studies developed later by scientists who disagreed with her conclusions. Slye developed a theory that cancer had a simple hereditary basis, and that this might be applied to prevention in humans; this, of course, proved to be untrue. We now understand that hereditary factors are important in some forms of cancer, but the role of heredity is complex. Maude Slye was awarded the AMA Gold Medal in 1914.[40]

FACTORS DETERMINANT TO SUCCESS

How did some of these early women make it when so many did not? Did they have anything in common? The backgrounds of the women who have achieved fame in the medical sciences are heterogeneous. They came from different ethnic, religious, and socioeconomic environments. However, is it coincidental that Sabin and Reed were classmates (first at Smith College and then at Johns Hopkins University), and that the Johns Hopkins medical program ultimately aborted Gertrude Stein, in her senior year, to a different sort of glory? In looking over the educational backgrounds of women who made it in medical science, one finds objective evidence that the type of collegiate and medical school experience of these women may have been a very important determinant of their academic productivity. I analyzed sixteen celebrated women doctors from the literature. The most prestigious and accomplished women scholars in the medical sciences attended women's colleges (9 of 16) and/or Johns Hopkins University School of Medicine (5 of 16).

The uniqueness of Johns Hopkins University arose from the terms of Mary Garrett's letter to the trustees,[41] which stipulated that in order to enjoy the donation which the school required to open its doors, women would have to be admitted "on the same terms as men." This resulted in substantial numbers of women in each class. There were fifteen women in a forty-two member class at Johns Hopkins during the year that Florence Sabin and Dorothy Reed matriculated. In the early twentieth century, nowhere else in the country was such a critical mass of women consistently present. However, Alice Hamilton was also one of thirteen women in a class at the University of Michigan. Tidball and Kistiakowsky have pointed out that women with undergraduate degrees from women's colleges may have more productive graduate careers than women from coeducational schools.[42] Analysis of the educational backgrounds of the most important women contributors in medical science suggests that the same factors may have been operative here. To the extent that critical mass is important, we should note that medical school admissions of women had risen to about 30 percent by 1976.[43] The Johns Hopkins percentage was, interestingly, in this range.

CICELY WILLIAMS

An outstanding discovery of worldwide significance was made by Cicely Williams (1893–   ), who spent twenty years working with natives in West Africa.[44] In 1931, she described a syndrome termed kwashiorkor (now termed protein-calorie malnutrition) among infants and young children in

Ghana.[45] As Darby has recently pointed out, "the importance of her early descriptions of this syndrome and related concepts cannot be overestimated."[46] Kwashiorkor "may well be the world's most appalling cause of early death and morbidity."

Cicely Williams was born in Jamaica in 1893, and after education in England and a diploma in tropical medicine and hygiene in 1929, she became the first woman to be appointed to the British Colonial Service in the Gold Coast. Her recognition of kwashiorkor as a distinct deficiency disease followed two years later and her findings were published in the *Gold Coast Colony Annual Medical Report* in 1931–32. In her next paper, which appeared in the *Archives of Diseases of Childhood* (in spite of reluctant editors), she suggested that an "amino acid or protein deficiency" might be responsible. It was a reasonable possibility, since kwashiorkor in Ga (Accram) means "the disease of the deposed baby when the next one is born," and is associated with weaning without adequate protein provision.[47] Writing in *Nutrition Review* in 1973 (on the occasion of the celebration of her eightieth birthday), she recalled receiving some "sharp rebukes." Most of these related to disagreement about kwashiorkor as an entity distinct from pellagra (niacin deficiency), and the verbal abuse extended to accusations of "carelessness and deceit."[48]

In the course of Dr. Williams's career, she produced more than fifty publications about nutritional problems and child health in developing countries. She spent part of the Second World War in a Japanese prison camp with her notes for a manuscript on rickets. Later, she became the first head of the Section of Maternal and Child Health of the World Health Organization. Ultimately, she was a full-time professor and consultant in international family health at Tulane University School of Public Health. Her philosophy is partially expressed in a quotation from a 1953 paper: "We worry a great deal about the persons we want to liberate from political tyranny, and we ignore those we could and should liberate from the tyrannies of dirt, ignorance, and hunger."[49]

MARY WALKER

In 1974, an obituary appeared in the *Lancet*[50] for Dr. Mary B. Walker (1892–1974; not to be confused with Dr. Mary E. Walker of Civil War fame, who dressed in men's clothes). The obituarian cited Dr. Walker's discovery of physostigmine as a treatment for myasthenia gravis in the 1930s as "the most important British contribution to therapeutic medicine up to that time." Myasthenia gravis is a relatively uncommon disease characterized by unusual fatigability of voluntary muscles. It is due to a lesion at the myoneural junction, and current evidence supports the character of the lesion as autoimmune. That is, the body manufactures antibodies against the postsynaptic acetylcholine receptor; the coating prevents the normal activation of the muscle by the nerve. In Walker's original communication

Mary B. Walker, M.D. Courtesy
of Pamela Furtek.

A. *Curve obtained from*
Myasthenic Muscle

B. *Curve obtained from*
normal Muscle

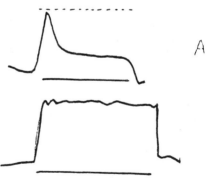

After an injection of
Prostigmin the Myasthenic
Muscle gives a Normal
Curve

Time marker $^1/10$
record

Frequency 900
per second

Diagram from Dr. Walker's doctoral thesis. Courtesy of Pamela Furtek.

in the *Lancet* in 1934 (an unassuming Letter to the Editor in which she thanked the medical superintendent of her hospital for permission to publish),[51] she compared the clinical features of myasthenia to the known effects of curare, and reasoned that physostigmine, a partial antagonist to curare, might benefit such patients. She reported a case of dramatic improvement with the drug, and wrote, "I think that this effect of physostigmine on myasthenia gravis is important . . . " since it supported the opinion that the fatigue was based on a myoneural problem, rather than being due to primary muscle disease. Her doctoral thesis, presented at Edinburgh University in 1937, was awarded a gold medal.[52] Later, she described a new disorder, periodic paralysis. Her obituarian puzzled why she should have been "relegated to such a lowly post at St. Francis Hospital." Mary Walker may have protested the system which denied her the opportunities to go on with her research (in fact, as her obituary indicated, she did protest, citing the poor treatment she received at one hospital), but the medical system did not encourage women, even when they made ground-breaking discoveries. Mary Walker had no mentors to nurture her unusual talents in scientific research. After her retirement she practiced medicine in Wigtown (Scotland), where she died at home at the age of eighty-two.

WOMEN IN PEDIATRICS

During the first half of this century, and not unexpectedly, some of the most outstanding conceptual clinical advances made by women were in pediatrics or closely related fields. This circumstance is undoubtedly related to the preferential distribution of women physicians in pediatrics which occurred early in the twentieth century and still occurs.

RUTH DARROW

An exciting advance in pediatric hematology occurred in the 1930s which would revolutionize the understanding of erythroblastosis foetalis (hemolytic disease of the newborn). Ruth Darrow (1895–1956) had personally suffered a series of stillbirths. In the *Archives of Pathology* in 1938 she provided a voluminous review of the clinical features, pathology, and familial aspects of erythroblastosis in the newborn, and analyzed all of the theories put forward to that time.[53] It was even unclear whether the basic pathogenesis resided in the mother, or in the infant. Then, as Nathan and Oski point out in *Hematology of Infancy and Childhood*, she "discarded all the current theories and concluded that the disease could only be explained as the result of maternal sensitization to an as yet unknown fetal antigen—a splendid example of the value of intelligent speculation."[54] Her hypothesis was subsequently confirmed and the unknown fetal antigen was the Rh factor. Today, immunization procedures have been developed to prevent the sensitization of Rh-negative mothers to Rh antigen present in fetuses.

MAUDE ABBOTT

Maude Abbott (1869–1940) received her training at McGill University, and published her first paper on functional heart murmurs (those not due to organic causes) in 1899. In the 1920s she presented a clinical classification of congenital heart diseases which distinguished between cyanotic (or blue-baby) and noncyanotic disorders, and this classification became the basis for the understanding of these diseases.[55] Abbott was applauded by her colleagues and students in extravagant terms for her abilities as a teacher and mentor of young people. Her work spanned the 1930s, and, at the time of her death in 1940, she left forty-eight contributions of journal articles and books relating to heart disease.

HELEN TAUSSIG

Helen Taussig (1898–1986), who attended Radcliffe College and Johns Hopkins Medical School, originated the notion for the "blue-baby operation," developed in 1945 as the Blalock-Taussig procedure (Blalock was the surgeon who performed it).[56] The procedure depends upon connecting the blood supply of the lungs to an arterial oxygen source, in children who developmentally lack the normal anatomical sources of such a supply. Blalock was elected to the National Academy of Sciences the following year, but Taussig did not receive her appointment as full professor until fourteen years after the development of the operation. Later, she was the first woman president of the American Heart Association, and she ultimately won her own election to the National Academy of Science.

HATTIE ALEXANDER

Another graduate of Johns Hopkins Medical School, Dr. Hattie Alexander (1901–1968), specialized in pediatrics and became famous for her development of an antiserum for *Hemophilus influenzae* meningitis.[57] She received the Oscar B. Hunter Memorial Award in 1961, and became president of the American Pediatric Society in 1964.

But women appeared to be absent from the biggest breakthrough of the 1940s, the discovery of antibiotics. Antibiotics are substances produced by living microorganisms which are capable of killing other microorganisms; the first was penicillin, produced by a certain mold, which could kill a variety of common disease-producing bacteria. Alexander Fleming, a British reseacher, discovered it, identified the mold, and followed through with the development of the process necessary to produce it. Brown has cited the use of penicillin by Elizabeth Stone "seventy years before Dr. Alexander Fleming discovered it." Stone treated victims of timber accidents in Teshigo, Wisconsin, and applied poultices made from moldy bread directly to the wound.[58] However, Stone was in good company with the Chinese of 3,000 years ago, and various male physicians over the ages, who used penicillin in a similar way. One of these was an American physician who, around the turn of the century, used mold but simply did not propagate his knowledge because he knew that it would be laughed at by his colleagues.[59] All of these people used mold which might or might not contain the active ingredient penicillin. The time was not right for the discovery of penicillin in any definitive sense. Until bacteriological techniques made it possible to identify the mold and exploit its possibilities in a systematic manner, its use was a matter of art, not science. In that sense Fleming was there on the right day, at the right time, with the right mindset.[60]

RECENT WOMEN IN MEDICAL SCIENCE

Over the past three decades, numerous women have made important contributions to the discipline; the ultimate significance of some of these will only be assessable over longer periods of time. During this century, women's contributions have been particularly impressive in biosocial spheres of medicine.

VIRGINIA APGAR

In 1953, a method of assessing the physical status of an infant at birth was developed.[61] The Apgar score, devised by Virginia Apgar (1909–1974), is so widely used that "every baby born in a modern hospital anywhere in the world is looked at first through the eyes of Virginia Apgar."[62] Apgar attended Mount Holyoke College and Columbia University, and went on to become professor of anaesthesia at New York Columbia-Presbyterian Medical Center. She was director of Medical Progress on Birth Defects for the National Foundation, March of Dimes. Her system for determining which infants need resuscitative efforts at birth uses a practical scoring method of heart rate, respiratory effort, muscle tone, reflex irritability, and color. It remains "the most simple and best way to evaluate the condition of an infant at birth." In addition to her development of the scoring system, she used it to compare the effects of various methods of anaesthesia on the scores of delivered babies, and showed that the method of anaesthesia used influenced the well-being of neonates.

THELMA B. DUNN

Thelma Dunn (1900–   ) became the leading authority in the world on the pathology of mouse tumors. She ultimately became the director of the Cancer Induction and Pathogenesis Section of the Pathology Laboratory at the National Cancer Institute. She received her M.D. degree from the University of Virginia Medical School in 1926, took postgraduate training at Bellevue, and interrupted her career until 1942 to have three children. After her retirement in 1970, she wrote a little-known book, *The Unseen Fight Against Cancer*, documenting historical milestones in cancer research.[63] Unable to find a publisher, she and her husband arranged for private publication.

KATHERINE SANFORD

Katherine K. Sanford (1915–) was the first person to clone a mammalian cell; that is, to isolate a single cell *in vitro* and allow it to propagate identical descendants. The recognition that this could be accomplished laid the groundwork for pure cell lines, with predictable metabolic and genetic properties. Cloning made possible the culturing of viruses, development of vaccines, study of metabolic disorders, and study of stem cells.

The technique initially used by Sanford proved cumbersome and hard to duplicate, but it opened the door for the development of easier ways. Sanford had been working in the National Cancer Institute for years, with Dr. Virginia Evans and a small group of tissue-culture workers who were trying to find predictable ways of invoking malignant transformation of mammalian cells *in vitro*. Prior to the development of a cloning technique, tissue and cell cultures were composed of a mixture of various body cells, some of which possessed the selective advantage of being able to outgrow others. Single cells separated out seemed to need other cells, or their products, to survive and grow.

Sanford considered whether a single cell might be able to survive and proliferate if it were kept in a closed environment in culture, where its own cellular products could not diffuse away from it into a large volume of medium. She developed micropipettes in which single cells could be picked up under the microscope, and then cultured a single cell in a fine microenvironment.[64] The successful outcome, a clone called L929 consisting of mouse fibroblasts, came after 929 attempts. The progeny were photographed emerging from the tube and the famous illustration has appeared on book covers and in books all over the world.

Sanford was born in Wellesley, and she and her two sisters attended Wellesley College. She received her doctorate in biology from Brown University, and then went to the National Cancer Institute, where she remained throughout her research career. She was one of the small group of women who emerged from the Brown University biology graduate program in the 1940s and continued on to distinguished careers.

BARBARA BAIN

Barbara Bain conceived and developed what we now term the MLC (mixed leukocyte culture) in the early 1960s. The technique is crucial to determining donor-recipient matches for organ or bone marrow transplantation. Her work formed the basis of her doctoral dissertation, and was done before the rapidly expanding research which has lead to our current understanding of human blood lymphocytes and their roles in immunologic processes.

Bain demonstrated that when the blood lymphocytes of one individual are mixed and cultured with lymphocytes of another individual, the lymphocytes become activated. The amount of activation can be measured by assessing the amount of DNA synthesis which occurs after five to six days *in vitro*. In addition, the cells show distinct morphologic changes which make them identifiable by microscopy. And the amount of activation is an index of the incompatibility of one set of cells with another. Thus, it can be used to predict the probability of tissue rejection by the donor (graft rejection), or of recipient cells by the transplanted cells (graft vs. host disease). Later refinements of the technique made it possible to do one-way measurements, that is, the cells of either person could be switched off so that they could not respond, and each set could then be tested separately against the other.

The work which initiated the recognition of the MLC in relationship to transplantation immunity was published in the journal *Blood* in 1964.[65] In 1979 it was published as a citation classic in *Current Contents*.[66] The paper had been cited more than five hundred times since 1964.

Bain says that she was "in the right place at the right time." A chance observation by Schrek and Donnelly[67] (not pursued by them) caught her attention and she rapidly established that the MLC phenomenon was due to complex genetic differences between donors. Bain also said that her important discovery was by no means the highlight of her scientific career.[68] Bain was born in Montreal and attended McGill University. At the affiliated Royal Victoria Hospital she carried out the work as a graduate student that led to the mixed leukocyte culture technique. But she was always, she says, much more oriented to teaching and zoology than to medical biology. She is now at the zoology department at the University of Western Ontario, where she is director of Electronmicroscopy and participates in a variety of teaching and research activities.

Mary Lyon

Although the existence of sex-linked diseases has been known since biblical times, the concept of how such diseases were genetically transferred awaited technologic developments which allowed visualization of human chromosomes. Hemophilia is a prototypic example of a sex-linked disorder. In this disease the affected males have a profound deficiency in a substance called Factor 8, which is necessary for blood clotting.

After it was established that females have two X chromosomes, while males have an X and a Y, it was soon appreciated that sex-linked meant X-linked; that is, the gene for the disease is present on one of the X chro-

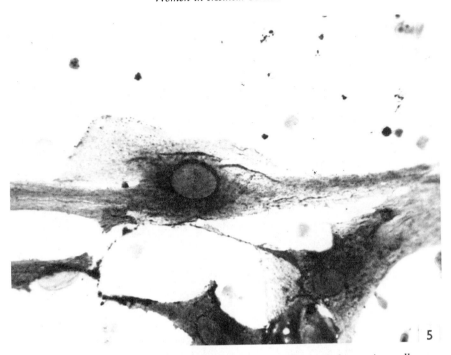

Large activated lymphocytes from a mixed leukocyte culture. A small unactivated lymphocyte is also present. Photo by P. Farnes.

mosomes of the mother. If that chromosome is passed on to the male child the disorder will occur. The mother who is the carrier for the trait generally appears normal, although occasionally there are partial expressions of the disease in the carrier woman.

This variation and expression of X-linked disorders in carriers is explained by the Lyon hypothesis, which was put forward by Mary Lyon in 1956. In brief, it proposes that since females (*or* males) only need the direction of one X chromosome in each cell, early in development there is a random inactivation of one X chromosome in each cell of females.[69] If all of the disease-bearing X's were by chance inactivated, there would be no discernible abnormality (by a laboratory test for Factor 8, for instance) in the carrier. If all of the normal X's were by chance inactivated, the disease would be expressed in the carrier. In the usual case some cells inactivate the normal chromosome, while others inactivate the disease-bearing X, and—as in the case of hemophilia—partial expressions of the deficiency in the carrier females can be demonstrated by assaying Factor 8. According to McKusic,[70] as of 1978 more than two hundred X-linked disorders had been identified. Included are a number of nerve and muscle syndromes, immunodeficiency states, and the more familiar color blindness.

Lyon developed her hypothesis from initial observations that female mice bearing two different X-linked coat-color genes showed mottled coat colors, with discrete patches of each color randomly arranged.[71] A number of controversies have emerged over the years about various details of the hypothesis, including the time frame of X-inactivation in fetal development. However, according to Ford, "there seems to be overwhelming evidence that the hypothesis is largely true for most cells in most tissues, and at most times."[72] Probably the most important practical consequence of the Lyon hypothesis in medical application has been its usefulness as a basis for devising screening studies of putative X-linked carrier states.

Mary Lyon was born in Norwich (England), and received her degrees from Girton College, Cambridge. In the early 1950s she was on the Medical Research Council (MRC) Scientific Staff, Institute of Animal Genetics in Edinburgh. Currently, she heads the Scientific Staff MRC Radiobiology Unit Genetics Section, Harwell.

LOSS, MOURNING, AND BURIAL IN MEDICAL SCIENCE

Women have undergone burial by the evolving medical establishment, from the time when herbal medicine was replaced by the more aggressive and, subsequently, more technological medical approach. As women gained access to the profession, they were isolated from the mainstream. Finally, women as well as men are subject to circumstantial effacement, which need not be gender-determined.

Certainly analysis of Nobel Prizes in Medicine and Physiology could be taken as a general indication of important conceptual advances which ultimately affected the field in some positive way. There are exceptions, as in the case of the award to the discoverer of prefrontal lobotomy. From 1901 until 1947, no woman was a Nobel laureate in these fields.

During this period, some of the major medical discoveries included mechanisms of immunity, elucidation of metabolic pathways, advances in neurophysiology, and the discovery of sulfa drugs and antibiotics. In 1947, Gerty Cori (whose work is detailed elsewhere in this volume) received the Nobel Prize with her husband, Carl, for work on the synthesis and degradation of glycogen. And during the same year she was abruptly promoted from research associate to professor of biochemistry at Washington University.

Did men do all of the work which led to Nobel awards? While there is no currently available documentation of the participation of women in Nobel research, it is clear that women have been involved in some cases, and that their activities have not been generally appreciated. For instance, Dr. Mari Krogh (1874–1943) was a co-investigator in the research of her husband (who received the Nobel Prize for work on capillary structure in

1920).[73] According to Bourkes, Forssman (who pioneered cardiac catheterization in the 1920s) had a wife who was a close professional colleague; however, his paper on the technique had no co-author.[74] It could well be that male Nobelists tend to marry collaborating wives. Zuckerman,[75] in a recent analysis of the professional activities of Nobel Prize winners, laid no emphasis on such collaboration.

The comments of one Nobel Prize wife, Ruth Hubbard, are worthy of note. Professor Hubbard, who is the wife of George Wald and who contributed work which was a crucial part of the discoveries for which he received the 1967 award in physiology and medicine, remarked, "I didn't win the Nobel Prize for George—a lot of us, including George, worked together. Laboratories should get prizes—not people."[76] She has also pointed out that once an individual is singled out by the prize, the work of the other members of the laboratory and their history in the field are erased, whether they are male, female, wives, or collaborators.

However many women underwent interment by the system, even more, ultimately, became lost by the effects of the special insulation of women from the medical mainstream. These women were, in effect, erased from the literature by their colleagues. An extraordinary compilation of *Women in Medicine: A Bibliography of the Literature on Women Physicians*, provides over four thousand references about women physicians from antiquity until 1977.[77] An unsuspected mode of burial becomes evident from review of this literature. Numerous references to important contributions of women are included. However, almost all of these citations appear in journals of extremely restricted distribution and availability. The most common were the *Journal of the American Medical Women's Association* and its progenitor, the *Women's Journal*. As we might expect, the majority of the citations were not concerned with groundbreaking contributions to medical science. Most of them *were* citations of women who had done "anything at all" at a scholarly level. Or (and of real social significance), the very important contributions of women physicians at the clinical level, especially in public health and missionary work, were lauded. Our early women physicians tended to be social pragmatists. They may have known about each other, but the rest of the world did not. In a banding together in exclusion from the system, women accidentally developed burial rites for each other.

Some significant losses relating to the evolution of medical science are circumstantial. They can occur for a wide variety of reasons, can involve women or men, and are probably more common than we might suppose. Indeed, with the current proliferation of medical journals, literature, conferences, and scientific output, we may expect such losses to increase in the future. After all, Gregor Mendel was almost lost, because of his obscure status and the circumstances under which he reported his genetic studies. Earlier in the development of medical science, geographic isolation contributed to burials. Communication channels were limited; opportunities for cross-fertilization of ideas and discoveries were minimal. Sometimes two people or even three independently reported the same thing, but only one achieved the recognition. Chediak[78] (in Cuba) and Higashi[79] (in Japan) both reported a new syndrome, and each believed it was unique; now we refer to the Chediak-Higashi syndrome. Many of the hyphenated syndromes so common in medicine arose in such circumstances. If Chediak or Higashi had been a woman, we can only conjecture how it would have turned out.

Hsu[80] provides an interesting example of circumstantial loss of a woman's discovery, and analyzes the reasons for it. Eleanor Slifer, in 1934, discovered that hypotonic solutions can spread metaphase chromosomes. This observation ultimately proved fundamental to the development of human and mammalian cytogenetics. Her observation (which was only of incidental significance to her study of grasshopper embryos) appeared in a widely read zoology journal, but it was not until almost twenty years later that it was rediscovered and the information finally applied to mammalian cells. Hsu explains the burial by the failure of the discoverer to realize the potential application to chromosome cytology, and the insularity of cytogeneticists, who simply failed to pick up the notion. It seems likely that the gender of the discoverer was irrelevant to the loss.

Once in a while, there is a burial which carries an element of humor with it. Hsu cites a self-burial of another woman. The rediscovery of hypotonicity came about through a laboratory error when a technician prepared a hypotonic solution by mistake. Hsu says, "Even today, I would love to give a peck on the cheek to that young lady who made an important contribution to cytogenetics," but no one would admit the error, and "this heroine must remain anonymous."

CONSEQUENCES OF BURIAL

If a woman makes an important scientific discovery, and it is lost or attributed to someone else, what are the real consequences (quite aside from the personal frustration of denial of credit)? I believe the only people who really consider this question are people who have been through the experience. However, it may be years later (if ever) that the scope of loss is appreciated by the woman. I will approach this problem from the other side of the mirror by asking: what happens when someone makes an important discovery in medical science and gets full credit in our system? If we look at it this way, we find out—by default, as it were—what there was to lose.

Supposing that two women scientists made a discovery in 1962 which, a few years later, emerge ˙ as an important contribution to immunology[81] (in fact, they did, and one of the women is the author of this chapter).

What would normally accrue to any man (or woman) if the discovery were wholly credited and its significance appreciated? Some of the following: developmental potentials for the area, applications for fellowships, institutional or community enthusiasm and support, invitations to lecture and participate in the network, new professional contacts and collaborations, grants, job offers, and pleas for job offers from young scientists. In the instance cited, none of these was forthcoming (except, oddly, pleas for job offers). The discovery was brushed off by many investigators, but a very astute government-supported group with enormous resources grabbed the ball and ran with it. There was no contest—after all, they had sixteen postdoctoral fellows sitting in a row waiting for new things to flow in. Within a year they were giving interviews to the medical media about their discovery. All we could do was mutter darkly (not even into our beards) while the other groups became the lecturers, the grantees, and, almost always, the ones cited. A few years ago our discovery was resurrected and proper priority was established when an account of the actual circumstances was included by T. C. Hsu in his book *Mammalian and Human Cytogenetics*.[82]

In any event, loss of citation *per se* could be considered as the least important consequence, when women go unrecognized. The greater loss is of the potential for academic development which is not nurtured properly in such women. What more might Mary Walker have achieved in neuromuscular research if her discovery of physostigmine and periodic paralysis had led to the kinds of opportunities listed above, instead of to the obscure post in St. Francis Hospital?

IS THERE A WOMEN'S MEDICINE?

As one analyzes the outstanding contributions of women in medical science, the social direction of many women physicians' professional lives becomes increasingly apparent. The following examples are prototypic.

MARIA MONTESSORI

Probably the first woman physician who profoundly changed some biosocial direction by her work in this century was Maria Montessori (1870–1952), who is usually not even listed in biographical resources as a physician; she is listed as an educator instead. Dr. Montessori was born in 1870 to an upper middle class Italian family and entered technical school at age thirteen intending to become an engineer. While she was taking courses in mathematics, physics, and natural sciences, she determined to change to medicine. She was ranked high in her class in 1896 when she became the first woman to graduate from the University of Rome Medical School. Her medical school experience had been characterized by isolation and she "dissected a cadaver alone." Montessori entered psychiatric research and undertook to select patients from insane asylums for treatment in the clinic.

She became interested in the educational theories of Itard, Sequin, and Rousseau; she was especially interested in the potentials for education of the mentally retarded. In 1900 she became the director of a practice demonstration school for the retarded. Her principles were directed toward "utilizing each individual's native potentials to develop along his natural line." She described her success as follows: "I succeeded in teaching a number of idiots from the asylum, both to read and to write so well that I was able to present them at a public school for an examination."[83] She began to apply her dramatically effective methodology to the education of normal children in 1907.

Her whole concept of children was different from the traditional one of her times. She stated, "Education is a natural process which develops spontaneously in the human being." She took the position that education was acquired by experience, not by listening. Children, she believed, should be observed, rather than probed.[84] Her methods rapidly achieved international reputation and the support of many important well-known, public figures. Early, both the Nazi regime in Germany and Mussolini's Fascist government in Italy became interested in these methods, but by the mid 1930s, both had second thoughts and the system was forbidden in both countries. She spent the war years in India, returning to London in 1946; she was invited back to Italy to reestablish her teaching methods in 1947. The Montessori method was introduced in the United States early in this century, but did not achieve the expected popularity. However, when it was resurrected in the late 1950s, it was much more favorably received. There are now over 3,000 Montessori schools in the United States. Dr. Montessori died in Holland in 1952.

MARGARET SANGER

My students have said, "Yes, but what if a woman had been involved in the development of The Pill? Wouldn't she have been concerned about all of the problems?" The Pill has been one of the most important and controversial developments of our time, and a woman was certainly involved.

Margaret Sanger's name has been synonymous with the history of birth control in the United States.[85] Sanger (1883–1966) was a nurse who began writing about sex education matters in 1912. On a trip to France in 1911 she discovered that there was widespread contraceptive information in France and on the continent. She spent the next years in Europe researching and visiting various clinics. Throughout her career Sanger single-mindedly focused all of her energies on the dissemination of birth control information in this country, which made her involvement fundamentally different from the various socialist and anarchist concerns in behalf of the same issue. Over the years there were the establishment of clinics (with government resistance to the clinics), the distribution of pamphlets (with similar resistance), indictments, hearings, and jail. Ultimately, her principles prevailed and she was instrumental in the founding of Planned Parenthood. According to the Seamans,[86] Margaret Sanger was introduced to Gregory Pincus of the Worcester Institute and promptly undertook to support him in research which would eventually lead to The Pill. Sanger has been much maligned for her elitist attitudes on eugenics, for taking "what might have been woman-controlled methods" and turning them over to "middle-men or doctors," and for failing to foresee the disruptive consequences which could attend hormonal manipulations of healthy young women. However, she could not be held responsible for the disgracefully inadequate pretesting of The Pill, for production methods which allowed estrogen contamination, or for the proceedings of the first Pill conference in 1962 where the earliest warnings of thrombotic complications were whitewashed.

Elisabeth Kübler-Ross

Elisabeth Kübler-Ross was born in 1926 in Zurich. At birth, she weighed barely 2 pounds. In reflecting on her later career in psychiatry and thanatology, she has wondered if her precarious neonatal period influenced her interest in developing new concepts relating to death. "After all," she writes, "I was not expected to live."

After the Second World War, during which she had been a hospital volunteer, she decided to study medicine. She obtained her M.D. in 1957 from the University of Zurich. After her marriage to E. R. Ross, she came to the United States, took postgraduate training, and specialized in psychiatry. In 1965 she became an assistant professor of psychiatry at the University of Chicago Medical School.

In the same year some theology students from the Chicago Theological Seminary came to her proposing a thesis on death, but there was little hard data on which to build their project. Thus began interviews of terminally ill patients, which met with considerable resistance from the physicians in the institution. Ultimately, however, the seminars became an accredited course. An outgrowth of these experiences was her first book, *On Death and Dying*,[87] in which a number of stages (of emotional dynamics) are identified which patients may experience when terminally ill.

Two other books followed, exploring feelings of the dying and developing ways in which responses of care-givers and families could begin to deal with these expressed feelings.[88] Kübler-Ross identified an area of medical care which had been virtually ignored by others, and addressed problems which had gone largely unappreciated by the medical establishment. She became the first physician to introduce a perspective of dying patients' personal needs, both for the medical establishment and for the public.

MARY CALDERONE

Mary Calderone was born in New York City in 1904; her father was the internationally known photographer Edward Stiken. In 1925 she graduated from Vassar College, where she had majored in chemistry. She had a strong interest in dramatics and tried acting for three years, after which she decided she was not good enough. An early marriage ended in divorce and she ultimately decided to study medicine. She obtained her M.D. in 1939 from the University of Rochester and her Master of Public Health degree in 1942 from Columbia University. At Columbia she met Dr. Frank Calderone and they were married in 1941. Her professional activities centered around issues relevant to public health. From 1953 until 1964 she was medical director of the Planned Parenthood Federation of America. Her attention was increasingly drawn to matters relating to sex education and she was part of the group to establish the Sex Information and Education Council of the United States (SIECUS) in 1964. She advocates a continuum of sex education beginning at an early age and integrating—with the biologic facts relating to reproduction—fundamental concerns with human relationships and responsibilities. She has become a pioneer in the development of responsible sex education for our children.[89]

WOMEN AS RECIPIENTS OF
MEDICAL SCIENCE

Women have been the recipients, at a level unparalleled in any other scientific discipline, of the results from research in medical science. Many of the contributions of women as scholars in medical science have had little to do with any unique perspectives brought by them because of their sex. Lymphatics, malnutrition, heart disease, getting born, are common to people, not just to women. So, I'll return to a question posed at the beginning of this chapter. Would women's participation in medical science investigation make any difference in the conformation of the discipline? Part of the answer is: they *were* there, and it often did not make a difference. The other part of the answer is: they *are* there now, and it remains to be seen what the results will be.

There are numerous areas we could define in which the perspectives of women could have influenced the things that happened in research and medical science. The thing about medical science which makes it different from the other sciences in this book (except, to some extent, biology) is that the impact of scholarly activity in this field is transmitted on a *personal* level to all people. I will consider two areas where women's perspectives have been and are uniquely important: mental health, and obstetrics and gynecology.

Freud said, early on in his career, "the position of women cannot be other than what it is: to be an adored sweetheart in youth, and a beloved wife in maturity."[90] Evading a lengthy analysis of the dominance of Freudianism in the evolution of the psychiatric discipline, and the well-recognized results of it on women, I'll simply pose a question. What if women had been in on all of his theories? What if women had been integral to his propositions about feminine psychology, and clitoral-vaginal orgasms (or the significance of each)? If they had, would his conclusions have been the same, and would the dogma have been carried over subsequent generations? One might imagine that the answer to this last would be a resounding no. But the answer is yes. Women *were* there, and I do not mean simply as patients. Disciples and co-authors were there in profusion. From Anna Freud (his daughter) to Karen Horney, Helene Deutsch, and Marie Bonaparte, women contributed their perspectives to the emergence of Freudianism and to its propagation. What were those perspectives?

Largely, they echoed his, with certain diversions and variations. For Marie Bonaparte, who was not allowed to study medicine for fear of jeopardizing her marriage potential, "the hole was the symbol of womanhood . . . frigidity was provoked by vital fear of penetration." Similarly, "To become truly a woman, the girl must grieve for and accept the loss of her penis." Daly cites the "outstanding more-freudian-than Freud women analysts [who] included Helene Deutsch and Marie Bonaparte."[91] She quotes an example from Deutsch: "the theory that I have long supported—according to which femininity is largely associated with passivity and masochism—has been confirmed in the course of years by clinical observation." Those are the words of a woman who was there.

It is an inescapable conclusion that both the qualitative validity of women's perspectives and the recognition of those perspectives depend upon the political orientation and wholeness of the women involved. If women disciples of Freud agreed with him that clitoral orgasm is a sign of immaturity, if women gynecologists have agreed that menstrual cramps are psychosomatic, then we have to conclude that the women who were there did not change the evolution of the discipline. We thus have to ask another key question: is women's representation on the scene any guarantee of a representation of women's perspectives? The answer historically has usually been no. The paralysis of female perspectives and experience is self-evident historically. More currently, we have to recognize that women in medicine (including psychiatry) have traditionally accepted the stereotypic, upper middle class male value system from which most of them arose. Most of the scholarly contributions from women in medical science are nongendered, in terms of scientific merit or application to women's lives. And even those which are gender-related may not reflect reality.

The rebirth of feminism in the 1960s has provided a different perspective about the nature of mental health for women. What had been considered as the stereotypic mentally healthy woman was questioned, as exemplified by the studies of the Brovermans.[92] The research of Masters and Johnson[93] provided entirely different facts about women's sexuality than those assumed by the Freudians; so did the Hite Report.[94] Feminist therapy has been introduced, and we can anticipate that feminist perspectives will continue to influence the discipline.

Throughout modern medical history, women as patients have been subject to the male-dominated medical establishment, and particular impact at the establishment-patient interface has been and is experienced in the areas of pregnancy (and its prevention), childbirth, and gynecologic diseases. Certain women's perspectives have been important in the development of medical science relating to these areas. Darrow's conceptualization of erythroblastosis foetalis, Sanger's drive for contraceptive development, and Apgar's objective criteria for newborns are examples.

Currently, what can we say about whether the presence of women in the discipline at the investigational level matters, or will matter? For example, it has been pointed out that one of the most critical areas of ethical concern for women in the field of contraceptive development is that the scientists, researchers, developers, physicians, drug company executives, and vendors of contraceptives will never have to subject themselves to the very pills, devices, implants, and injections they are promoting. Of course they won't, unless they are women, and unless they are *concerned women*.

Would more women in medical science lead to different priorities, different utilization of investigational resources, or changes in attitudes of the medical establishment toward such matters as dysmenorrhea, styles of childbirth, male contraceptive development, or hysterectomy? Perhaps. And perhaps not. There is a popular notion that in medicine (in contrast to physics, or astronomy, or mathematics, for example) more women would invoke profound changes in the discipline. While this is possible, it is by no means necessarily the case. The number of women in medical science is less important than the political perspectives of these women and men in the discipline. As we have seen, the most well-known achievements by women in medical science have been in fields which were of minimal interest to men.

NOTES

1. P. Farnes and E. W. Schweers. Women and health care: A model for women's studies. *J.A.M.W.A.* 35:182, 1980.

2. G. Lerner. New approaches to the study of women in American history. *J. Soc. Hist.* 4:333, 1969.

3. I. Koprowski. Women in ancient medicine. *J.A.M.W.A.* 31:453, 1978.

4. S. M. Stuard. Dame Trot. *Signs* 1:537, 1975.

5. R. H. Major, *A History of Medicine*. Springfield, IL: Charles C. Thomas, 1954.

6. H. E. Sigerist. *The Great Doctors.* New York: Dover Press, 1971.

7. F. A. Marti-Ibanez. *Prelude to Medical History.* New York: MD Publications, 1961.

8. W. Withering. *An Account of the Foxglove.* Birmingham, England: M. Swimey, 1785.

9. Major, *A History of Medicine.*

10. S. L. Chaff, R. Haimbach, C. Fenichel, and N. B. Woodside. *Women in Medicine: A Bibliography of the Literature on Women Physicians.* Metuchen, NJ: The Scarecrow Press, Groliet Educational Corp., 1977.

11. B. Griggs. *Green Pharmacy: A History of Herbal Medicine.* New York: The Viking Press, 1982.

12. Ibid., p. 113.

13. Ibid., p. 124.

14. Ibid.

15. B. Ehrenreich and D. English. *Witches, Midwives and Nurses.* Glass Mountain Pamphlet no. 1. Old Westbury, NY: The Feminist Press, 1973.

16. G. Pickering. *Creative Malady.* New York: Delta Books, Dell, 1974.

17. I. B. Cohen. Florence Nightingale. *Scientific American* 250:128–137, 1984.

18. L. Monteiro. On separate roads: Florence Nightingale and Elizabeth Blackwell. *Signs* 9:520, 1984.

19. Cohen, Florence Nightingale.

20. B. Ehrenreich and D. English. *For Her Own Good: 150 Years of the Experts' Advice to Women.* Garden City, NY: Anchor Press, Doubleday, 1978. p. 57.

21. Ibid.

22. R. Truax. *The Doctors Jacobi.* Boston: Little, Brown, and Co., 1952.

23. Chaff et al., *Women in Medicine.*

24. R. Satran. Augusta Dejerine-Klumpke. *Ann. Int. Med.* 80:260, 1974.

25. C. Spieler. *Women in Medicine.* New York: Josiah Macy Jr. Foundation, 1976.

26. R. F. Hume. *Great Women of Medicine.* New York: Random House, 1964.

27. F. R. Sabin. Studies on living human blood cells. *Bull. Johns Hopkins Hosp.* 24:277, 1923.

28. L. Kass and B. Schmitzer. *Monocytes, Monocytosis, and Monocytic Leukemia.* Springfield, IL: Charles C. Thomas, 1973.

29. M. K. Phelan. *The Story of Dr. Florence Sabin: Probing the Unknown.* New York: Thomas Y. Crowell Co., 1969.

30. H. S. Kaplan. *Hodgkin's Disease.* Cambridge, MA: Harvard University Press, 1972.

31. D. Zucker-Franklin et al. *Atlas of Blood Cells, Function and Pathology.* Philadelphia: Lea & Febiger, 1981.

32. D. Reed-Mendenhall. Milk: The indispensable food for children. Washington, DC: Bureau of Publications #163. Government Printing Office, 1926.

33. Chaff et al., *Women in Medicine.*

34. W. R. Slaight. *Alice Hamilton: First Lady of Industrial Medicine.* Ann Arbor, MI: University Microfilms, 1974.

35. J. Addams. *Twenty Years at Hull-House.* New York: MacMillan Co., 1910.

36. A. Hamilton and H. L. Hardy. *Industrial Toxicology.* Acton, MA: Publishing Sciences Group Inc., 1974.

37. B. D. Davis. *Microbiology.* New York: Lippincott, Harper & Row Medical Division, 1968.

38. W. Ashby. Determination of length of life of transfusion blood corpuscles in man. *J. Exp. Med.* 29:267, 1919.

39. *In* M. M. Wintrobe. *Blood, Pure and Eloquent.* New York: McGraw-Hill Book Co., 1980.

40. "Sketch of Maude Slye," University of Chicago Archives, April 10, 1936.

41. "The conditions of Miss Garrett's Gift to the Medical School of Johns Hopkins University. *Boston Med. Surg. Journal* 128:71, 1893.

42. E. M. Tidball and V. Kistiakowsky. Baccalaureate origins of American scientists and scholars. *Science* 193:646, 1976.

43. E. Crovitz. Women entering medical college: The challenge continues. *J.A.M.W.A.* 35:291, 1980.

44. V. Gollanez. *Cicely A. Dally: The Story of a Doctor.* London, 1968.

45. C. D. Williams. Deficiency Diseases in Infants. Appendix E (Health Branch), Gold Coast Colony Annual Medical Report, 1931/32, p. 93.

46. W. J. Darby. Cicely D. Williams: Her life and influence. *Nutr. Rev.* 31:392, 1973.

47. C. D. Williams. Kwashiorkor. *J.A.M.A.* 153:1280, 1953.

48. C. D. Williams. The Story of Kwashiorkor. *Nutr. Rev.* 31:334–40, 1973.

49. Williams, Kwashiorkor.

50. Obituary: "Dr. Mary Walker," *Lancet* 2:1582, 1974.

51. M. B. Walker. Treatment of Myasthenia gravis with physostigmine. *Lancet* 1:1200, 1934.

52. M. B. Walker. A contribution to the study of Myasthenia Gravis. University of Edinburgh Library, 1934.

53. R. R. Darrow. Icterus gravis (Erythroblastosis) neonatorum. *Arch. Path.* 25:378, 1938.

54. D. G. Nathan and F. A. Oski. *Hematology of Infancy and Childhood.* Philadelphia: W. B. Saunders, 1974.

55. P. D. White. Maude E. Abbott, 1869–1940. *Am. Heart J.* 23:567, 1942.

56. H. B. Taussig. Diagnosis of tetralogy of Fallot and indications for operation. *J. Thoracic Surg.* 16:241, 1947.

57. H. E. Alexander. Guides to optimal therapy in bacterial meningitis. *J.A.M.A.* 152:662, 1953.

58. M. Hartman and L. W. Bonner. *Clio's Consciousness Raised.* New York: Harper and Row, Colophon Books, 1974.

59. L. Kavaler. *Mushrooms, Molds, and Miracles.* New York: John Day Co., 1965.

60. W. H. Hughes. *Alexander Fleming and Penicillin.* New York: Crane, Russak & Co., 1977.

61. V. Apgar. A proposal for a new method of evaluation of the newborn infant. *Curr. Res. Anesth. Analg.* 32:260, 1953.

62. V. Apgar and J. Beck. *Is My Baby All Right?* Trident Press, 1972.

63. T. B. Dunn. The unseen fight against cancer. Privately published, 1975.

64. K. K. Sanford, R. M. Merwin, G. L. Hobbs, M. C. Fioramonti, and W. R. Earle. Studies on the difference in sarcoma-producing capacity of two lines of mouse cells derived *in vitro* from one cell. *J. Nat. Cancer Inst.* 20:121, 1958.

65. B. Bain, M. R. Vas, and L. Lowenstein. The development of large immature mononuclear cells in mixed leukocyte cultures. *Blood* 23:108, 1964.

66. Citation Classic, Current Contents, March 12, 1979, p. 14.

67. R. Schrek and W. J. Donnelly. Differences between lymphocytes of leukemic and non-leukemic patients with respect to morphologic features, motility, and sensitivity to guinea pig serum. *Blood* 18:561, 1961.

68. B. Bain. Personal communication.

69. M. F. Lyon. Gene action in the X-chromosome of the mouse (Mus musculus). *Nature* 190:372, 1961.

70. V. A. McKusick. *Mendelian Inheritance in Man.* 5th ed. Baltimore: Johns Hopkins University Press, 1978.

71. J. S. Thompson and M. W. Thompson. *Genetics in Medicine.* 3rd ed. Philadelphia: W. B. Saunders Co., 1980.

72. E. H. R. Ford. *Human Chromosomes*. New York: Academic Press, 1973.

73. T. L. Bourkes. *Nobel Prize Winners in Medicine and Physiology. 1960–1965*. New York: Abelard-Schuman, Harper & Row, 1966.

74. W. Forssman. Die Londierung des rechten Herzens. *Klin. Wochenschinit* 8:2085, 1929.

75. H. Zuckerman. *Scientific Elite: Nobel Laureates in the United States*. New York: The Free Press, Macmillan Publishing Co., 1977.

76. R. Hubbard. Personal communication.

77. Chaff et al., *Women in Medicine*.

78. M. M. Chediak. Nouvell anomalie leucocytaire de caractere constitutional et familial. *Rev. Haemat*. 7:362, 1954.

79. O. Higashi. Congenital abnormality of peroxidase granules—a case of "congenital gigantism of peroxidase granules," a preliminary report. Tohoku. *J. Exp. Med*. 58:246, 1953; 59:315, 1954.

80. T. C. Hsu. *Human and Mammalian Cytogenetics: An Historical Perspective*. New York: Springer-Verlag, 1979.

81. P. Farnes, B. E. Barker, L. E. Brownhill, and H. Fanger. Mitogenic activity in Phytolacca americana (Pokeweed). *Lancet* 1964.

82. Hsu. *Human and Mammalian Cytogenetics*.

83. M. Montessori. The work of Dr. Maria Montessori. *Woman's Med. J*. 21:205, 1911.

84. E. G. Hainstock. *The Essential Montessori*. New York: Mentor Books, New American Library, 1978.

85. M. Sanger. *Margaret Sanger–An Autobiography*. New York: Dover Publications, 1938.

86. B. Seaman and G. Seaman. *Women and the Crisis in Sex Hormones*. New York: Bantam Books, 1977.

87. E. Kübler-Ross. *On Death and Dying*. New York: MacMillan, 1969.

88. E. Kübler-Ross. *Death—The Final Stage of Growth*. Englewood Cliffs, NJ: Prentice-Hall, 1975.

89. M. S. Calderone. *In* Successful women in the sciences: An analysis of determinants. *Ann. N.Y. Acad. Sci*. 208:47, 1973.

90. R. W. Clark. *Freud: The Man and the Cause—A Biography*. New York: Random House, 1980.

91. M. Daly. *Gyn/Ecology*. Boston: Beacon Press, 1978.

92. D. Broverman, E. Klaiber, Y. Kobayashi, and W. Vogel. Roles of activation and inhibition in sex differences in cognitive abilities. *Psychological Review* 75:23–50, 1968.

93. W. Masters and V. Johnson. *Human Sexual Inadequacy*. Boston: Little, Brown, 1970.

94. S. Hite. *The Hite Report: A Nationwide Study of Female Sexuality*. New York: MacMillan, 1976.

JANE A. MILLER

Women in Chemistry

Chemistry is the study of matter and how it changes. It is the science central to the understanding of almost all scientific disciplines, for it is of fundamental importance to the study of genetics, physiology, archaeology, agriculture, metallurgy, nutrition, and many other fields of twentieth-century science.[1]

Chemistry began as a practical discipline, carried on by bakers and brewers, smelters and refiners of metals, potters and glassmakers. Artisans, using chemical processes, prepared leather and medicines and followed recipes for making dyes, perfumes, cosmetics, and cleaning agents.[2] There is little evidence of any theoretical basis for early chemistry, and it is very likely that any orally transmitted theory used by practitioners has been lost.[3]

It was the Greek philosophers who first exhibited a concern for the nature of matter and who began the search for the fundamental substance from which everything is made. The theory of Thales (early sixth century B.C.E.), who declared that water was *the* elemental substance, led to that of Empedocles, who, during the fifth century B.C.E., taught the four-element theory (earth, air, fire, and water), which was adopted in medicine as the four humours (blood, phlegm, yellow bile, and black bile). Aristotle described the "qualities" or properties of the four elements and added a basic element, the quintessence.

The ancient philosophers also considered changes in matter and whether matter existed in continuous or discontinuous form. Democritus, a student of Thales,[4] conceived of matter as separate particles or atoms, a theory which was later adopted by the Epicurians. The atomic theory was, however, generally considered to be of little importance,[5] and the dominant theory of chemistry became the idea that forms of matter could be transformed one into another.

In China the philosophers worked with five elements (earth, wood, fire, metal, and water) and two contraries or opposites which determined the structure and properties of matter.[6] The Chinese also believed that matter could easily change from one form to another and that a ferment could be found which would cause this transformation.

The melding of theory and practice into the science of chemistry probably occurred in Alexandria, Egypt[7] during the early centuries of our era.

Here the philosophies of the Greeks could interact with those of the Arab world and practical chemistry could fuse with astrology and magic. From this intermingling, alchemy would arise as the dominant theory and practice of chemistry for many centuries.

Women have been involved in the study and practice of chemistry from its very beginnings and have made significant contributions to this "central science." The first chemists whose names we know are women, the Mesopotamian perfumers Tapputi-Belatekallim and (   )ninu.[8] On tablets dating from 1200 B.C.E., they are identified as a mistress of a household and in charge of manufacturing perfume, and the author of an ancient text on perfumery. Levey[9] notes that most Mesopotamian chemical apparatus appears to come from kitchens or to have evolved from kitchen utensils. "Women were active in various branches of chemical technology until Alexandrian times when they achieved important positions in alchemy."[10]

Although women played an important role in alchemy, we find few references to their contributions after the scientific revolution of Galileo and Newton. During the seventeenth and eighteenth centuries, women served as helpers of chemists and popularizers of chemistry, but few were able to take part in making the exciting discoveries which followed Lavoisier's chemical revolution at the end of the eighteenth century.

Participation of women in chemistry, as in many other scientific areas, was limited by their lack of access to higher education, particularly doctoral study, and to laboratories in industry or at the better universities. Women, however, found subdisciplines of chemistry which were hospitable to them, or invented areas of science in which they could work. In the United States particularly, women turned to the study of biochemistry and to chemical research in departments of home economics.

This chapter will discuss women's contributions to early chemistry and will provide information on the careers of some outstanding women chemists. These include women who advanced the fields of organic chemistry, agricultural chemistry, chromatography, histochemistry, and particularly biochemistry—women whose contributions have been important in building the vast structure of chemistry.

ALCHEMY

Chemistry as a science begins with alchemy.[11] This mysterious discipline, allied with the occult, dependent on the technical achievements of artisans, was practiced by philosophers, religious and secular scholars, and mystics, as well as charlatans. Alchemy was part of the cultures of China, India, Greece, Rome, and Egypt. Western alchemy probably began in Hellenistic Egypt, was strongly influenced by Arabic writings, and was widely practiced into the seventeenth century. Based upon a belief in the continuity of matter, and therefore its interconvertability, alchemy attracted some of

Alchymia. From Thurneysser's *Quinta Essentia*, 1570.

the world's greatest thinkers, including Isaac Newton, who only became disillusioned with the art in 1693.[12] A basic concept of alchemy was the combination of the male-female principles, the Yin-Yang of China and the King-Queen pair of the West. Women were important to alchemy both as founders and adepts and as providers of an essential ingredient thought to ensure its success.

Isis, the Egyptian goddess whose worship grew from that of the great Earth Mother, is considered one of the mythological deities who gave birth to the science of chemistry. Isis was the bringer of wisdom and writing to the Earth. She taught humans the practice of agriculture and weaving, and instituted marriage and the art of medicine.[13] She was the goddess of the flooded Nile, of the Egyptian Delta, the Kemet (black earth), which most scholars believe to be the origin of the word "chemistry."[14]

Early writers of Greek manuscripts on alchemy mention two women who are considered among the founders of the art, Maria the Jewess, sister of Moses, and Cleopatra. Although some consider these women to be mythological, as their names suggest, their writings and the laboratory apparatus which they purportedly invented were widely used.

There are many references to the work of Maria the Jewess[15] (early third

Isis, the mythological founder of chemistry. With permission of The Oriental Institute, University of Chicago.

century C.E.): Zosimus, in the third century, gives the first exposition of her accomplishments, and Michael Maier, in the seventeenth century, included her as one of the twelve sages of alchemy. Zosimus states that Maria was the first to prepare copper burnt with sulfur, the "first material" for the preparation of gold. She taught that the "Great Work" could only be prepared in the early spring and that God had given its secret exclusively to the Hebrews. Maria believed that all matter is basically one, and that success in making gold will come when parts are joined: "One becomes two, two becomes three, and by means of the third the fourth achieves unity; thus two are but one."[16] There is, in her writings, an analogy between humankind and the metals: "Join the male and the female and you will find that which is sought."[17]

The serpent Ourboros, an Egyptian emblem of the cosmos, of eternity, appears both in abstract and direct forms in Cleopatra's manuscripts on gold making. In the middle of the serpent there is the inscription "The one is all." Adapted from A. Coudert, *Alchemy: The Philosopher's Stone*.

As influential as her philosophy was, it is the laboratory apparatus with which Maria is credited which makes her an important figure. She built and described many ovens and apparatus for cooking and distilling. The most famous is the water bath, a double vessel, the outer part filled with water while the inner contained the substance to be heated. Her association with the water bath or double boiler is acknowledged today by the names given this vessel, *bain Marie* in French and *Marienbad* in German. Maria also provided the oldest description of stills, particularly the complex *tribikos* with three tubes made of copper, a little thicker than that of a pastry cook's copper frying pan, to keep the distillate liquids separate. The dome was likened to a woman's breast because of its shape "and condensed vapors were drained away through the three tubes into flasks and not through the nipple-like projection left on top."[18] Maria also invented the *kerotakis*, a sealed container in which metal shavings were exposed to the action of vapors.

Michael Maier, a seventeenth-century alchemist, names four women who had discovered the secret of producing the philosophers' stone: Maria, Taphuntia, Medera, and Cleopatra.[19] Cleopatra was both a philosopher and a practical experimentalist. Her manuscript the *Chrysopoe* (Goldmaking) was a major source for alchemists and her dialogue with the philosophers was vividly written, drawing on metaphors of the womb and treating nature symbolically.[20] According to Lindsay her discourse was "the most imaginative and deeply-felt document left by alchemists." It conveyed a "deep sense of life and its possibilities, a delight in the beauty of the earth," and a conviction that one had the power to blend harmoniously with nature. In her dialogue she used the imagery of conception and child-birth and saw scientific quest not as "an abstract search for knowledge" but "as a means of enriching human life."[21] The earliest shorthand for the elements is attributed to Cleopatra.[22]

In later centuries women were cited as important co-workers with well-known alchemists. In the fourteenth century Nicholas Flamel's wife, Per-renelle, was his partner in making precious metals from mercury. He writes "I have made [gold] three times, with the help of Perrenelle, who under-stood it as well as I, because she helped me in my operations, and without doubt, if she would have enterprised to have done it alone, she had attained to the end and perfection thereof."[23]

A seventeenth century Dutch physician, John Frederick Helvetius, de-scribes the preparation of gold by his wife in his book the *Golden Calf*. After she was given a small portion of a stone with which gold could reputedly be made, she succeeded, within a quarter of an hour, in totally transmuting a mass of lead into the best and finest gold.[24]

Needham's story[25] of Chhêng Wei's wife shows that women were prac-ticing alchemists in China. Chhêng Wei, a gentleman of the Han Imperial Court, was trying to make gold, but without success. "One day his wife went to see him just as he was fanning the charcoal to increase the heating of a reaction vessel in which there was mercury. She said, 'Let me show you what I can do,' and taking a small amount of some substance from her pouch she threw it into the vessel, and they saw that the contents had turned to silver." In spite of bribes and beatings, Chhêng Wei was unable to extract the secret from his spouse. Finally "she went mad, rushed out naked, smeared herself with mud, and died."[26] (It has been suggested that this description of her death might indicate mercury poisoning.)

These few examples of the involvement of women in alchemy show that women, perhaps as representative of the female principle, were im-portant to the art. Western alchemical writings make many comparisons of the steps in the production of gold to women's work: cooking, roasting, washing clothes. Although we see little evidence of women acting alone, there are several illustrations which picture men and women working to-gether as practitioners of the Black Art.[27]

SEVENTEENTH-CENTURY WOMEN
OF LEARNING

During the seventeenth century chemistry began to diverge from its al-chemical roots. A group of practical chemists inspired by the iatrochemists (chemists allied with medicine) espoused new ideas and carried out new reactions, not to discover the philosophers' stone, but to study the nature of matter. Many of these chemists were pharmacists who had ready access to laboratories, others were physicians or self-taught chemists, such as

Johann Rudolph Glauber (1604–1670), who wished to make Germany economically self-sufficient, or Robert Boyle (1635–1703), the amateur scientist who wished to apply the new mechanical philosophy to chemical reactions. Although much of seventeenth-century chemistry was based on the Paracelsian elemental theory, which recognized three "principles" (salt, sulfur, and mercury), the concept of atoms (discontinuous matter) was taking hold.[28]

MARIE MEURDRAC

During the mid-seventeenth century books about science written especially for ladies began to appear. The invention of the telescope and the microscope had inspired women of the upper classes to study natural science, and to write for their peers.[29] Probably the first book on chemistry written by a woman was published in 1666.[30] Marie Meurdrac (dates unknown), a Frenchwoman, divided her book, *La Chimie Charitable et facile, en faveur des dames*, into six parts: 1) Principles and operations, including apparatus; 2) The properties of simples [elements], their preparation, the method of extracting their salts, etc.; 3) Animals; 4) Metals; 5) Method of making compound medicine [rosemary is extolled as a universal antidote]; and 6) For the Ladies, in which there is a discussion of everything capable of preserving and increasing beauty.

Madame Meurdrac explains that she has been careful not to go beyond her knowledge, that everything she writes is true and that all the remedies have been tested. In her foreword she states that she began the book solely for her own satisfaction and to retain the knowledge she had acquired through long work and through "various oft-repeated experiments." For two years she debated with herself over the publication. "I objected to myself that it was not the profession of a lady to teach, that she should remain silent, listen and learn, without displaying her own knowledge; that it was not her station to offer a work to the public and that a reputation gained thereby is not ordinarily to her advantage since men always scorn and blame the product of a woman's wit [mind]." However, Madame Meurdrac declares that "minds have no sex and that if the minds of women were cultivated like those of men, and that if as much time and energy were used to instruct the minds of the former, they would equal those of the latter. . . . "

WOMEN AND THE CHEMICAL REVOLUTION

Newton's concept of attraction between bodies and the view of the corpuscular nature of matter were the most influential theories of the seventeenth and eighteenth centuries. Shapes and kinds of particles were considered fundamental to reaction, and tables of affinities were prepared and included in the new dictionaries and encyclopedias of the period.[31] As Leicester notes, "Practical chemistry greatly increased the knowledge of elements and compounds; quantitative methods came to be accepted as essential to chemical investigation; and the whole new field of gasses was opened up. The combination of all these factors made possible the foundation of modern chemistry by Lavoisier and the French school."[32]

Lavoisier, through the use of ingenious experiments, verified the law of conservation of mass, elucidated the nature of combustion and respiration and the role of oxygen in these phenomena, and defined and identified elements in modern terms.[33] This revolution in the understanding of chemistry led to the extraordinary growth of the discipline in the nineteenth and twentieth centuries.

By the time Antoine Lavoisier brought about the chemical revolution in the last decades of the eighteenth century, women's role in chemistry had generally become that of helper or of popularizer of the discipline. These activities, however, were indispensable in helping to spread the "new" chemistry and in stimulating its growth.

MARIE ANNE PIERRETTE PAULZE

Lavoisier's wife, Marie Anne Pierrette Paulze (1758–1836), is the best known of these helpmates.[34] She is described as a lively, sensible lady, well versed in her husband's chemical theories, which she could discuss with the greatest scientists of the time. Her salon was the gathering place for scientists, not only from France but from throughout Europe. She translated works needed by her husband, including the writings of Priestley, Cavendish, and Henry, as well as Kirwin's *Essay on Phlogiston*, in which she pointed out scientific errors in the text. Madame Lavoisier prepared the plates for the illustrations in the *Traite de Chemie* and edited and published her husband's *Memoires*. She also took an active part in Lavoisier's experiments on the chemical nature of respiration. Without her aid, much information about her husband's revolutionary ideas might have been lost.[35]

JANE HALDEMOND MARCET

By far the foremost popularizer of chemistry was Jane Haldemond Marcet (1769–1858), whose *Conversations on Chemistry*, first published in 1805, went through sixteen editions and was often pirated.[36] The daughter of a Swiss family living in London, she married Alexander Marcet, a physician and chemist. Jane Marcet, like many women of her day, attended the lectures of Davy at the Royal Institution and met many other scientists at evening gatherings. Dr. Marcet encouraged his wife's interest in chemistry and her authorship of the *Conversations*. Mrs. Marcet insisted on an experimental approach to chemistry, and the conversations between Mrs. B, the teacher, and her students, Caroline and Emily, were well illustrated with woodcuts of appropriate apparatus and were constantly updated as the ideas of the "new" chemistry spread and as discoveries, such as gaslight, changed the face of Europe.

Conversation on Chemistry was so popular that one hundred and sixty thousand copies were sold in the United States before 1853. But its greatest influence was on a thirteen-year-old boy who read the book as an apprentice to a bookbinder. Michael Faraday later wrote that after doing the experiments in Jane Marcet's book, "I felt that I had got hold of an Anchor of chemical knowledge, and clung fast to it." Faraday praised Mrs. Marcet as "one able to convey the truth and principle of those boundless fields of knowledge which concern natural things, to the young, untaught and inquiring minds."[37]

AFTER THE CHEMICAL REVOLUTION

During the nineteenth century women had little opportunity to contribute to the explosive growth of chemistry which took place after the chemical revolution. While the English, the French, and especially the German scientists were laying the foundation for modern organic chemistry, discovering its reactions, the tetravalent nature of the carbon atom, the ring nature of benzene, and the phenomena of isomerism; while physical chemistry became a separate part of chemistry; and while the identification and understanding of the molecules which made up biochemistry were being pursued, most laboratories and universities were closed to women. All the one hundred and forty chemists who attended the influential Congress of Karlsruhe in 1860 were males,[38] and although some women were allowed to work in laboratories, it was only in Switzerland that women interested in chemistry were able to obtain the doctoral degree.[39] Women came to Swiss universities from as far away as Russia[40] and America. When Helen Abbott traveled from Philadelphia to Europe in the late 1800s to continue her scientific studies, she found that "women students abounded" at the University of Zurich.[41]

Although Helen Abbott Michael (1857–1904) did not stay at Zurich, her work in plant chemistry remains an example of what a very few women could achieve without a university education. Abbott was born in Philadelphia into a well-to-do family and for a time studied medicine at the Women's Medical College. An injury prevented the continuation of her studies and she began private research.

In 1883, on hearing that several children had died after eating wild parsnip roots, Abbott began analyzing the chemicals present in plants. From 1884 to 1887 she worked under Henry Trimble at the Philadelphia

College of Pharmacy, where she isolated a "new wax" from "*octilla*" and published nine papers in plant chemistry. The trustees invited Abbott to lecture to the students and provided new research laboratories (including space for research by women) for the continuation of her work. Although she was able to isolate few pure compounds, Abbott's work, particularly that on saponins, provided valuable information to pharmacists of the late nineteenth century.

Abbott became interested in plant evolution and proposed that the manufacture of chemicals in plants was in response to plant functions and structure. She suggested that these secondary chemicals might be used as a chemical classification of plants. Alston and Turner credit Abbott with the first plan for biochemical classification.[42]

In 1887 Abbott decided to study chemistry at Tufts College under Arthur Michael. They were married in 1888 and settled on the Isle of Wight, where Mrs. Michael continued to do research in organic chemistry in the laboratory there, publishing four papers in German journals. Later Helen Michael returned to Boston and received an M.D. degree in 1903 from Tufts. She practiced at a free hospital, which she established, but died from an infection in the next year.[43]

Another outstanding, largely self-taught chemist was Agnes Pockels (1862–1935), who was a founder of surface chemistry.[44] Pockels was born in Venice, where her father, an officer in the Royal Austrian army, was stationed. In 1871 Captain Pockels retired and settled in Brunswick in Lower Saxony. Pockels attended the Municipal High School for Girls but, because women were not accepted into institutions of higher education and because of family duties, she remained in her parents' house, where her pioneering research was done on the kitchen table. Her brother Friedrich, as a student at the University of Göttingen, helped Pockels in her search for knowledge.

The research on surface films dates from observations made about 1881. By 1882 Pockels had invented the slide trough, "a method of extending or reducing the surface area of water by means of a wire or metal strip placed over it. . . . "[45] In 1891 Pockels wrote to Lord Rayleigh about her work, and he had her communication published in *Nature* in the following year. In this letter Pockels describes her trough, which was capable of "attaining any condition of tension which is at all possible" without the possibility of "currents breaking through,"[46] and noted that a contaminated water surface can exist in two conditions, one of normal water, the other of a monolayer of oil. Pockels also invented the method for producing a clean surface by sweeping it with the trough barrier, and the method of applying water-insoluble compounds to the water surface by dissolving them in benzene and allowing the solvent to evaporate. With her surface film balance she was able to calculate the thickness of a surface film at the moment it formed in what is now called a stable monolayer. This minimum area, about 20Å per molecule, is known as the Pockels' point.[47]

Pockels made many observations on the effects of contaminants on the surface tension of organic substances and of the interfacial tension between water and liquids immiscible with water. She found that a monolayer could be detected by dusting with a fine powder such as lycopodium, thereby discovering the ring method of determining surface tension, and she identified hysteresis (the time lag occurring when there is a change of forces) in the contraction and expansion of films. Giles and Forrester note that Pockels suggested that osmotic pressure effects are possible in certain surface films.[48]

Pockels's work was recognized and developed by other scientists: Rayleigh adopted her methods in his research, and Langmuir, who received the Nobel Prize in 1932 for investigating monolayers, improved the Pockels' trough and extended her pioneering research.[49] (See Physics chapter.)

A Russian woman, Yulua Vsevolodovna Lermontova (1846–1919), was one of the few females able to receive a doctorate from a German university.[50] In 1869 Lermontova joined her friend the mathematician Sofya Vasillievna Kovalerskaya in Heidelberg to study there for two years because Russian universities were closed to women. Although they could not register, they were allowed to attend lectures, and Lermontova learned qualitative and quantitative analysis and did research in the separation of the metals in the platinum family. The two women then went to Berlin, where Lermontova published an article on her research on diphenene (4,4'-diaminohydrazo-benzene) done in Hofmann's laboratory.[51] Although neither Berlin nor Heidelberg would grant them degrees, the University of Göttingen could award doctorates in absentia to foreigners. Lermontova received a Doctor of Philosophy in Chemistry in 1874, after being examined on her thesis, "The Study of Methylene Compounds."

On returning to Russia, Lermontova spent one year in the Moscow University Laboratory of V. V. Markownikov. She prepared 1,3-dibromo-propane and, with Markownikov, synthesized glutaric acid, a dibasic organic acid. She then moved to St. Petersburg to work with A.M. Butlerov, studying the action of tertiary butyl iodide on isobutylene in the presence of metallic oxides. She was able to synthesize two new hydrocarbons, C_8H_{16} (a precursor to isooctane) and $C_{12}H_{24}$. Her research included the catalytic synthesis of the dimer and trimer of isobutylene and of 2-butyne, published in 1881. Lermontova was called back to Moscow because of family pressures and joined Markownikov in his research on petroleum. She designed an apparatus for the continuous treatment of petroleum during the high pressure splitting of hydrocarbons using metal catalysts. Lermontova was a member of the Russian Chemical Society and the Russian Technical Society.

However, in 1886, because of family duties, she retired to the family farm near Moscow and her research in chemistry ceased.

As more universities outside of Switzerland opened their doors to women, female chemists earned doctoral degrees in Germany (from 1908),[52] in Paris, and in England (from the 1870s). M. Carey Thomas set up a Ph.D. program for women at Bryn Mawr, and the University of Chicago, from its founding in 1891, accepted both sexes on equal terms in both graduate and undergraduate colleges. The University of Pennsylvania, Cornell, Yale, and the University of Illinois were among the schools of higher education from which women earned doctoral degrees.[53]

Even with growing admission to graduate schools women had difficulty in finding employment as research chemists. Few were accepted in industry and fewer still on the regular chemistry faculties at Ph.D.-granting institutions.[54] Women, therefore, found opportunities in fields outside the normal disciplines of organic, inorganic, and physical chemistry and concentrated on research in agricultural chemistry, home economics, or nutrition. At women's colleges, beginning at Mount Holyoke, women faculty members and students formed research groups, which have been productive and influential. Women also made names for themselves in nuclear chemistry by having a part in the discovery of five elements: Marie Curie (polonium and radium), Lise Meitner (protactinium), Ida Tacke Noddack (rhenium), and Marguerite Perey (francium).[55] Biochemistry and crystallography have been fields particularly open to research by women and many excellent chemists have found opportunities there. Women also played an important part in the development of chromatography.

WOMEN CHEMISTS AS SPECIALISTS

ORGANIC CHEMISTRY

By the mid-nineteenth century chemical knowledge had become so vast that division of the science into specialities became necessary.[56] The most fruitful research area and the one which attracted most chemists was organic chemistry, the study of carbon compounds. Organic chemists were beginning to understand the arrangement of atoms in molecules and the nature of the reactions which compounds could undergo. They were able to prepare large numbers of compounds, both those which occur in nature and those which were previously unknown. Organic chemistry was especially valuable in encouraging the growth of the chemical industry, and the success of many companies was based upon the synthesis of such compounds as aspirin, Novocain, and saccharin.

Chemistry laboratory at Mount Holyoke College. With permission, Mount Holyoke College Library/Archives.

An influential organic chemist, Emma Perry Carr (1880–1972), studied at the University of Chicago with Alexander Smith and Jules Stieglitz, receiving a Ph.D. in 1910.[57] Although she returned to the faculty at Mount Holyoke College, a women's college with a liberal arts emphasis, she was able to continue carrying out research in chemistry, which, she acknowledged, had added immeasurably to the joy of teaching.[58] By involving excellent women faculty, undergraduates, and teaching assistants who were pursuing master's degrees, Carr was able to develop a unique group research situation. Allied with Carr in the early years were Mary Lura Sherrill (1888–1968), who also studied with Stieglitz,[59] and Lucy W. Pickett (1904–   ), a physical chemist who received a Ph.D. from the University of Illinois.[60]

Carr had noticed that while foreign journals were publishing papers on the relationship between ultraviolet absorption spectra and structures of organic molecules, no comparable work was being done in the United States. She decided to concentrate on the spectra of the hydrocarbons, and work on cyclopropane derivatives was begun in 1913. Because of the complexity of these compounds, Dr. Sherrill chose a single group, the carbon-

carbon double bond, for subsequent study. The Mount Holyoke group prepared and purified large numbers of hydrocarbons, both straight chain and cyclic, with single and multiple double bonds. The compounds were prepared in such a degree of purity that, several years later, the spectra were found to be identical with the standards of the American Petroleum Institute.[61] Carr felt that the most important result of the work was the identification of two different types of electronic bands in the spectrum of the mono-olefins.[62] A better theoretical interpretation of the energy relationships of small unsaturated compounds was made possible by this work and was used by Mulliken and others in the development of basic chemical theory.[63] After Pickett joined the faculty, the studies were extended to the near and far ultraviolet.

As was typical of many women chemists, Sherrill's research during the two world wars was defense-related: first at the Edgewood Arsenal, where she was part of a group attempting a practical synthesis of a nontoxic irritant gas, and, during the forties, when she worked on the synthesis of anti-malarial drugs with the Mount Holyoke group.[64]

The research projects at Mount Holyoke attracted grants to the department, providing needed instrumentation and study leaves. This group research also stimulated the interest of women students in chemistry and in chemical research so that, between 1920 and 1980, there were ninety-three Mount Holyoke graduates who later received Ph.D.'s in the discipline.[65]

Gertrude B. Elion (1918–   ) is another example of a woman whose determination to be a chemist, along with her innate ability, allowed her to succeed in industry. Elion is an outstanding organic chemist, pharmacologist, and administrator and director of research in industry.

After Elion's graduation, summa cum laude, from Hunter College in 1937, fifteen schools rejected her application for a graduate assistantship because she was a woman. For several years she worked at marginal laboratory jobs (once without a salary), taught in New York City high schools, and attended New York University part-time. Dr. Charles Frey, who interviewed her for a research position in the job-scarce pre–World War II days, recalls talking to a young, enthusiastic woman "who wanted to become a great research scientist." He remembers her disappointment when all he could offer her was a laboratory dishwashing job for which she was overqualified.[66] Fortunately, her great potential was realized at Burroughs Wellcome, where she was employed in 1944 to work with nucleic acids.

Elion is renowned for her chemotherapy research. She has synthesized and studied drugs which are used to treat leukemia and to ensure successful organ transplants. As a result of her metabolic studies, a new approach to treating gout and hyperuricemia was found, in which the enzyme xanthine oxidase is inhibited and uric acid formation is blocked in the last step of the metabolic pathway. The drug which she synthesized was allopurinol.

One of the first drugs which Elion prepared is 6-mercaptopurine, which is still a drug of choice used in the chemotherapy of children with acute leukemia. Her immunosuppressant, azathioprine (Imuran), was used in the first heart transplant.[67] She is an acknowledged leader in the field of purine antimetabolites for the treatment of cancer. In 1988, Elion shared the Nobel Prize in medicine with her collaborator George Hitchings.

AGRICULTURAL CHEMISTRY

During the nineteenth century and the early years of the twentieth, chemists became increasingly important to the agricultural community. In state experiment stations and in the laboratories founded by the Department of Agriculture, chemists labored to solve the problems specific to farming and the use of farm products. Women chemists in the United States were a part of this movement and were often able to find in these laboratories an opportunity for research, denied elsewhere.

Probably the earliest American woman to receive a Ph.D. in chemistry was Rachel Lloyd (1839–1900) whose interest in the chemistry of sugar beets arose during her studies in Switzerland.[68] Lloyd's husband died in 1865 and, after a stay in Europe, she accepted a teaching position at a private girls' school in Kentucky. During the years 1876–1884, she attended the Harvard Summer School, doing research with C. F. Mabery on acrylic acid derivatives. Lloyd was the first woman to publish in the *American Chemical Journal*. She then went to Zurich, and was awarded a doctorate in 1886. Her dissertation research with V. Merz was in organic chemistry, concerning the high-temperature conversion of phenols to aromatic amines.

Lloyd accepted an appointment as an associate professor of analytical chemistry at the University of Nebraska in 1887 and as assistant chemist at the Nebraska Agricultural Experimental Station. She was promoted to full professor in 1888, and, as the chemistry department had only two members, she carried a heavy teaching load. At the experimental station, Lloyd and the director, H. H. Nicholson (who was the other member of the university department), planned tests to determine whether successful sugar beet factories could be established in Nebraska. They distributed seeds, arranged for experimental plots to be planted, and analyzed questionnaires concerning climate and soil conditions, costs, and yields. In addition Lloyd analyzed beets from each plot for specific gravity, total solids, percent of sucrose and reducing sugars, and the "coefficient of purity," a measure of the ease of refining white sugar.[69]

The favorable results of these experiments caused two sugar factories to be built in Nebraska. The chemists continued research in the production of better seed, the effects of storage, and use of by-products as stock-feed, which led to the publication of twelve bulletins in the "Sugar Beet Series." In 1977 Nebraska ranked eighth in the nation in sugar beet production.

Another woman active in agricultural chemistry is Allene R. Jeanes, (1906–   ), who spent thirty-five years in research at the U.S. Department of Agriculture's Northern Regional Research Laboratory (NRRL) at Peoria, Illinois.[70] Jeanes was born in Waco, Texas, and received the A.B. in chemistry from Baylor University and the M.S. in organic chemistry from the University of California–Berkeley. For five years she taught on the faculty of Athens College in Alabama, and then in 1935 she entered the graduate program at the University of Illinois, where she studied under Roger Adams. She was awarded a Ph.D. in 1938 and accepted one of the first Corn Industries Research Foundation Fellowships at the National Institutes of Health. With Horace S. Isbell she developed a new technique of periodate oxidation of starches.

Dr. Jeanes become a member of the staff of the NRRL three months after it opened and initially studied the nature and structural role of the branch points in starch. In a successful attempt to isolate isomaltose, she also became interested in dextrans, especially D-gluans produced by microbes from sucrose and based upon the a $(1 \rightarrow 6)$ linkage. Because of the Korean War and an increased incidence of serum hepatitis, scientists were searching for an acceptable blood-plasma substitute, and Dr. Jeanes and her group were able to find a chemical dextran which was used successfully to expand plasma volume. A fermentation process to produce this dextran was developed using a medium containing sucrose, yeast extract, potassium phosphate, and small amounts of other inorganic salts. Dr. Jeanes and her group isolated and characterized over one hundred different dextrans which have had great value in research, especially in immunology and immunochemistry. Additional refinement of the periodate oxidation method made possible the rapid differentiation and characterization of these dextrans. In addition, Jeanes's isomaltose was used as a standard for dextran metabolism studies.

Dr. Jeanes also explored other types of microbial polysaccharides which might come from the carbohydrates of cereal grains. This led to the discovery in 1962 of xanthan, an industrially important polysaccharide gum.[71]

Dr. Jeanes was honored for her work when she received the U.S. Civil Service Commission's Federal Woman's Award, the Garvan medal of the American Chemical Society, and the USDA Distinguished Service Award.

BIOCHEMISTRY

Biochemistry was born from agricultural chemistry and physiological chemistry,[72] from problems encountered on farms and in the practice of medicine. It is concerned with the physiology of living systems, both plant and animal. In the twentieth century, as scientific methodology and understanding progressed, it became necessary to concentrate research more and more on the molecular level. Biochemistry deals with the close-ups: how molecules, atoms, and particles interact with one another to synthesize and degrade important chemical compounds; how energy is formed, stored, and released; how pathways mediating these events are controlled; and how perturbations on these systems affect the organism.

Women scientists have made some extraordinary contributions to biochemical research and the scope of these contributions has been diverse. In this section we will detail discoveries of new methodologies, of visionary research, of the definition of new metabolic pathways, of pharmacological developments, and of new concepts for understanding the biochemical basis of diseases. The biochemical contributions span species and organ systems. They include bacteria, plants and animals, enzymes, and organic molecules. Because biochemistry has been the branch of chemistry most open to women, it would be impossible to cover all their contributions. This small sample, however, will show the importance of the work of women to this field.

The most renowned of these women biochemists is Gerty Theresa Cori (1896–1957), who, with her husband Carl, received the Nobel prize in medicine and physiology in 1947 for the elucidation of the metabolism of the essential carbohydrate glycogen. Glycogen forms the body's storage pool for glucose, the immediate prime energy-producing molecule used by cells. Glycogen is a polymer; that is, it is made up of successive units of glucose molecules chained together by a variety of bonds. The ultimate structure of glycogen is a branching molecule. Branches are placed onto the glycogen molecule by certain enzymes and removed by others; in addition, still other enzymes act on the main chain structure.

By identifying metabolic intermediates, and the enzymes involved, the Coris established how glycogen was synthesized and under what conditions. Gerty Cori isolated and synthesized a key molecule in glycogen metabolism, glucose-1-phosphate.[73] The Coris crystallized the principal enzyme, phosphorylase, which mediates glycogen breakdown to glucose.[74] They established the relationships between liver and muscle glycogen and blood glucose and lactic acid, now known as the Cori cycle.[75] In 1939 they carried out the first test-tube synthesis of glycogen.[76]

This work led to an examination of glycogen storage diseases. These diseases had been recognized since the 1920s and cases described by individuals became known as von Gierke's disease, Pompe's disease, or McArdle's disease. All of these disorders exhibit some sort of abnormality in the way glycogen is formed, stored, or released.

Gerty Cori developed the conceptual classification of these glycogen storage diseases. She postulated that Type I was due to glucose-6-

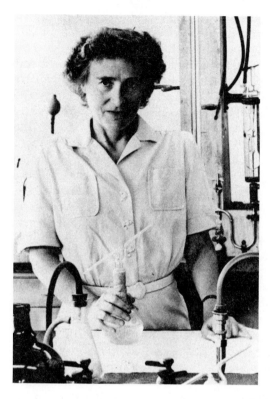

Gerty Cori. Courtesy of the Archives of Washington University School of Medicine.

phosphatase deficiency,[77] and Type III, due to deficiency in the debranching enzyme, amylo-1,6-glucosidase.[78] The former is now known as Type IGSD, and the latter as Type III. This was the first demonstration that a defect in an enzyme was the cause of a human genetic disease.[79] Herman Kalckar stated in 1958, "The demonstration that an aberrant architecture of a polymer is rooted in a well-defined enzyme defect still remains an unmatched scientific achievement."[80] Subsequently a number of other specific enzymatic defects have been described in the disease spectrum, including deficiency of the enzyme phosphorylase.

Gerty Cori (née Radnitz) was trained as a physician, and graduated from the German University of Prague in 1920. She and Carl were married in August of the same year and emigrated to the United States in 1922 to work at the New York State Institute for the Study of Malignant Diseases. During this period Gerty and Carl were advised to end their research collaboration, because it would be detrimental to his career.[81]

In 1931 they moved to Washington University in St. Louis; Carl as chairman of the Department of Pharmacology, Gerty as a research associate,[82] a usually dead-end position typically allotted to wife-collaborators throughout the scientific establishment. Gerty was finally promoted to Professor of Biochemistry in 1947, the year of the Nobel award. Although it is "impossible to separate the contributions of Gerty from those of Carl Cori, as they always worked in close collaboration ever since their first joint publication in 1923"[83] and shared a common bibliography, the honors awarded the pair differed. Gerty was elected to the National Academy of Sciences eight years after her husband (1948). Carl alone received the Albert Lasker Award of the American Public Health Association and the prestigious Willard Gibbs Medal of the American Chemical Society. Gerty did, however, receive the Garvan Medal in 1948 from the A.C.S., the society's award for excellence to a woman chemist.[84]

An early Swiss biochemist who was also interested in enzymes was Gertrud Woker (1878–1968), a pioneer in the study of catalysis. Woker was born in Berne, Switzerland, and attended the University of Berne, where her maternal grandfather and her father were professors.[85] She received the Ph.D. in 1903 under von Konstanecki with a dissertation on the synthesis of 3,4 dioxyflavone. After further study in Germany, Dr. Woker returned to Berne and synthesized flavanone and flavone in von Konstanecki's laboratory. These compounds exist in nature as yellow dyes and in stalks and stems of many plants. They are the active ingredients in many folk medicines and are important for normal plant growth and defense against infection and injury. Gertrud Woker was then invited to carry out physicochemical research at the Berne medical clinic. She became a member of the science faculty in 1907 and her *habilitation* lecture concerned the problems of catalysis, an interest which became the driving force in her research for the remainder of her life.

Dr. Woker was appointed director of the Institute for Physical and Chemical Biology at the university in 1911 and predictably selected biological catalysts (enzymes) for her own studies. In 1913–14 she found that extracts from fungi brought about decomposition of hydrogen peroxide, without an expected peroxidase reaction, showing that a secondary peroxide acts as carrier for both the catalase and peroxidase reactions by liberating oxygen to oxidizable substances (peroxidase reaction) and destroying excess H_2O_2 (catalase reaction).

This study led to a general examination of oxido-reductases (enzymes which can cause oxidation and reduction), which can remove hydrogen as a dehydrase or act as a catalase by releasing oxygen from H_2O_2. Dr. Woker and her group discovered that ascorbic acid (vitamin C) could take the place of an enzyme in raw milk and could provide redox reactions in boiled milk. A microchemical method for the determination of ascorbic acid was developed using its redox formation of furfural.

A study of the reactions of furfural and other aromatic aldehydes followed, as did an investigation of the reactions of quinone (an organic oxidizing reagent) with albumin, amino acids, vitamins, hormones, etc. Dr.

Woker has also studied the chemical basis of assimilation of carbon dioxide, the mechanism of urea formation, cocarboxylase as an intermediate catalyst in nerve muscle exchange, and the process of narcosis.

Early in her career Dr. Woker was asked to write the section on catalysis for *Die chemische Analyse,* and in 1910 her first book on catalysis in analytical chemistry was published. This was followed by volumes on inorganic catalysts in 1915, on hydrolyzing biological catalysts in 1924, and on respiration catalysts in 1931.

Dr. Woker was a strong advocate for peace and women's rights. Her book *Der kommende Gift und Brandkrieg* (The Coming War of Poison and Fire), first published in 1931, went through many editions and was publicly burned in Nazi Germany.

Another pioneer biochemist was the American Agnes Fay Morgan (1884–1968).[86] After receiving her Ph.D. from the University of Chicago in 1914, she chose to follow the lead of Ellen Richards and pursue her career in the field of home economics.[87] She went to the University of California at Berkeley, where she taught the first scientific courses in human nutrition to be offered there. She became professor of household science in 1923 and when the Department of Home Economics was formed in 1938, she became its first chairperson. She also had an appointment as a biochemist at the Agricultural Experiment Station.[88]

Work from Morgan's laboratory covered a wide range of chemical research relating to nutrition. Her group carried out analyses on natural and processed foods,[89] discovered that overcooking changed the nutritional values of meat and other proteins, and observed the effect of sulfur dioxide (used as a preservative) on vitamin C and thiamine.[90]

Dr. Morgan's most important research was done in her study of vitamins, particularly the effect of vitamin deficiencies on animals. She observed the lower absorption of total glucose, the lower liver glycogen count, and the lower intestinal absorption of glucose in animals with a deficiency of ascorbic acid,[91] and that in scurvy, blood glycogen and sugar levels were elevated[92] and there was a reduction of body fat.[93] In 1939, she and Sims reported hemorrhage and atrophy in rats deficient in a "filtrate factor," later identified as pantothenic acid,[94] and studied the relationship between this vitamin and adrenal cortex hormones.[95] Morgan's vitamin D studies included a comparison of the hypervitaminosis induced by irradiated ergosterol and fish liver oil concentrates. She also showed that massive doses of vitamin A could decrease the morbidity as well as the skeletal resorption produced by vitamin D deficiency in rats.[96]

Morgan directed a large laboratory which provided important information on nutritional requirements; however, it was her pioneering work in the biochemistry of vitamins which had the greatest influence by providing the basis for further study of these important compounds.

Morgan's most outstanding student was Icie Macy Hoobler (1892–1984). This biochemist, whose research was primarily in nutrition, has been described as one of the best physiological chemists of the first half of the twentieth century.[97] Born on a farm in Missouri, she attended a female academy and Randolph-Macon Woman's College. There she met Mary Sherrill, who encouraged Hoobler to continue her studies at the University of Chicago.[98] Jules Stieglitz, her adviser, urged her to go to graduate school, but thinking that she needed to do some teaching first, he found her a position as laboratory assistant at the University of Colorado. While at Boulder, Hoobler became interested in physiological chemistry and prepared to go to Yale to study with Lafayette Mendel.[99] There, because they were not allowed to teach in the laboratories, women were provided with research assistantships, which greatly helped their academic careers.[100] Hoobler was assigned a project important to America in 1918, the possible toxicity and food value of cotton seeds, which had been substituted for wheat flour during the war. She found that the illness observed in animals which were fed cotton seeds was due neither to starvation nor to vitamin deficiency but to gossypol, a poison present in the seeds.

Mendel was anxious for Hoobler to work in chemistry related to the health sciences, because he believed these were fields "to which women scientists have much to contribute."[101] He challenged her to consider the analysis of human milk and to study the nutrition of mothers and children, a project which became a life-long endeavor.

Hoobler's first position was at West Pennsylvania Hospital, where she carried out analyses of fetuses for calcium and magnesium and studied the comparative compositions of the bloods of mothers and their newborn infants. After spending some time at the University of California with Agnes Fay Morgan, in 1923 Hoobler went to the Merrill-Palmer School for Motherhood and Child Development in Detroit and then to the Children's Fund of Michigan Laboratory, where she remained until her retirement in 1954. She was one of a handful of women in the United States to be director of a research laboratory.[102]

During her long tenure in Detroit, Hoobler and her group (often as many as sixty scientists)[103] studied the effect of nutrition on both mother and child. Through exhaustive research, she determined the composition of human milk and the variability of its composition[104] and the nutritional requirements of both pregnant and nursing mothers and of infants and children.[105] She showed that malnutrition in women, even before pregnancy, had a great effect upon infant health[106] and that calories were a factor in children's growth.[107] The research group was instrumental in showing the need for vitamin D and in encouraging the irradiation of milk.[108] In addition, Hoobler studied amino acids in foods and the standardization of vitamins B and C.[109] She published papers on the composition of blood cells, bone development, gastrointestinal studies, and the effect of iodine on calcium metabolism.[110] Her research had far-reaching effects upon the determination of nutritional requirements for women and children, because studies had previously been done only on adult males[111] and there were many nutritional myths. Hoobler's group, by presenting chemical evidence, was able to change the diets of the women and children of America.

Dr. Hoobler was also active in professional organizations. She was the first woman to chair a division of the American Chemical Society and to chair the local Detroit section of the society. She was active in organizing the Women's Committee of the A.C.S. and in establishing the Women's Award, later known as the Garvan Medal, which she received in 1946.[112]

The chemistry of the life processes was also the focus of the research of the Englishwoman Marjory Stephenson (1885–1948), who was the foremost authority on bacterial metabolism during the early part of this century. Although trained as a biochemist, she was inspired by her mentor, F. G. Hopkins, to begin a study of bacteria.[113] Throughout her career, she was able to combine several different disciplines, always with an emphasis on living systems. Stephenson's classic volume on bacterial metabolism was part of the prestigious "Monographs on Biochemistry" series.[114]

Stephenson attended Newnham College from 1903 to 1906, and later was awarded a Beit Memorial Fellowship for medical research. Her work was interrupted for four years during World War I when she served with the British Red Cross, first in France and then in Salonika. During the 1920s she concentrated on bacterial metabolism and, in 1929, after receiving a series of annual grants, she was appointed to the permanent staff of the Medical Research Council. She was also a Reader in chemical microbiology in the University of Cambridge. She became the first woman in the biological sciences to be elected a Fellow of the Royal Society, in 1945. She was president of the Society of General Microbiology at the time of her death in 1948.[115]

In 1930 Stephenson published her book, *Bacterial Metabolism*, "which has gone into three editions and is still a standard work."[116] In 1948 she stated, "Bacterial metabolism is now such a wide study that it is no longer convenient for one person to attempt to cope with all its branches. . . ."[117] Nevertheless, because of her prominence in the field, the third edition had no co-authors.

Stephenson was concerned, in the early days of her bacterial research, with enzymes and how bacteria adapted to changes in their environment. In 1928, she separated lactic dehydrogenase from *E. coli* and the gonococcus.[118] This key enzyme plays a fundamental role in carbohydrate metabolism and is used in clinical medicine as an index of tissue destruction and inflamation. In 1929, she discovered a microbe which was responsible for a stench which developed in the Ouse River each summer. Stephenson found that it was a new enzyme, hydrogenase, which was changing the sulfate ion in the river to hydrogen sulfide.[119] Subsequent studies identified the properties of hydrogenase, a bacterial enzyme activating molecular hydrogen,[120] and the hydrogenylases, bacterial enzymes liberating molecular hydrogen.[121] Enzymology in the 1930s was a crude science, but the efforts of the early investigators began to illuminate key points which allowed the concept of cycles, such as the Cori cycle, to be introduced. Stephenson, a woman scientist working outside the traditions of one discipline (biochemistry), was able to bring about an innovative integration of several disciplines to carry out pioneering research in bacterial metabolism.

Sarah Ratner's research with amino acids has been fundamental to the understanding of many mysteries of protein metabolism. Her early work on the condensation of cysteine with formaldehyde served as a model system for the later determination of the structure of penicillin. Ratner (1903–) pioneered in the use of nitrogen isotopes to follow metabolic changes in body proteins and amino acids. In the mid-1940s, her studies of the role of enzymes in amino acid metabolism incorporated one of the earliest successful attempts to use cell-free systems. A description of the course of urea synthesis on an enzymatic level and of the biosynthesis of arginine resulted from this research. During the investigation of the urea cycle, Ratner discovered the function of a new enzyme and isolated and identified a key intermediate, arginosuccinic acid. Dr. Ratner developed a sensitive color test to study the formation of this acid in brain tissue. This led to a better understanding of the nature of a defect which causes mental retardation in children and is associated with large amounts of arginosuccinic acid in urine and cerebrospinal fluid.[122]

Although Ratner had excellent training as a biochemist (B.A. and M.A. at Cornell and Ph.D. at Columbia) and joined the biochemistry department at Columbia's College of Physicians and Surgeons in 1930, it was not until 1946 that she became an assistant professor at New York University's College of Medicine, becoming an adjunct associate professor in 1953. From 1957 she was at the Research Institute of the Board of Public Health of the City of New York. Dr. Ratner has been appointed to the editorial boards of the *Journal of Biological Chemistry* and *Analytical Biochemistry*. She has received the prestigious Neuberg and Garvan medals and has been elected to the National Academy of Sciences.[123]

HISTOCHEMISTRY

Another area in which women have made important contributions is histochemistry, a technique in which chemically specific reagents are used to demonstrate specific chemical properties of tissues. The advantage of this technique is the localization of some particular chemical property to particular cell populations within a tissue, and the localization of that chemical property to some specific area within individual cells, a determination which could not be made by a chemical analysis of whole tissues.

Historically, according to Pearse,[124] histochemistry emerged as a science in the first half of the nineteenth century. Its development has been and is dependent upon the evolution of chemistry and of biochemistry and upon the newer disciplines of biophysics and molecular biology.

Two women scientists won lasting renown as investigators in histochemistry: Maude Menten, who was responsible for the discovery of the azo-dye technique, and Helen Wendler Deane, who applied the science extensively and critically to the characterization of mammalian tissues.

Maude Menten (1879–1960) was one of the most versatile, innovative investigators in chemistry in the first part of this century. She received her B.A. in 1904 and her M.D. in 1911 from the University of Toronto. She was, successively, a demonstrator of physiology in MacCallum's laboratory at the University of Toronto, a research fellow at the Rockefeller Institute, and a research fellow at Western Reserve University. While she was studying with Michaelis in Berlin, the Michaelis-Menten equation was developed.[125] This equation, which was fundamental to the interpretation of how an enzyme reacts upon its substrate, gives an expression for the rate of an enzyme reaction. It is assumed that the reaction takes place in two steps: 1) the formation of a complex between the enzyme (E) and the substrate (S), followed by 2) the decomposition of the complex into the product (P) and the regenerated enzyme.

$$1) \quad E + S \rightleftarrows ES \qquad 2) \quad ES \rightleftharpoons P + E$$

The rate of the reaction is proportional to the concentration of the complex (ES) formed. The Michaelis-Menten expression, derived from this, is:

$$v = \frac{V \, [S]}{Km + [S]}$$

where v is the velocity of the reaction, V is its maximal velocity, and Km is the Michaelis constant.[126]

Ultimately, Menten earned a Ph.D. in biochemistry at the University of Chicago, and became a professor on the faculty of the University of Pittsburgh School of Medicine. She also served as chief pathologist at the Children's Hospital in Pittsburgh. She was known for her efforts in behalf of sick children and as an inspiring teacher.

In histochemistry, her publication in 1944 (with Junge and Green) of a

Maude Menten. Courtesy of Dr. G. H. Fetterman, University of Pittsburgh.

new technique for the demonstration of the enzyme alkaline phosphatase ushered in the new azo-dye method.[127] Later, Menten and Janouch confirmed the practical application of the method in biological investigation by showing that decreases in the enzyme related directly to the degree of renal damage in an experimental model.[128]

Prior to the development of azo-dye methods, demonstration of enzymes within cells depended upon cumbersome procedures involving the use of calcium and/or cobalt salts, and results were ambiguous. One of the principal uses of azo-dye coupling techniques in medicine today is in the diagnosis of different forms of leukemia, which is essential in determining what treatment regimes are indicated for individual patients. In addition, azo-dye techniques became important experimental tools in many areas of biology.

In 1944, Menten and co-workers were the first to use another important technique, electrophoresis, to solve a biochemical problem.[129] In 1940 almost nothing was known at the molecular level about the constitution of the red blood cell hemoglobins (the red protein pigments which carry oxygen). It was perceived, however, that the "F" hemoglobin of the mammalian fetus was biochemically different from the "A" hemoglobin of adult red blood cells. Menten succeeded in the separation of the two hemoglobins, thereby permitting the study of their different chemical properties.

The principle of electrophoretic separation depends upon the fact that molecules travel across an electric field at different rates. These rates are governed by the size and the cumulative exposed charges of the proteins. Some years later, Linus Pauling used the method of hemoglobin electrophoresis to show that in sickle cell disease the "S" molecule was different from the "A" molecule. This discovery ushered in the now-famous concept of molecular disease.

In contrast to Menten, who applied pioneering new approaches to problems in various widely separated disciplines, Helen Wendler Deane (1917–1966) applied approaches via histochemistry (among other scientific tools) to interdisciplinary concerns. Her accomplishments in this arena catalyzed the introduction of histochemistry into a broader biologic base at a time when numerous methods were being discovered but when applications to "real life" were yet to be explored. Deane possessed an enormous body of scientific knowledge which spanned anatomy, biochemistry, and cell biology, and she used histochemistry as an integrating tool in more than seventy of her publications. In *The Adrenocortical Hormones*,[130] anatomy, chemistry, and physiology are combined to lead to a better understanding of the adrenal glands.

In her studies of the adrenal glands, Deane identified the kinds of cells which line the adrenal vessels[131] and determined the distribution of ascorbic acid in the cortical tissues.[132] In vitamin studies she mapped the changes both in the cells of various zones and in adrenal hormone secretions when there was a deficiency in pantothenic acid[133] and depletion of pyridoxine[134] and thiamine.[135] She showed that sodium deprivation resulted in marked histological alteration in the adrenal cortex and investigated the effect of the sodium/potassium ratio on cortical secretions,[136] as well as the relationship between the hormone deoxycorticosterone and sodium retention.[137]

Also noted among Deane's significant contributions were her studies with Manfred Karnovsky on the nature of the fat-droplets formed in hormone-secreting tissues. These droplets had been assumed to be hormonally related substances, when in fact they are aldehydes (byproducts not in the mainstream).[138]

Senator McCarthy's witch hunts of the 1950s found Deane, who was "deeply humanitarian" and "guided by a genuine sense of fairness," accused and then cleared while she held an appointment at Harvard. The appointment "was allowed to expire." Ultimately, she obtained a position at the Albert Einstein School of Medicine. Greep, her obituarian, stated that "in passing this way, she contributed much to the advancement of truth and to the welfare of mankind to which she was earnestly dedicated."[139]

CHROMATOGRAPHY

The invention of new instruments and techniques during the twentieth century has made possible much of the continuing rapid growth of chemistry and has added immeasurably to our understanding of the nature of matter. Mass spectrometry, nuclear magnetic resonance techniques, X-ray crystallography, and chromatography are among the important tools of modern chemists and biochemists. Women are seldom mentioned as agents in the development of these techniques, except in the case of crystallography and chromatography. The unique contributions of two women were vital to the development of chromatography, and the work of one of these scientists was of such a fundamental nature that she has been called the "mother of chromatography."[140]

Chromatography is a powerful technique in which the substances present in a mixture are separated by different rates of travel down a column or over a "flat bed." The chromatographic process was first described in a publication in 1906 by Michael S. Tswett (the "father of chromatography"), a Russian botanist. He coined the name from the Greek words meaning "color writing" because he could see, on a chalk column, the separation of the brilliantly colored plant pigments from leaves.[141] For over forty-five years little use was made of this technique, but after World War II chromatography was reborn. Among the scientists who were responsible for the rebirth were Erika Cremer, a physical chemist who laid the foundation of modern gas adsorption chromatography, and Marie S. Schraiber, a pharmacist who, with N. A. Izmailov, developed thin-layer chromatography.[142]

Erika Cremer (1900–   ) was born in Munich into a distinguished German scientific family. Her great-grandfather, grandfather, father, and two brothers were scientists and university professors.[143] Cremer received her Ph.D. from the University of Berlin in physical chemistry in 1924. For many years she was unable to obtain a permanent appointment because she was a women. Finally, during World War II Cremer was given a temporary appointment as Dozent at the University of Innsbruck in the Tyrol. It was assumed that she would relinquish the position when the war ended and the men returned; however, in 1945 she was appointed head of the Institute of Physical Chemistry at Innsbruck, because of her obvious talent and capabilities. In spite of her position as head of an institute, she was not promoted to professor at the university for six years.

As in the case of Lise Meitner, the turmoil in Europe in the thirties and forties played a large role in Erika Cremer's career, both helping and hindering its development. World War II certainly gave her the opportunity to demonstrate her unique abilities as a scientist and an administrator. However, because of conditions at the end of the conflict, her first chromatography paper, which set forth the basic theoretical principles for gas-solid chromatography, was never published. This short note on the theoretical relationship of chromatographic characteristics to their adsorption energies was set for publication in early 1945 in *Naturwissenschaften*; however, the journal suspended publication after the Nov./Dec. issue and the proofs disappeared.[144] It was not until 1976 that the paper was published in *Chromatographia* because of its historical importance.[145]

Even after an air raid damaged the university and all research was suspended, Dr. Cremer with her first graduate student, using any spare parts they could find, built the first modern gas chromatograph and research on this instrument continued during the immediate postwar years. Erika Cremer was, therefore, the first to develop gas chromatography as an analytical technique for both qualitative and quantitative analysis and for physicochemical measurements. More importantly, she developed much of the fundamental theory of the method, introducing many of the basic terms used routinely in chromatography. She continued to carry out research in adsorption studies and various aspects of gas chromatography, especially in the development of substance-selective detectors.

From 1940, Professor Cremer had been reporting her results at various scientific meetings, but the general reaction was "fairly skeptical and negative," especially from organic chemists. Finally, in 1951, her theoretical and experimental work was published.[146]

The development of chromatographic instrumentation has played a major role in many scientific "breakthroughs" in the past three decades and the gas chromatograph is today indispensable to the organic chemist.

We know relatively little about the other woman, Marie Semenova Schraiber (1904–   ), who contributed to the rebirth of chromatography.[147] A native Ukrainian, she received her graduate degree in 1939 at the Ukrainian Institute for Experimental Pharmacy under Nikolai A. E. Izmailov. Izmailov was professor and head of the physical chemistry department of Kharkhov State University as well as head of the physical chemistry laboratory at the institute. Schraiber was appointed a senior researcher in the laboratories of analytical chemistry there and, for the next forty years, she was active in the field of pharmaceutical analysis, publishing papers on such techniques as complexometry, nonaqueous titrations, and chromatography. With Izmailov, she pioneered the development of the technique of thin-layer chromatography (TLC), their first publication appearing in 1938. That paper, entitled "A Spot Chromatographic Method of Analysis and Its Application in Pharmacy,"[148] marked the invention of this important research technique. In many texts it is Egon Stahl, not the Russian scientists, who is credited with first developing TLC. Stahl, however, recognized the priority of the Russians' work and, in 1962, wrote in his classic volume, "I dedicate my book to Prof. N. Izmailov and M. Schraiber, pioneers of thin-layer chromatography."[149] Today thousands of papers have been written on TLC and it is used in almost every field of the physical and biological sciences, in clinical chemistry, and in quality control in industry.

Although Dr. Izmailov was recognized for his scientific work and awarded the Mendeléev Prize of the Academy of Sciences of the USSR, the only record we have of recognition for Dr. Schraiber was when, on her seventieth birthday, "the Collective of the Kharkhov Chemistry and Pharmacy Institute honored Dr. Schraiber on her distinguished career of forty years in analytical research."[150] Dr. Schraiber remained a senior researcher throughout her long scientific career.

CONCLUSION

Women have had a part in the development of chemistry from prehistoric times, when their household tasks and utensils became the techniques and apparatus of early chemistry. Women played an important role in the alchemy, serving as founders and contributors to the "Black Art" and as inventors of methods and tools, and providing a presence which was important to a practice based upon the theory of opposites. Although the volume written by Marie Meurdrac[151] indicates that there were women studying chemistry in the seventeenth century outside of the alchemical tradition, women served only as assistants or popularizers of the discipline during the exciting early days of modern chemistry.

During the first seventy-five years of the nineteenth century, we find few women chemists because the doors of institutions of higher education were largely closed to them. Gradually, however, women were able to attend graduate schools and find places in laboratories to carry out research in chemistry. At least one woman worked in her kitchen, while others inspired the formation of laboratories for their sex.

Successful women chemists in the first half of the twentieth century had to overcome many barriers, which impeded their careers. These women produced innovative solutions, however, which allowed them to satisfy the urge to discover more about the nature of matter. If organic chemistry laboratories were closed to them, they worked in nutrition or in agricultural chemistry. They discovered that they were welcome in the field of biochemistry: they established group research in women's colleges. Although few were able to gain high academic positions, their achievements were of inestimable importance to the development of chemistry.

These women chemists, only a small number of whom could be included in this chapter, won many awards, headed large laboratories, laid the foundation for techniques which changed the practice of chemistry, discovered the cause of diseases, and carried out research which was fundamental to our understanding of the reactions of matter. Their contributions make complete the story of the history of chemistry, for they have surveyed frontiers of the "boundless horizons" of science.[152]

NOTES

1. George B. Kauffman and H. Harry Szmant, eds., *The Central Science: Essays on the Uses of Chemistry* (Fort Worth: Texas Christian University Press, 1984); Theodore L. Brown and H. Eugene LeMay, Jr., *Chemistry: The Central Science*, 2nd ed. (Englewood Cliffs, NJ: Prentice-Hall, 1981), pp. xvii–xviii.

2. Eduard Farber, *The Evolution of Chemistry*, 2nd ed. (New York: The Ronald Press, 1969), p. 16.

3. Robert P. Multhauf, *The Origins of Chemistry* (New York: Neale Watson Academic Publications, 1967), p. 35.

4. Ibid., p. 43.

5. Ibid., p. 54.

6. Farber, *The Evolution of Chemistry*, pp. 28–29.

7. J. R. Partington, *A Short History of Chemistry* (London: Macmillan & Co., 1951), p. 19.

8. Martin Levey, *Chemistry and Chemical Technology in Ancient Mesopotamia* (Amsterdam: Elsevier, 1959), p. 143.

9. Martin Levey, "Some Chemical Apparatus in Ancient Mesopotamia," *Journal of Chemical Education* 32(1955):180–183.

10. Ibid., p. 182.

11. Sources for alchemy include: C. A. Burland, *The Arts of the Alchemists* (New York: Macmillan, 1967); M. Caron, *The Alchemists* (New York: Grove Press, 1961); Allison Coudert, *Alchemy, the Philosopher's Stone* (Boulder,CO: Shambhala, 1980); Henry M. Leicester, *The Historical Background of Chemistry* (New York: Dover Publications, 1956); Jack Lindsay, *The Origins of Alchemy in Graeco-Roman Egypt* (New York: Barnes & Noble, 1970); Ronald Pearsall, *The Alchemists* (London: Weidenfeld & Nicolson, n.d.); John Read, *Prelude to Chemistry, an Outline of Alchemy* (Cambridge, MA: M.I.T. Press, 1966); John Maxson Stillman, *The Story of Alchemy and Early Chemistry* (New York: Dover Publications, 1960); F. Sherwood Taylor, *The Alchemists* (New York: Henry Schuman, 1949).

12. Richard S. Westfall, *Never at Rest, a Biography of Isaac Newton* (Cambridge: Cambridge University Press, 1980), pp. 530–531.

13. *Egyptian Mythology* (New York: Tudor Publishing Co., 1965), p. 57.

14. Partington, *A Short History of Chemistry*, p. 20.

15. Raphael Patai, "Maria the Jewess—Founding Mother of Alchemy," *Ambix* 29(1982):177–197.

16. Ibid., p. 182.

17. Ibid., p. 183.

18. Burland, *The Arts of the Alchemists*, p. 22.

19. Patai, "Maria the Jewess," p. 191.

20. Lindsay, *The Origins of Alchemy in Graeco-Roman Egypt*, pp. 256–257.

21. Ibid., 260.

22. Taylor, *The Alchemists*, p. 51.

23. Read, *Prelude to Chemistry*, p. 66.

24. Coudert, *Alchemy, the Philosopher's Stone*, pp. 49–52.

25. Ibid., pp. 60–61.

26. Joseph Needham, *Science and Civilization in China* (Cambridge: Cambridge University Press, 1956), Vol. 3. Pt. 3, p. 38.

27. Burland, *The Arts of the Alchemists*, pp. 190–197.

28. Leicester, *The Historical Background of Chemistry*, p. 103.

29. G. C. Meyer, *Science for English Women, 1650–1760, the Telescope, the Microscope and the Feminine Mind* (Berkeley: University of California Press, 1955).

30. Lloyd O. Bishop and Wills DeLoach, "Marie Meurdrac, First Lady of Chemistry?" *Journal of Chemical Education* 47(1970):448.

31. Leicester, *The Historical Background of Chemistry*, pp. 126–128.

32. Ibid., p. 119.

33. Aaron J. Ihde, *The Development of Modern Chemistry* (New York: Harper & Row, 1964), pp. 57–88.

34. Denis I. Duveen, "Madame Lavoisier," *Chymia* 4(1953):13.

35. For another view of Mme. Lavoisier's contribution, see Margaret Alic, *Hypatia's Heritage* (Boston: Beacon Press, 1986).

36. Eva Armstrong, "Jane Marcet and Her Conversations on Chemistry." *Journal of Chemical Education* 15(1938):53.

37. Ibid., p. 56.

38. Clara De Milt, "Carl Weltzein and the Congress of Karlsruhe," *Chymia* 1(1948):153.

39. Ann Tracey Tarbell and D. Stanley Tarbell, "Helen Abbott Michael: Pioneer in Plant Chemistry," *Journal of Chemical Education* 59(1982):548–549.

40. Vera Bogdanovskara Popova (1868?–1897) received a Doctor of Science degree in 1892 from the Unviersity of Geneva for work on dibenzylketone. She returned to Russia, taught for a time at St. Petersburg Women's College, and was killed in an explosion in her private laboratory in 1897. Eleanor S. Elder and Sue-Dee Lazzerini, "The Deadly Outcome of Chance—Vera Estaf'evna Bogdanoskara," *Journal of Chemical Education* 56(1979):251.

41. Helen Abbott Michael, *Studies in Plant and Organic Chemistry and Literary Papers* (Cambrdige, MA: Riverside Press, 1907), p. 73.

42. R. E. Alston and B. L. Turner, *Biochemical Systematics* (Englewood Cliffs, NJ: Prentice-Hall, 1963), pp. 45, 46, 67.

43. Tarbell and Tarbell, "Helen Abbott Michael," p. 548.

44. C. H. Giles and S. D. Forrester, "The Origins of the Surface Film Balance," *Chemistry and Industry* 1971(1971):43.

45. Ibid., p. 47.

46. Ibid., p. 43.

47. M. Elizabeth Derrick, "Agnes Pockels, 1862–1935," *Journal of Chemical Education* 59(1982):1030.

48. Giles and Forrester, "The Origins of the Surface Film Balance," p. 46.

49. Ibid., p. 52.

50. Charlene Steinberg, "Yula Vsevolododovna Lermontova (1846–1919)," *Journal of Chemical Education* 60(1983):757.

51. In 1887, when Abbott inquired about attending Hofmann's lectures in Berlin she was told it would have to be in secret "since women are not permitted in the auditorium, nor to work in the rooms with men students." Helen Abbott Michael, *Studies in Plant and Organic Chemistry*, p. 46.

52. This date, which is universally accepted, may be late. Paul A. Jones has found that Ina Milroy received a Ph.D. from Berlin in 1904. (Private communication.)

53. Margaret Rossiter, *Women Scientists in America, Struggles and Strategies to 1940* (Baltimore: Johns Hopkins Press, 1982), pp. 29–50, chaps. 7 and 9.

54. There were, of course, exceptions. Rachel Lloyd had an appointment at the University of Nebraska, Dorothy Nightingale (1902–   ) was professor at the

University of Missouri, and Leonora Bilger, at the University of Hawaii. Hazel Bishop conceived her formula for "indelible lipstick" while doing research for the military during World War II.

55. Mary Elvira Weeks and Henry M. Leicester, *Discovery of the Elements*, 7th ed. (Easton, PA: Journal of Chemical Education, 1968), pp. 778–780, 781–783, 792, 823–827, 838–839.

56. Ihde, *The Development of Modern Chemistry*, p. 285.

57. Edward R. Atkinson, "Emma Perry Carr," in Wyndham D. Miles, ed., *American Chemists and Chemical Engineers* (Washington, D.C.: American Chemical Society, 1976), pp. 66–67.

58. Emma Perry Carr, "Research in a Liberal Arts College," *Journal of Chemical Education* 34(1957):468–469.

59. Edward R. Anderson, "Mary Lura Sherrill," in Miles, *American Chemists and Chemical Engineers*, p. 436; see also *Chemical and Engineering News* 46(51), (1968):57.

60. *Chemical and Engineering News* 35 (17), (1957):70.

61. Mary L. Sherrill, "Group Research in a Small Department," *Journal of Chemical Education* 34(1957):466–468.

62. Carr, "Research in a Liberal Arts College," p. 469.

63. Atkinson, "Emma Perry Carr," p. 67.

64. Sherrill, "Group Research in a Small Department," p. 467.

65. Alfred E. Hall, "Baccalaureate Origins of Doctoral Recipients in Chemistry," *Journal of Chemical Education* 62(1985):406.

66. "Garvan Medal to Gertrude B. Elion," *Chemical and Engineering News* 46(3), (1968):65.

67. Nina Matheny Roscher and Phillip L. Ammons, "Early Women Chemists of the Northeast," *Journal of the Washington Academy of Sciences* 71(4), (1981):181.

68. Ann T. Tarbell and D. Stanley Tarbell, "Rachel Lloyd (1839–1900): American Chemist," *Journal of Chemical Education* 59(1982):743–744.

69. Ibid., p. 744.

70. Paul A. Sanford, "Allene R. Jeanes," *Carbohydrate Research* 66(1978):3–5.

71. "Research of Tap: The Dextran Story," *Chemical and Engineering News* 32(1956):1983.

72. Ihde, *The Development of Modern Chemistry*, p. 418.

73. C. F. Cori, S. P. Colowick, and G. T. Cori, "The Isolation and Synthesis of Glucose-l-phosphoric Acid," *Journal of Biological Chemistry* 121(1937):465.

74. C. F. Cori, G. T. Cori, and A. A. Green, "Crystalline muscle phosphorylase III. Kinetics," *Journal of Biological Chemistry* 151(1943):39.

75. C. F. Cori and G. T. Cori, "Glycogen Formation in the Liver from d- and l-Lactic Acid," *Journal of Biological Chemistry* 81(1929):389.

76. G. T. Cori and C. F. Cori, "Crystalline muscle phosphorylase IV. Formation of glycogen," *Journal of Biological Chemistry* 151(1943):57.

77. G. T. Cori and C. F. Cori, "Glucose-6-phosphatase of the Liver in Glycogen Storage Disease," *Journal of Biological Chemistry* 199(1952):661.

78. G. T. Cori, "Glycogen Structure and Enzyme Deficiencies in Glycogen Storage Disease," *Harvey Lectures*, 1952–53 p. 145.

79. John Parascandola, "Gerty Cori, 1895–1957," *Radcliffe Quarterly* 66(4), (1980):11.

80. Herman Kalckar, *Science* 128(3314), (1958):17.

81. Parascandola, "Gerty Cori," p. 12.

82. Bernardo Houssay, "Carl F. and Gerty T. Cori," *Biochemica et Biophysica Acta* 20(1956):11.

83. Severo Ochoa, "Gerty T. Cori, Biochemist," *Science* 128(3314), (1958):16.

84. Houssay, "Carl F. and Gerty T. Cori," p. 15.

85. Ralph Oesper, "Gertrude Woker," *Journal of Chemical Education* 30(1953):435.

86. Mel Gorman, "Agnes Fay Morgan," in Miles, *American Chemists and Chemical Engineers*, pp. 348–349.

87. Rossiter, *Women Scientists in America*, p. 78.

88. *Chemical and Engineering News* 46(33), (1968):158.

89. Gorman, "Agnes Fay Morgan," p. 348. See also W. H. Sebrell and Robert S. Harris, eds., *The Vitamins, Chemistry, Physiology, Pathology and Methods*, 2nd ed., vol. 2, p. 28, and Paul Gyorgy and W. N. Pearson, eds., *loc. cit.*, vol. 6, p. 215 (New York: Academic Press, 1968).

90. Gorman, "Agnes Fay Morgan," p. 348.

91. Sebrell and Harris, *The Vitamins*, vol. 1, pp. 445–446.

92. Agnes Fay Morgan, "The Effect of Vitamin Deficiencies on Adrenocortical Function," *Vitamins and Hormones* 9(1951):161.

93. Sebrell and Harris, *The Vitamins*, vol. 1, p. 450.

94. Morgan, "The Effect of Vitamin Deficiencies," p. 184.

95. Ibid., p. 189.

96. Edmund R. Yendt et al., "Clinical Aspects of Vitamin D," in H. F. DeLuca and J. W. Suttie, eds., *The Fat Soluble Vitamins*, 6th ed. (Madison: University of Wisconsin Press, 1970), p. 149.

97. Margaret A. Cavanaugh, "Contributions of Icie Macy Hoobler to Chemistry," paper presented at the 178th National Meeting of the American Chemical Society, September 1979.

98. Icie Macy Hoobler, *Boundless Horizons: Portrait of a Pioneer Woman Scientist* (Smithtown, NY: Exposition Press, 1982), p. 40.

99. Oral proceedings of the Symposium on Women in Chemistry, American Chemical Society, September 1979.

100. Cavanaugh, "Contributions of Icie Macy Hoobler," p. 5.

101. Hoobler, *Boundless Horizons*, p. 72.

102. Cavanaugh, "Contributions of Icie Macy Hoobler," p. 9.

103. Ibid., p. 10.

104. Icie G. Macy, "Composition of Human Colostrum and Milk," *American Journal of Diseases of Children* 78(1948):589.

105. Icie G. Macy and Helen A. Hunscher, "An Evaluation of Maternal Nitrogen and Mineral Needs During Embryonic and Infant Development," *American Journal of Obstetrics and Gynecology* 27(1934):878.

106. Icie G. Hoobler and Harold C. Mack, "Implications of Nutrition in the Life Cycle of Women," *American Journal of Obstetrics and Gynecology* 68(1954):131.

107. Icie G. Macy and H. A. Hunscher, "Calories—a Limiting Factor in the Growth of Children," *Journal of Nutrition* 45(1951):189.

108. Hoobler, *Boundless Horizons*, p. 98.

109. Cavanaugh, "Contributions of Icie Macy Hoobler," p. 10.

110. Ibid., p. 10.

111. Virginia A. Beal, *Nutrition in the Life Span* (New York: John Wiley & Sons, 1980), p. 7.

112. Cavanaugh, "Contributions of Icie Macy Hoobler," pp. 12–15.

113. Robert E. Kohler, "Innovation in Normal Science, Bacterial Physiology," *Isis* 76(1985):162–181.

114. Marjory Stephenson, *Bacterial Metabolism*, 3rd ed. (London: Longmans, Green and Co., 1949).

115. D. Needham, "Women in Cambridge Biochemistry," in Derek Richter, ed. *Women Scientists, the Road to Liberation* (London: MacMillan, 1982), pp. 161–162.

116. Ibid., p. 162.

117. Stephenson, *Bacterial Metabolism*, p. vii.

118. M. Stephenson, "On Lactic Dehydrogenase: a Cell-free Enzyme Preparation Obtained from Bacteria," *Biochemical Journal* 22(1928):605.

119. Kohler, "Innovation in Normal Science," p. 175.

120. M. Stephenson and L. H. Strickland, "Hydrogenase, a Bacterial Enzyme Activating Molecular Hydrogen. I. The Properties of the Enzyme," *Biochemical Journal* 25(1931):205.

121. M. Stephenson and L. H. Strickland, "Hydrogenases. Bacterial Enzymes Liberating Molecular Hydrogen," *Biochemical Journal* 36(1932):712.

122. *Chemical and Engineering News* 39(14), (1963):103. A woman biochemist, Grace Medes (1886–1967) was responsible for the discovery and study of another metabolic defect, tyrosinosis (prevention of the oxidation of tyrosine to 2,5 dihydroxyl phenylpyruvic acid). *Chemical and Engineering News* 33(15), (1955):1505.

123. *American Men and Women of Science*, Jacques Cottell Press, ed., 15th ed. (New York: R. R. Bowker, 1982), vol. 6, p. 61.

124. A. G. E. Pearse, *Histochemistry, Theoretical and Applied* (Boston: Little, Brown and Co., 1960).

125. L. Michaelis and M. L. Menten, *Biochem. Z.* 49(1913):333.

126. Keith J. Laidler, *Introduction to the Chemistry of Enzymes* (New York: McGraw-Hill Book Company, 1954), pp. 18–25.

127. M. L. Menten, J. Junge, and M. H. Green, "A Coupling Histochemical Azo-dye Test for Alkaline Phosphatase in the Kidney," *Journal of Biological Chemistry* 153(1944):471.

128. M. L. Menten and J. Janouch, "Changes in Alkaline Phosphatase in Kidney Following Renal Damage with Alloxan," *Proceedings of the Society of Experimental Biology and Medicine* 63(1946):33.

129. M.A. Aadersch, D. A. Wilson, and M. L. Menten, "Sedimentation Constants and Electrophoretic Mobilization of Adult and Fetal Carbonylhemoglobin," *Journal of Biological Chemistry* 151(1944):301.

130. H. W. Deane, *The Adrenocortical Hormones—Their Origin, Chemistry, Physiology, and Pharmacology* (Berlin: Springer-Verlag, 1962).

131. Albert B. Eisenstein, ed., *The Adrenal Cortex* (Boston: Little, Brown & Co., 1965), p. 23.

132. Ibid., p. 316.

133. Ibid., p. 321.

134. Ibid., p. 324.

135. Frank Hartman and Katharine A. Brownell, *The Adrenal Gland* (Philadelphia: Lea and Febiger, 1949), p. 282.

136. Eisenstein, *The Adrenal Cortex*, p. 332.

137. Hartman and Brownell, *The Adrenal Gland*, p. 69.

138. M. L. Karnowsky and H. W. Deane, "Aldehyde Formation in the Lipid Droplets of the Adrenal Cortex during Fixation, as Demonstrated Chemically and Biochemically," *Journal of Histochemistry and Cytochemistry* 3(1955):85.

139. Roy O. Greep, "Professor Helen Wendler (Deane) Markham," *Journal of Histochemistry and Cytochemistry* 14(1966):881–883.

140. L. S. Ettre and A. Zlatkis, "75 Years of Chromatography—A Historical Dialogue," *Journal of Chromatography Library* (Amsterdam: Elsevier Scientific Publishing Co., 1979), pp. 17, 21–30.

141. Ibid., pp. 483–490.

142. Ibid., pp. 413–447.

143. L. S. Ettre, *Chromatographia* 8(7), (1975):309.

144. L. S. Ettre, *Chromatographia* 9(8), (1976):363–366.

145. E. Cremer, "Dr. Cremer's first unpublished paper on gas chromatography: On the Migration Speed of Zones in Chromatographic Analysis. How We Started Work in Gas Adsorption Chromatography," *Chromatographia* 9(8), (1976):365.

146. E. Cremer and F. Prior, *Zeitschrift für Electrochemie*, 55(1951):66–70; E. Cremer and R. Muller, *Zeitschrift für Electrochemie*, 55(1951):217–220; E. Cremer and R. Muller, *Mikrochem/Mikrochim Acta*, 36/37(1951):553–560.

147. Ettre and Zlatkis, "75 Years of Chromatography," pp. 413–417.

148. N. A. Izmailov and M. S. Schraiber, *Farmatsiya* 3(1938):1.

149. E. Stahl, ed., *Dunnschichtchromatographia* (Berlin: Springer Verlag, 1962).

150. Ettre and Zlatkis, "75 Years of Chromatography," p. 413.

151. Marie Meurdrac, *La Chimie charitable et facile, en faveur des dames*, 2nd ed. (Lyon: Chez Jean Baptiste Deville, 1680).

152. Hoobler, *Boundless Horizons*.

Parts of this chapter have appeared in: Jane A. Miller and Virginia F. McConnell, "A Brief History of Chemistry," an educational tape published by the American Chemical Society, 1982; and Jane A. Miller, "Daughters of Isis: Women in Chemistry," *2YC Distillate* 3(3), (1985):1. This chapter also benefited from manuscripts prepared by Phyllis Brown and Melanie Ianuzzi, and Patricia Farnes. Virginia Fenner McConnell provided valuable bibliographic and factual material.

MAUREEN M. JULIAN

Women in Crystallography

The Science of crystallography has been accused of being overrun with women and has been likened to "intellectual knitting." The two premises implied by these assertions concern both the number of female crystallographers and the nature of crystallography itself. Some consider crystallography to be a service rather than a discipline because of the many materials which can be studied and the specialized techniques that are necessary. There is also a feeling that the discipline is overrun with women. I have asked both crystallographers and people familiar with crystallography what percentage of crystallographers they thought were women. The reply always lies between 40 and 60 percent. A male reviewer for the profile of Rosalind Franklin told me that I had no cause for complaint because, of all of the physical sciences, crystallography is "much too full of women." The state of affairs is somewhat similar to the baseball world when a few blacks made it to the top causing many people to consider this proof of the breaking of the color line when in fact it represented exploitation of a few geniuses.

WHAT IS CRYSTALLOGRAPHY?

Crystallography includes structure analysis of crystals, semicrystalline materials, and glasses, the study of crystal growth, diffraction theory, surface chemistry and physics, solid state reactions, molecular orbital theory, electron density distribution, neutron and electron diffraction, diffraction by synchrontron radiation, optics, mathematics, symmetry, twinning (one form of crystal growth error), the production and analysis of X-rays, computer programming, equipment development, X-ray photography, X-ray spectroscopy, the history of crystallography, and crystallography in science education.

The basic experimental technique of X-ray diffraction consists of mounting a sample in the path of a beam (usually an X-ray or a neutron beam), recording the positions and intensities of the diffracted beams, and using this information to determine the molecular structure of the material under consideration. To do this the mathematics of Fourier transforms, group theory, Patterson functions, and the like are used. Clever calculating analogs were devised by early crystallographers. However, when several thousand pieces of data are collected, the long and tedious computations require computers. Starting in the 1960s crystallographers were among the first and most intensive users of computers.

Diffraction patterns are also used as fingerprints for identification of crystalline materials. In the past few years, thousands of diffraction patterns have been indexed in computers. These files can be searched and matched for identification of unknown substances.

Crystallography is an interdisciplinary field in which chemists, biochemists, physicists, geologists, engineers, and mathematicians all belong to an active International Union of Crystallography. The diversity among crystallographers is due in part to the great variety of materials that yield information by crystallographic techniques: diamonds, DNA, penicillin, insulin, vitamin B-12, frog venoms, polypeptides, kidney stones, plant viruses, and coal. Women have made major contributions to the elucidation of every one of these materials. Crystallography is such an interdisciplinary science that Kathleen Lonsdale once likened the subject to cuckoos found nesting in various departments.

WOMEN IN CRYSTALLOGRAPHY

I counted the number of women crystallographers in the name index of P. P. Ewald's definitive historical work, *Fifty Years of X-ray Diffraction*,[1] and found that of the over 1,200 names listed only 3 percent were of women. The women can be identified because most of their names are preceded by a "Miss" or a "Mrs." following a protocol of title not demanded of male scientists. In the 1976 membership directory of the American Crystallographic Association, a count of the first 1,200 names showed that 8 percent were women. For comparison, in the 1977 membership directory of the American Physical Society, of the first 1,200 names 2 percent were women.

The 1981 *World Directory of Crystallographers*[2] permits further analysis with a data breakdown by country (Table 1). Between 1976 and 1981 there was an overall gain in women crystallographers, and an increase in the United States from 8 percent to 11 percent. The international figure for women in crystallography in 1981 was 14 percent of a population of 8,174 entries in the *Directory*. Thus the concentration of women crystallographers is seven times as great as the concentration of women physicists. Nonetheless, the idea that crystallography is a woman's field is certainly erroneous, since 86 percent of the world's crystallographers are men.

If the proportion of women in crystallography is no greater than 14 percent and if the impression still exists that the field is saturated with women, then what is the reason for this impression? Why does this small percentage of women seem so large? Have women's contributions in crys-

TABLE 1.
Country-by-country breakdown of total crystallographic population and percent of women crystallographers

Country	Total Population of Crystallographers	Percent Women
Thailand	28	50
Chile	43	37
Portugal	20	35
Russia	657	32
Poland	153	32
Argentina	28	32
Bulgaria	74	27
Hungary	71	25
Romania	32	25
Philippines	8	25
France	348	21
Mexico	39	20
Korea	15	20
Venezuela	10	20
Ireland	5	20
Sweden	206	19
Brazil	108	19
Denmark	62	19
Italy	276	18
Egypt & Arab Republic	77	18
Colombia	28	18
Yugoslavia	128	17
Czechoslovakia	91	16
German Democratic Republic	376	15
Indonesia	13	15
Canada	165	14
Spain	113	14
Ivory Coast	8	12
United States	1,622	11
Israel	71	11
Tunisia	18	11
United Kingdom	651	10
India	340	10
Belgium	87	10
Greece	82	10
South Africa	80	10
New Zealand	49	10
Nigeria	9	10
Switzerland	133	7

Table 1 (continued)

Country	Total Population of Crystallographers	Percent Women
Austria	90	7
Pakistan	31	6
Germany, Federal Republic of	455	5
Australia	200	5
Malaysia	20	5
Bangladesh	25	4
Turkey	60	3
Japan	516	2
Bolivia	12	0
Ghana	8	0
Kenya	4	0

tallography been so pervasive as to make these women appear to be more numerous than they are?

Before looking at the history of women's contributions to crystallography, I would like to examine why women go into crystallography.

Since X-ray crystallography was born during the second decade of this century, and only developed after World War I, the field is relatively new and perhaps more free of traditional prejudices. The early appearance and welcome of a few brilliant women into the laboratories of the Braggs was crucial (see below). Many of the co-workers of these women, especially J. D. Bernal (1901–1971), later carried on the tradition of inviting women students and colleagues into their own laboratories. The first influential women crystallographers were Kathleen Lonsdale (1903–1971), a student and later colleague of the elder Bragg, and Dorothy Crowfoot Hodgkin (b. 1910), who had worked with J. D. Bernal. In turn, Lonsdale's and Hodgkin's own laboratories included many male and female graduate students and visiting scholars.

Another reason women are more evident participants in crystallography is an interesting condition inherent in at least one important part of crystallography known as structural analysis. This is the solution of the three-dimensional jigsaw puzzle in which each atom of the crystal is precisely located. Once this geometry is correctly solved, the crystallographer receives credit. Any further work on that substance is "refinement" and is referenced to the original work. This differs from other sciences, where early work may be completely supplanted by "slightly" more sophisticated techniques, which may measure an effect to one more significant place or at a slightly higher or lower temperature, or provide some other technical variation.

At first the crystallographic calculations were done by hand. Arnold

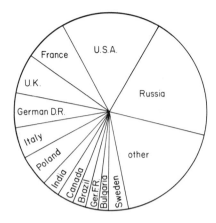

Distribution, by country, of women crystallographers. The number of women crystallographers is just over 1,000.

Beevers and Henry Lipson invented a series of strips which were aids to the calculations and these were widely used until the early sixties, when computers took over. This author remembers a big wooden box at Cornell University filled with hundreds of these strips, each about a half-inch wide and six inches long with numbers running along the edges. With the advent of modern electronics, many mathematically inclined women were attracted by the complex mathematics and extensive computing necessary for crystallographic research. Often these women, who would be turned away by prejudice from physics and mathematics, could practice a form of these disciplines in crystallography under the mantle of chemistry or geology. Indeed many crystallographers often leave crystallography for higher-paying computing careers.

In the early days of crystallography, the X-ray tubes were hand-blown and the high-voltage setups necessary to run them were dangerous and cumbersome. Now X-ray equipment is so expensive and difficult to maintain in good running condition that expert technicians, rather than crystallographers, are required to operate and repair the high-voltage and high-vacuum systems. This tends to decrease the male cultural advantage which assumes that men are better at fixing equipment than women. Unfortunately, offsetting this effect is the reluctance to allow women to head laboratories with large financial investments.

Anne Sayre, author of *Rosalind Franklin and DNA*[119] and wife of crystallographer David Sayre, wrote to me:

There is something in the ancient history of crystallography that is hard to isolate but nevertheless was there, that I can best describe as modesty. I have often wondered how much the Braggs were responsible for the unaggressive low-key friendly atmosphere that long prevailed in the field (and no longer seems to very much). Somehow the first and second and a few of the third generation crystallographers consistently conveyed an impression of working for pleasure, for the sheer joy of it—the idea of competition didn't seem to emerge very strongly until the 1960s or so. Uncompetitive societies tend to be good for women.

The Communist countries have a higher percentage of women crystallographers than do the non-Communist countries. Anne Sayre pointed out in a letter to me that both J. D. Bernal of England and M. Mathieu in France were card-carrying Communists. Their idealism included equality for women in the laboratory. This is borne out in the country-by-country breakdown of crystallographers (Table 1). Marie Przybylska, a Polish-born, Scotland-trained Canadian crystallographer, thought that the economic necessity of two incomes to the family in the Communist countries brought with it a new and better economic and technical equality.

Another common, but unflattering, opinion of how women first entered crystallography was stated in F. H. Portugal and J. S. Cohen's book, *A Century of DNA*:

Since the high speed computer had not yet been invented, the business of calculating data was a very laborious occupation and smart fellows who could find other things to do would generally do them, unless they were absolutely dedicated to the business of X-ray crystallography. Is it possible that these first class women got to be X-ray crystallographers because they were willing to do this work. [p. 267]

EARLY HISTORY OF CRYSTALLOGRAPHY

Wilhelm Conrad Roentgen (1845–1923) discovered X-rays in 1895 and took the first photograph showing bones in the living hand. Although he searched for experimental evidence of diffraction of X-rays, it was not until 1912, under the inspiration of Max Laue (1879–1960), that experimentalists Walter Friedrich and Paul Knipping recorded diffraction spots from a beautiful blue crystal of copper sulfate pentahydrate.

In England during the summer of 1913, Laue's paper generated much excitement for Professor William Henry Bragg (1862–1942) of the University of Leeds and his son, William Lawrence Bragg (1890–1971), who was on vacation from graduate studies at Cambridge University. Father and son had many enthusiastic scientific discussions. When young Bragg returned to Cambridge, he examined X-ray reflections from a crystal of mica and discovered the relationship connecting the wavelength of the incident X-ray beam, the periodicity of the crystal, and the locations of the spots in the diffraction pattern. This relationship is now known as Bragg's Law. Back at Leeds, his father invented the Bragg X-ray spectrometer, which is used to collect accurate X-ray data. Common salt, sodium chloride, was the first structure successfully analyzed. The brilliant complementary talents of the experimentalist father and the theoretician son placed England in the forefront of crystallographic development.

World War I interrupted almost all pure research. After the war, the Braggs divided the crystal world between them. The father, at University College, London, chose the organic structures and quartz, and his son, at the University of Manchester, took the rest of the inorganic substances.

ASSOCIATES OF WILLIAM HENRY BRAGG

The story of women in crystallography began after World War I when Nobel laureate William Henry Bragg started to establish his postwar laboratory at University College, London.[19,64] He was unhappy there and in 1923 moved to the Royal Institution. In the original group of twelve scientific workers that went with him, three were women: I. Ellie Knaggs, G. Mocatta, and Kathleen Yardley (later Lonsdale). I. Ellie Knaggs came from Cambridge University with a wide knowledge of mineralogical and optical methods.[64,78,79]

Later, other women joined the Bragg team. Ida Woodward, a mathematician, studied porphine and a whole range of transformations of single crystals of potassium nitrate.[81,142] Natalie C. B. Allen, from Australia, tested the sensitivity of various photographic plates to X-rays in 1919 and later worked on organics such as benzil.[5,6] Thora C. Marwick, from New Zealand, studied strychnine and cellulose.[11,111,112] Lucy Pickett, from the United States, worked on such organic compounds as the isomers of pentene, and from 1927 to 1937 Helen S. Gilchrist helped prepare long chained compounds with Berta Karlick of Austria.[45,46,78,115] M. E. Bowland made measurements of the diamagnetic anisotropy of various crystals in 1932, and C. F. Elam (later Tipper) worked on metals, especially aluminum and copper.[37,78]

Through the years, more than seventy scientific workers were associated with William Bragg at the Royal Institution. Five of these became Fellows of the Royal Society and established schools of crystallography of their own in British universities. Each in turn was responsible for the future inclusion of women into crystallography (Table 2). They were William Thomas Astbury (1898–1961), at Manchester; John Desmond Bernal, first at Cambridge and then at Birkbeck College; Kathleen Yardley Lonsdale, at University College, London; John Monteath Robertson (b. 1900), at Glascow; and Ernest Gordon Cox (b. 1906), at Birmingham and later Leeds.

While still working at the Royal Institution, W. T. Astbury collaborated with Kathleen Yardley on the space group tables or symmetry arrangements in three-dimensional space. The symmetry of the structure of every crystal can be described by one of the 230 space groups. The tables, which were published in 1924 in the *Philosophical Transactions of the Royal Society*,[8] were so helpful that an unusual second printing had to be made. Although

Table 2.

Genealogical table of women in crystallography whose professional lives have been directly influenced by the Braggs. * denotes female.

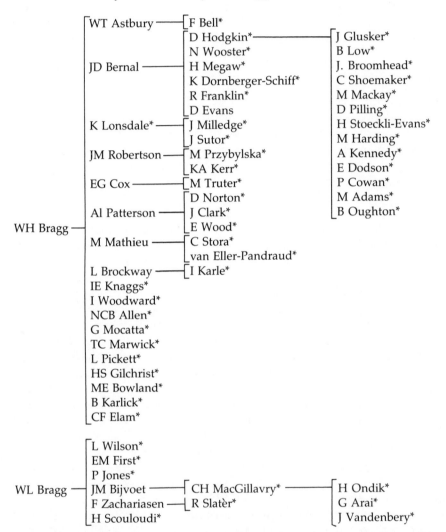

the mathematical results were well established, the Astbury-Yardley tables were easily read and therefore used by practicing crystallographers. In 1928, Astbury went to Leeds, where he studied natural fibers, such as wool and whalebone. A Cambridge graduate research scientist, Florence Bell (later Sawyer), worked with him on the X-ray patterns of proteins.[9] Some of their photographs laid the foundation for the later DNA studies.

Kathleen Lonsdale proved the planarity of the benzene ring,[82–84] edited the *International X-ray Tables*,[92,95,97] and had her own school of crystallography at University College, London.

Perhaps the largest number of women crystallographers was associated with J. D. Bernal (Table 2). The most famous, Dorothy Crowfoot Hodgkin (see profile), was with him at Cambridge University, where, in 1934, she began taking pictures of crystals of vitamin B-1, vitamin D, several of the sex hormones, and the protein pepsin. Nora Martin,[143] who later married crystallographer W. A. Wooster, was Bernal's first graduate student. She studied halogen compounds and brookite, a polymorph of titanium oxide. Bernal and Helen D. Megaw published a classical study of hydrogen bonding in metal hydroxides.[13] The hydrogen bond is an intermolecular bond where hydrogen acts as a bridge. Professor Lipson wrote to me that he believed this important work introduced the hydrogen bond. Helen Megaw's study of the titanates influenced the theory of ferroelectricity. A ferroelectric substance has a spontaneous electric polarization which can be reversed in an electric field. Recently (1982), when G. V. Gibbs of Virginia Polytechnic Institute and State University wanted to use Megaw's crystal structure textbook, *Crystal Structures: A Working Approach*,[113] in his X-ray course, he found, unfortunately, that the book was out of print.

Kate Dornberger-Schiff a refugee from Göttingen, joined Bernal at the beginning of World War II, when he first arrived at Birkbeck. Fearing that Birkbeck might be bombed, Schiff and others were evacuated, first to Bristol, and then to Oxford, where Schiff helped Dorothy Hodgkin with the calculations and interpretations of some three-dimensional Patterson maps of insulin. Schiff's work with Bernal was on silicate chemistry. She also published a study of the structure of hemoglobin of patients with sickle-cell anemia.[33,34] She introduced a systematic treatment of order-disorder structures,[35] especially those with regular layers stacked in irregular ways; disorder refers to the degree to which the crystal departs from its regular growth. The structure of purpurogallin[36] was an application of this theory. In 1966 Schiff became head of the Institute for Crystal Structure Research in Berlin. She died in 1981 just before she was to present a paper at the Twelfth International Crystallographic Congress.

Rosalind Franklin, after her unhappy period with Maurice Wilkins on the DNA project at King's College, London, came to Birkbeck in 1953 to work with Bernal on plant viruses (see profile).

John Monteath Robertson, another of the William Bragg "descendants," had a number of women crystallographers at his school in Glascow. Among them were Maria Przybylska,[116–118] now a senior research officer at the National Research Council of Canada in Ottawa, who worked on the analysis of complex natural product structures, and K. Ann Kerr,[21,77] an outstanding Canadian crystallographer at the University of Calgary, who is interested in biologically active compounds.

E. Gordon Cox from Bragg's laboratory worked with and later married Mary Truter.[24,135–136] She started in Leeds in 1947 and was made a Reader in 1960. While in Leeds she wrote 53 papers on coordination chemistry and the comparison of uncomplexed and complexed molecules. Complex ions are polyatomic ions to which one or more atoms or groups of atoms are bonded. Compounds containing complex ions are called coordination compounds. Truter's first crystal contained the entity responsible for aromatic nitration, nitronium perchlorate. The twelve-year gap between papers[24,135] arose because the two independent oxygen atoms in the perchlorate ion appear to have different numbers of electrons. On chemical grounds this seemed highly improbable, and only when allowance was made for anisotropic vibration of the constituent atoms in the crystal was the problem resolved. This was one of the first compounds to have been subjected to full molecular trimethylplatinum beta-diketone complexes.[136]

In 1966 Truter changed from university teaching to full-time research when she became deputy director of the newly formed Agricultural Research Council Unit in Structure Chemistry. She established the unit and in 1973, when it was transferred to Rothamsted Experimental Station as the Molecular Structures Department, she was appointed head. This work was aimed at understanding, at a molecular level, the discrimination shown by biological systems between chemically similar alkali and alkaline earth metal cations. So far Professor Truter has published 82 papers from this group.

The influence of W. H. Bragg was monumental. Almost everyone who worked with him at the Royal Institution was influenced by this contact. A. L. Patterson, a New Zealander who was educated at McGill in Canada, was in residence at the Royal Institution from 1924 to 1926. He set up an X-ray laboratory at Bryn Mawr, where he began teaching in 1936. One of his students was Dorita A. Norton, who received her Ph.D. in 1958. At Roswell Park Memorial Institute she began her work on the steroid hormones and authored or coauthored some fifty scientific papers. She was also a professor in the Department of Biophysical Sciences at the State University of New York at Buffalo and directed summer programs in X-ray crystallography for gifted high school students. In 1967 Dorita Norton and her staff transferred to the Medical Foundation of Buffalo, where, during the last three years of her life, she established a center for molecular endocrinology.

Patterson introduced Joan Robertson Clark to crystallography when he hired her to work at the Institute for Cancer Research. In 1956, at his recommendation she went to the U.S. Geological Survey and two years later she got her Ph.D. from Johns Hopkins University. Prior to that she worked on the Manhattan Project and in industry. She elucidated the crystal chemistry of the borates as exemplified in her work on hydrogen bonding in hydrated borates (1977). In 1969 she was a co-investigator of the Apollo lunar samples.

Another graduate of Bryn Mawr, although not a student of Patterson's, was Elizabeth Armstrong (Wood). She got to know Patterson later in connection with their work for the National Committee for Crystallography. Elizabeth Armstrong received her Ph.D. in 1939 in geology and chemistry under E. H. Watson. Her dissertation was on "Mylonization of Hybrid Rocks near Philadelphia, Pa." She taught geology and mineralogy at Bryn Mawr and Barnard College (1934–1943). Until 1967 she spent most of her professional life at Bell Telephone Laboratories on such projects as the irradiation coloring in quartz[7] and the phase transitions in silicon.[138] She was a member of the National Committee for Crystallography and in 1957 she was president of the American Crystallographic Association. She wrote *Crystal Orientation Manual*[139] and two popular science books, *Crystals and Light*[140] and *Science for the Airplane Passenger*.[141]

Marcel Mathieu was at the Royal Institution with Professor Bragg from 1925 to 1926. One of Mathieu's first students at the Sorbonne was Cecile Stora,[127] who later succeeded him there. Stora used X-ray diffraction to study organic dyes, derivatives of triphenylmethane and of indigo, with Mrs. van Eller-Pandraud. Another student of Mathieu's was Jacques Mering, who introduced the brilliant physical chemist Rosalind Franklin to X-ray diffraction, which she applied to the nearly crystalline charcoals.

Later, from 1938 to 1939, Lawrence O. Brockway, from the University of Michigan, visited the Royal Institution and worked with Bragg. Brockway's student and collaborator, Isabella Karle (see profile), is now at the Naval Research Laboratory in Washington, D.C., where she was awarded the 1976 American Chemical Society Garvan Medal for her work in molecular structures. She, her husband Jerome Karle, and others have developed a new mathematical technique called "direct methods" which has revolutionized crystallography. He has primarily been the theoretician and she the experimentalist.[65–70] Presently, the overwhelming majority of structures are solved using the "direct methods" technique.

The crystallographic study of organic compounds can be traced to the influence of William Henry Bragg. Women played an essential role in this development. Kathleen Lonsdale proved the planarity of the benzene ring. At the Royal Institution laboratory women worked on prophine, benzil, pentane, strychnine, cellulose, and also inorganic compounds such as potassium nitrate, and metals, primarily aluminum and copper. Dorothy Hodgkin began her association with Bernal when they collaborated on their early vitamin studies, several sex hormones, and the pepsin work. Helen Megaw elucidated hydrogen bonding; Rosalind Franklin spent her last years on plant viruses; and Mary Truter studied nitronium perchlorate. In the United States Dorita Norton worked on steroids, Joan Clark on the borates, Elizabeth Wood on silicon, and Isabella Karle on direct methods applied to such organics as the peptides. Thus women worked in every area of organic crystallography and have made consequential contributions to the field.

The history that William Henry Bragg began in crystallography of the organic world, his son, William Lawrence Bragg, began for the crystallography of the mineral world.

ASSOCIATES OF WILLIAM LAWRENCE BRAGG

William Lawrence Bragg came to the University of Manchester after World War I as a successor to Ernest Rutherford. Lawrence Bragg held a physics professorship and R. W. James helped him set up the X-ray laboratory using the original X-ray spectrometer that Bragg's father had made at Leeds before the war. W. L. Bragg's first research student was Lucy Wilson, from Wellesley College in Massachusetts, in 1923–24. The next year Dorothy Haworth joined his laboratory.

Continuing Lawrence Bragg's original interest in NaCl, common table salt, R. W. James and Elsie M. Firth (later Taylor) measured the intensities of rock salt from room temperature down to the temperature of liquid air to see the effects of the thermal vibrations of the atoms.[54] This was the first experimental verification of the Third Law of Thermodynamics, as well as a direct manifestation of the basic principle of quantum mechanics. Many scientists in the early 1900s expected energy, not entropy, to go to zero at complete rest (absolute zero degrees Kelvin). The atomic vibrations of the cooled rock salt were much larger than could be accounted for by thermal energy; hence the atomic vibrations were a measure of zero point energy.

J. M. Bijvoet spent a brief but intensive postdoctoral period with Lawrence Bragg at Manchester. Bijvoet's student, colleague, and successor at the University of Amsterdam was Caroline H. MacGillavry, who studied the properties of aliphatic dicarboxylic acids for odd and even numbers of carbon atoms. For even-numbered compounds, she found that the packing of the carboxylic groups causes a planar molecule, while the odd-numbered compounds were twisted molecules. Later, she made another study on the steric hindrance of vitamin-A-related compounds. She showed that the oxides of sulfur and phosphorus were structurally parallel to the silicates. She also made disorder and twinning studies; disorder is the degree to which a crystal departs from its regular geometry, and twinning results from "errors" in crystal growth. MacGillavry studied single crystals of the alloy Cu_3Au to develop these order-disorder concepts.[103–105] When she became the first woman elected to the Royal Netherlands Academy of Science, she commented, "but I am not the first one to deserve it." MacGillavry was coauthor of the book *Roentgenanalyse von Krystallen*,[15] the standard X-ray crystallographic text in the Netherlands. Her interest in X-rays was exhaustive: from absorption studies[102] to the mathematical techniques of "direct methods" and the derivation of the Harker-Kasper inequalities.[106,114] These mathematical inequalities are related to the fact that the electron density can never be negative and that it is near zero except for isolated resolved peaks at the atomic positions.

Caroline MacGillavry had a small, active group that often had visitors from abroad. Among them was Helen Ondik, now at the National Bureau of Standards in Washington, D.C., and coauthor of *Crystal Data*. Professor MacGillavry influenced the careers of many people by arranging stays abroad; for example, Gerda Wessel Arai is now at Zenith Radio Corporation in Chicago and Joke M. Vandenbery is now at Bell Laboratories. Caroline MacGillavry worked with Kathleen Lonsdale on Volume III of the *International X-ray Tables*.[97] In her book *Fantasy and Symmetry: The Periodic Drawings of M. C. Escher*[107] MacGillavry shares her delight in the excitement of the magic of Escher's patterns.

After Sagrario Martinez Carrera received her doctorate from the University of Madrid in 1954, she became a research associate at the Consejo Superior de Investigaciones Cientificas at the Instituto de Quimica Fisica in Madrid.[109] Funds were in such short supply that the students even made their own Beevers-Lipson strips by hand. To learn crystallographic computing and how to work with three-dimensional data, she went to Amsterdam to the laboratory of Caroline MacGillavry.[110] Later, in 1961, Sagrario Carrera went to Pittsburgh with G. A. Jeffrey to work on the IBM 7070. She described this work in a letter to me:

> Now we have two single crystal diffractometers, a Univac 1108 computer, our people have been abroad and about thirty papers have been published from our laboratory. This is enough to be happy with for a while, but I will not stop because I would like one day to do the structure of a big molecule like a protein. . . . but Spain is a poor country and there is little money for research.

When Clara Brink (later Shoemaker) was preparing her Ph.D. with A. E. van Arkel at Leiden University, she came to work with MacGillavry because then there was no crystallographer in van Arkel's laboratory. Clara Brink's work involved the crystal structures of complexes of monovalent ions.[16] Upon receiving her Ph.D. she obtained a fellowship from the International Federation of University Women and went to work with Dorothy Hodgkin in Oxford on vitamin B-12.[48] MacGillavry suggested to Clara Brink that she work at MIT with David P. Shoemaker on sigma-phase related structures.[17] Brink arrived at MIT in 1953 and married Shoemaker in 1955. They continued to work as a team on sigma-phase structures, which they later described as "tetrahedrally close-packed."[123-125] The Shoemakers then went to Oregon State University in Corvallis, she as a research associate professor and he as a full professor. Soon she had published fifty papers, many in collaboration with her husband.

F. W. H. Zachariasen also spent time with Lawrence Bragg at Manchester. His student Rose C. L. Mooney (afterwards Slater) taught crystallography from 1926 to 1952 at a women's college, Sophie Newcombe, at Tulane University. She was interested in the electronic structure of the halogens, especially iodine. In 1952 she became a senior physicist at the National Bureau of Standards, until 1956, when she joined her husband, physicist John Slater, at MIT, where she was a research scientist.[126]

In 1954, when Lawrence Bragg came to the Royal Institution in his father's footsteps, he was joined by Helen Scouloudi.[18,120] She was originally trained as a physicist in Athens, Greece, and then took her Ph.D. at Birkbeck College of the University of London, in protein crystallography with C. H. Carlisle where she worked on ribonuclease. At the Royal Institution she worked on hemoproteins, especially on myoglobin. She is presently in the Zoology Department at Oxford University.

ASSOCIATES AROUND THE WORLD

Although everyone in crystallography owes a debt to the Braggs, there are many whose connections are more vague. Schools of crystallography arose in the United States; for example, there was Martin Buerger at MIT. The Templetons came to crystallography through a circuitous route by their work in absorption and synchrotron radiation. Also, the lineage of crystallographers in countries such as Japan, Russia, Argentina, and Yugoslavia is difficult to trace.

Table 3 is a summary of the country, name, and general area of crystallographic expertise of several women who are difficult to fit into the Bragg schema. This section will describe the contributions of these women.

Martin Buerger, a leading mineralogist, developed the precession camera in 1944 and wrote five classic crystallographic texts. Gabrielle Hamburger received the first Ph.D. in crystallography from MIT from a committee formed for the purpose under Buerger in 1948. Her thesis was on the structure of tourmaline.[30] To complete her Ph.D. requirements, Hamburger enrolled in a course in optical crystallography taught by Jose Donnay, a professor of crystallography and mineralogy at The Johns Hopkins University. They were married in 1949 and collaborated on many papers. They undertook a series of theoretical studies on one-dimensional crystals,[28,29] which they used as a wedding announcement as shown by the name changes between the first and second papers. They elucidated the structures of many minerals; an example is their work on the four minerals bastnaesite, parisite, roentgenite, and synchisite,[31] which form intergrowths in pairs. This work explained the complex morphology of these minerals. One of the minerals, roentgenite,[32] was discovered by Gabrielle during this study. She named it after "Wilhelm Konrad Roentgen (1845–1923) who discovered X-rays because X-ray methods alone proved its existence and established its formula." The Donnays have two sons, one of whom is the godson of Kathleen Lonsdale. For twenty years Gabrielle Donnay was a staff member in crystallography at the Geophysical Laboratory of the Carnegie Institute in Washington, D.C. In 1971, she became a professor of crystallography at McGill University in Montreal.

TABLE 3.

Women in crystallography whose connections with the Braggs are not as direct as those in Table 2

Country	Name	Topic
Argentina	M Rodriguez	optical crystallography
Canada	G Donnay	chemical crystallography
England	O Kennard	structural information retrieval
	D Wrinch	mathematics and crystallography
Japan	M Takagi	X-ray topography
Netherlands	A Vos	interatomic interactions
Russia	V Iveronova	scattering
	M Neklyudova	plastic deformation of crystals
	S Grzhimailo	crystal optics
	V Konstantinova	ferroelectrics
United States	L Templeton	absorption and synchrotron radiation
Yugoslavia	K Kranjc	small angle scattering

One of her students, Suzanne Fortier, after spending seven years at Buffalo with Herbert Hauptman on "direct methods," became a member of the chemistry faculty at Queens University, Ontario.

I met Liselotte K. Templeton, a research scientist at Berkeley, and her husband, David Templeton, a professor at Berkeley, at the 1982 spring meeting of the American Crystallographic Association at Gaithersburg, Maryland. Nepotism rules were always a great problem for them. Their interest in crystallography arose from their work in absorption and thus they are not connected historically to the Braggs in the same way that others in this chapter are. Lieselotte kindly wrote the following for me:

> It was really my interest in computer programming and the subsequent re-working of the analytical absorption program (AGNOST) now called ABSOR that got me seriously involved in crystallography. ABSOR was useful in helping me solve several crystal structures of heavy-element compounds, but it was particularly important in recent work with my husband on studying anomalous dispersion at absorption edges with synchrotron radiation. Our measurements with compounds of cesium[131] and several rare earth elements[132] demonstrated the exceptionally large effects which occur at L absorption edges and which are very promising for helping to solve the phase problem for protein and other macromolecular structures.
>
> We also have used the polarized nature of synchrotron radiation to demonstrate X-ray dichroism in anisotropic molecules such as vanadyl,[133] uranyl[134] and bromate ions and to measure for the first time in diffraction experiments the polarized anomalous scattering which comes with this dichroism. We are now very busy trying to learn more about these polarization effects and their possible applications in structure research.

In Argentina, Lic Maria Angelica Rodriguez de Benyacar has been fondly called "la cristalografa madre," the mother of cyrstallographers, for the present generation. She is head of the crystallography division at Comision Nacional de Energia Atomica in Buenos Aires. One of her studies was the relationship between optical properties such as indices of refraction and crystal structure.[108]

Aafje Vos, a leading Dutch crystallographer, received her Ph.D. in chemistry under E. H. Wiebenga from the University of Groningen, where she is now. Her professional interests are in accurate structure analysis and interatomic interactions.

In 1951 Olga Kennard set up a small crystallographic laboratory at the National Institute for Medical Research, Mill Hill, London. This work was generally in the area of applied biological standarization and crystallographic documentation. Presently she directs the Cambridge Crystallographic Data Centre and has published many structures of small biological molecules.[75,76]

Another British crystallographer was mathematician Dorothy Wrinch,[3,62] who applied mathematical techniques to proteins (see profile). She emigrated to the United States and spent many years at Smith College. She also wrote a widely used book on Fourier analysis.[150]

Mieko Kubo Takagi, who was born in Tokyo in 1919, is the leading woman crystallographer in Japan. Until 1946 a woman could not get into an advanced physics program. Mieko Kubo was an "irregular student" at Osaka University from 1941 to 1943. However, Seishi Kikuchi (of the Kikuchi pattern in electron diffraction) was very kind and treated Kubo as a regular student. He also gave her a chance to do research in electron diffraction. In a letter to me, she wrote: "I was completely attracted by the beautiful diffraction patterns on the flourescent screen and I worked hard at Prof. Kikuchi's lab." She then joined S. Miyake at the Kobayashi Institute of Physical Research and moved with him to the Tokyo Institute of Technology in 1949. That year she presented her doctoral thesis to Osaka University. During this time she worked on electron diffraction studies of liquid-solid transitions of thin metal films.[129] She felt that X-ray topography would be a very powerful tool for research of ferroelectrics, and in 1961 she began a year's study with A. R. Lang at the University of Bristol, England, on X-ray topographic studies on diamond.[130] Although she retired as professor at Tokyo Institute of Technology in 1979, she has been active as a temporary lecturer at Ochanomizu University and Aoyama University. In 1949, she married crystallographer S. Takagi; they have two daughters. She also wrote to me: "With the help of my husband and housekeepers I have been working continuously for 35 years." At present, she is organizing a movement to encourage and help Japanese women scientists.

Katarina Kranjc was the first woman in Yugoslavia to use X-rays. She was born in Zagreb in 1915 and graduated with a degree in physics in 1936.

At first she taught in several high schools and then became a demonstrator in physics on the veterinary faculty. Her interest in crystallography began in 1948 when she heard Drago Grdenic, a chemist, describe his postgraduate studies with A. I. Kitaigorodskii in Moscow just after World War II. Grdenic, using the benzene ring as an example, explained in general terms how to determine a structure. From Abbe's theory she could see the power in the method. Unfortunately she was not able to get a research position until Professor Paic arrived from Paris to set up a laboratory and then invited applications. Kranjc was given the choice between spectroscopy or X-ray diffraction and she chose the latter. To study small-angle X-ray scattering,[80] she constructed her own X-ray apparatus. Although she was burnt as she improvised a curved monochromator without any shielding, she kept her enthusiasm, which was inversely proportional to her experience. Later, Professor Kranjc described her research as always limited to topics for which she had the necessary equipment. She wrote to me that

Only now do I realize that I have chosen themes where nice pictures are seen and nice pictures are associated with single crystals. Ferroelectric and antiferroelectric domains in Berg-Barrett micrographs, the sodium dendrites grown in sodium chloride crystals, the lead dendrite grown by electrolysis are surely nice. Now I am working with a group which investigates the rapidly quenched alloys (from the melt) which are far from single crystals. However, I find interest in electron micrographs where a single crystal can be seen however the grains are small; I compute the contrast around the precipitates due to lattice distortions, and the Moire fringes which appear on the precipitates. As computer programming became my hobby, I find pleasure even in small-angle scattering although there are no nice pictures.

Katarina Kranjc was the first woman to receive a Ph.D. in Yugoslavia in physics. From Zagreb, crystallography spread throughout the universities of Yugoslavia.

Despite the large number of Russian women scientists it is difficult to learn much about them. I wrote individual letters to fifteen women in various parts of Russia and never received a reply. As far as I have been able to ascertain, the women that will be discussed below are mainly involved in areas related to weapons research.

At the time of Laue's discovery there were two schools of crystallography: E. S. Fedorov directed the Mining School at Petersburg, and G. V. Wulf did likewise at the People's University at Moscow. Wulf independently discovered what is known as Bragg's Law in 1913 [*Phys. Zs.* 14:217]. After World War I there were several crystallographic groups working in Moscow. N. V. Belov worked in the silicates, G. S. Zhdanov in ferroelectrics, G. B. Bokij in complex compounds with multiple bonds, and A.

I. Kitaigorodskii with organics. In 1963 Belov wrote in *Fifty Years of X-ray Diffraction*:[1]

> Only this year I was exceedingly happy and felt myself rewarded by the discovery by one of my pupils (she is a mineralogist) that there exists a dimorphous silicate which forms two types hitherto regarded as exclusive in one and the same substance: epididymite is a chain silicate while eudidymite, having the same formula, is a layer or net silicate (Phyllosilicate). [p. 521]

Note that the name of the woman student was never mentioned. Even now I am unable to locate the name of this crystallographer.

Valentina Ivanovna Iveronova, professor at the Department of Solid-State Physics of the physics faculty of Moscow University, is one of the leading crystallographic chemists in the USSR. Her book with G. P. Revkevich is a standard reference for the dynamic theory of X-ray scattering. In 1953, when the physics department moved to new quarters in the Lenin Hills, she organized the curriculum to double the number of students and at the same time improve the quality of teaching. She edited a textbook, *General Physics Practical Course*, which became the chief text for physics faculties in the USSR. As a Communist, she was active in public affairs and has served at three conventions as the deputy for the Krasnopresnensk Regional Council of Deputies of Workers. Her awards include the Order of Lenin and the Badge of Honor.

Marina Viktorovna Klassen-Neklyudova, born in January 1904, is one of the foremost specialists in the physics of strength and the plasticity of materials. She was in the first class of the Physiocotechnical Institute in Leningrad. In 1932 she organized the Laboratory of Plastic Deformation of Crystals and in 1936 she defended one of the first doctoral disserations in the Soviet Union. She was the first to study the mechanical properties of synthetic ruby crystals, which are very hard and brittle at ordinary temperatures; she showed that these crystals are capable of plastic deformation. Her research has been applied to corundum crystals and preparation of fine industrial stones, wire dies, ruby lasers, and high-pressure lamps. She was the first to use the optical polarization method to observe ferroelectric domain twins in Rochelle salt; these played an important part in the development of the present ideas concerning ferroelectric crystals. She is the author of over ninety publications, of which ten are monographs and symposia. Her awards include the Order of the Red Banner of Labor and the Badge of Honor.

Sof'ya Vladimirovna Grum-Grzhimailo (1908–1969), of the Institute of Crystallography, specialized in experimental crystal optics. She and her students demonstrated that the valence and coordination of impurity centers in crystals can be determined from their optical spectra. She showed that light absorption of crystals containing impurity ions of the iron group is due principally to electronic processes in ions with complete d shells. Studies of the absorption spectra of various crystals as a function of temperature, down to that of liquid helium, enabled her to discover a change in the form of the broad spectra band, determine the vibration structure of these bands, and elucidate the spectral ultrastructure produced by purely electronic transitions in the impurity ions. She authored over a hundred papers and was active in popularizing the crystallographic sciences to a broad range of workers through the *Znanie* (Knowledge) Society. Her other interests included archaeology and art.

Varvara Pavlovna Konstantinova (1907–1976) was born to a large family in Leningrad. Her parents died early, leaving the care of the children to herself and her older brothers. Along with most of her brothers and sisters, who became eminent engineers and physicists, Varvara entered Leningrad Polytechnic Institute. Before World War II she studied the structure, formation, and adhesive strength of soils. Then she was evacuated to Kazam with her husband and small children until the end of the war. Upon returning to Moscow she worked on the piezoelectric textures of Rochelle salt. She developed methods for growing perfect crystals of a large number of ferroelectrics; she studied their domains, dislocation structures, and mechanical properties. After her death the manuscript of her latest article was found on her desk.

Other Russian scientists include Serafima Ilyinishna Berkhin (b. 1919), who is now at the Institute of Geology, Mineralogy and Petrography in Moscow and is working in clay minerals. The following women are all from the Institute of Crystallography in Moscow: Leonila Bavrilovna Chentsova (b. 1902) is interested in crystal optics; Marina Alexandrovna Chernysheva (b. 1911) studies the mechanical properties of crystals; Nina Yuryevna Ikornikova (b. 1911) is interested in hydrothermal synthesis of crystals and the physical and the chemical studies of solutions; Anastasiya Arsentyevna Popova (b. 1916) studies crystal growth from melt; and Lyudmila Ivanovna Tatarinova (b. 1903) works in synthetic polypeptide structures. At the Moscow Institute of Steel and Alloys, Marianna Petrovna Shaskol'skaya (b. 1913) works on crystal physics and plastic deformation. Mariya Ivanovna Zakharova (b. 1904) is at Moscow State University studying phase transformations. These women, all born before 1920, indicate the great depth and scope of the work of women Russian crystallographers.

The influence of the Braggs on the science of crystallography is monumental. William Henry Bragg presided over the organics at the Royal Institution in London while his son, William Lawrence Bragg, had the minerals and metals at Manchester. There were groups of crystallographers throughout the world who received inspiration from the Braggs but who were not directly connected to them. That women played important parts in crystallography is, in large measure, due to the tradition begun by the

Braggs. A popular myth that women dominated crystallography turns out not to be true. Nonetheless their influence has been significant and we must assume that the apparent "overpopulation" of women in the field is due, in part, to their extraordinary success.

PROFILES

Now that the contributions of many women crystallographers, representing a cross section of the spectrum of crystallographic studies, have been described, I would like to narrow the field and examine a few careers more closely. The next section will focus on the biographies of five women crystallographers. Kathleen Lonsdale, Rosalind Franklin, Dorothy Wrinch, Isabella Karle, and Dorothy Hodgkin have been chosen because of their significant contributions.[61] These women studied a variety of materials, had quite different personalities, and coped with their life situations in distinct ways.

Kathleen Lonsdale was the first woman crystallographer to attain a worldwide reputation. She was vice-president of the International Union of Crystallography from 1960 to 1966 and president in 1966. Her emphasis was on a thorough study of small molecules. One of her most dramatic discoveries was her definitive proof that the benzene ring is flat and hexagonal.[82–84]

The second crystallographer whom I have chosen had her life marred with almost unendurable professional antagonism and tragically cut short by death at thirty-seven. Rosalind Franklin, although British-born, learned her crystallography in France and applied the techniques first to coal, then to DNA, and finally to plant viruses. Her work represents an attack on substances that were on the borderlines of crystallinity.

Dorothy Wrinch was a Renaissance scholar whose depth, vision, leadership, and energy were resented rather than respected. Her work spanned mathematics, philosophy, sociology, chemistry, physics, and biology. She dedicated her life to the solution of one of nature's most important and challenging mysteries, the architecture of protein molecules. While her theories were not implemented in nature in quite the way she anticipated, they were a critical stimulus to research. Nevertheless, she herself remained an outsider and a restless, disappointed renegade.

Isabella Karle, born in 1921, a year and a half after Rosalind Franklin, is an experimentalist, and her husband, Jerome Karle, is a theoretician. He and others worked out a mathematical theory which greatly extended the scope of crystallography, and Isabella applied it and made it popular and accessible to the crystallographic world. Her applications are fascinating and range from frog venom and poisonous mushrooms to DNA breakdown and transport across membranes. Isabella Karle was president of the American Crystallographic Association in 1976.

Dorothy Hodgkin was awarded the Nobel Prize for her work on structural studies on molecules of biological importance. She remembers reading Kathleen Lonsdale's papers on hexamethylbenzene in graduate school, being impressed by their clarity, and immediately incorporating them into a paper she was writing.[51] Each new increase in computing capabilities was brilliantly exploited by Dorothy Hodgkin. Her three outstanding contributions are on penicillin, vitamin B-12 and insulin, each work sequentially increasing in complexity and permanently changing crystallography.

KATHLEEN LONSDALE

Kathleen Yardley was born on January 28, 1903, in Newbridge, Ireland, just south of Dublin.[51,55,56,58,59] She was the youngest of ten children. Her father, the postmaster for the British garrison stationed in Ireland, was both an agnostic and a heavy drinker. When Kathleen was five years old, her parents separated. Mrs. Yardley took the children who were still at home to Seven Kings, a small town east of London which turned out to be on the zeppelin route during the World War I blitz.

At sixteen Kathleen Yardley enrolled in Bedford College, a small women's college of the University of London. Three years later William H. Bragg was among her B.Sc. oral examiners. She headed the university list with the highest marks in ten years. Bragg asked her, "How did you manage to do so well?" Answering his own question, he offered her a position in his laboratory at University College, London. The following year she moved with him to the Royal Institution. For her master's dissertation she studied the structure of succinic acid and related compounds.[152] In *Fifty Years of X-ray Diffraction*,[1] Kathleen Yardley told a story about Sir Alfred Yarrow, who had endowed several research fellowships at the Royal Institution. He had a theory that intelligence was inherited on the maternal side. Inconsistently, he argued that women should not be allowed in the laboratory because they only leave to get married. Kathleen asked "where his intelligent mothers would come from if only those with no professions were allowed to marry" [p. 414].

During this period Kathleen Yardley and William Astbury began their collaboration on a set of 230 space group tables, mathematical descriptions of the crystal symmetries, which became an indispensable tool of (especially British) crystallographers.[8] In 1936, to avoid error, her handwritten structure factor tables were photo-litho-printed from the original manuscript. For her study of the ethane derivatives,[153] she was awarded the prestigious Doctor of Science degree.

On 27 August 1927 she married fellow student Thomas Lonsdale. They took up residence in Leeds, where Thomas was working for the British Silk Research Association. At night he set up experiments to complete his doctoral thesis on the torsional strengths of metals. Kathleen Lonsdale was at the University of Leeds working on the hexamethylbenzene study, the first experimental proof of the planarity of the benzene ring. Although she continually kept Bragg informed of her progress, the work was done independently and was "communicated" by Professor R. Whiddington, who had welcomed her to the University of Leeds. Christopher K. Ingold, a professor of organic chemistry at Leeds, offered her some crystals of hexamethylbenzene, which, unlike other known benzene derivatives, proved to have only one molecule to the unit cell. Surmounting incredible odds Kathleen Lonsdale solved the first crystal structure of an aromatic compound and definitively proved that the benzene ring is hexagonal and flat.[82–84] Although this information upset some of Bragg's own theories of a "puckered" benzene ring, he rejoiced at her brilliant findings.

Between 1929 and 1934 their three children, Jane, Nan, and Stephen, were born. In 1930 Thomas Lonsdale's job in Leeds collapsed and the family returned to London, where he was able to get a permanent position at the Testing Station of the Experimental Roads Department in the Ministry of Transport at Harmondsworth. Kathleen Lonsdale worked mostly at home on the structure factor tables. In 1930 they bought a used typewriter with an extra long platen for five pounds to type the long formulas.

In 1934 Kathleen Lonsdale returned to the Royal Institution and remained with Bragg until his death in 1942. At first, after the disappointing news that no X-ray set was available, she focused her energies on a big, old electromagnet. Since the diamagnetic susceptibilities of aromatic compounds are greater perpendicular to the aromatic ring than in the same plane, she was able to show that the orbitals representing the sigma electrons were of atomic dimensions and that the orbitals representing pi electrons were of molecular dimensions. This experimental work established the reality of sigma and pi electrons and their representation by molecular orbitals.[85,86,95]

Kathleen Lonsdale became interested in thermal vibrations as a result of observing some diffuse (or non-Bragg) spots on an X-ray diffraction photo of benzil. Diamonds also have diffuse reflections, which led to her work in natural and man-made diamonds.[87–89] She applied the techniques of divergent X-ray beams and obtained precise lattice constants for diamond, showing that the distance between carbon atoms was 1.54451 angstroms with a variation of only 0.00014 angstroms in different diamonds.[90] In 1966 a rare wurtzite, or hexagonal, form of meteoritic diamond was named "lonsdaleite."[101] In a letter to Clifford Frondel of Harvard University, who suggested the name, she said, "It makes me feel both proud and rather humble that it shall be called lonsdaleite. Certainly the name seems appropriate since the mineral only occurs in very small quantities (perhaps rare would be too flattering) and is generally rather mixed up!"

Three years after William H. Bragg's death, Kathleen Lonsdale founded her own crystallography department at University College, London, where she had begun almost a quarter of a century before with Bragg. After living on year-to-year grants until she was forty-three years old, she finally got her first permanent position. A special chair was created for her in 1949 at the initiative of Christopher K. Ingold. At first most of her time was taken up as editor-in-chief of the *International X-ray Tables*.[92,95,97] She wrote a popular textbook, *Crystals and X-rays*.[91]

In 1949 Judith Grenville-Wells (later Milledge) came from South Africa to study diamonds. At first, in exchange for room and board at the Lonsdales, Judith did some of the secretarial work on the *International X-ray Tables*. Judith Milledge became Kathleen's scientific colleague and devoted friend. Judith Milledge and Kathleen Lonsdale's studies together included natural and artificial diamonds,[96] minerals at high pressures and high temperatures, and the mechanisms of solid state reactions. The most important example of the latter is the detection of the X-ray diffraction patterns of the intermediate products in the conversion of the photo-oxide of anthracene into a mixed crystal of anthrone and anthraquinone.[98] Eventually Judith Milledge became Kathleen Lonsdale's literary executrix and continued the diamond and solid-state reaction work at University College.

After seeing a collection of body stones—kidney, bladder, and gall stones—Kathleen Lonsdale began an extensive chemical and demographic study.[99] (She loved to exhibit an X-ray diffraction photograph of Napoleon III's bladder stone.) D. June Sutor, a New Zealander who earned her first Ph.D. in her native land and whose second Ph.D. was from Cambridge University, worked with Professor Lonsdale on this project and eventually took over and extended it.[100] Before coming to University College, June Sutor had done many structures, one of which was caffeine.[128]

From her youth, when Kathleen Lonsdale lived under the threat of the zeppelins, until her death in 1971, she felt the horrors of war and worked, wrote, and lectured for peace. Thomas and Kathleen became Quakers in 1936. They sheltered refugees in their home during World War II. Kathleen Lonsdale chose to go to jail for a month as a pacifist in 1943 rather than register for war duties. In the summer of that year she gave her first international paper at the Institute for Advanced Studies in Dublin, Ireland. Eamon de Valera, the founder of the Institute and the Prime Minister of Ireland, also had prison experiences.[59] Later Kathleen Lonsdale became interested in prison reform and actively served on many prison boards. On a scientific trip to Russia in 1951 she asked to visit a prison. Upon leaving, she noticed her interpreter grinning; Kathleen Lonsdale asked why. "The prison governor wants to know how it is such a nice lady knows so much about prisons," was the reply.

In 1956, upset by extensive nuclear testing in Russia, the United States, and Great Britain, Kathleen Lonsdale interrupted other work and in six weeks wrote her book *Is Peace Possible?*[93] In the foreword she stated that the book was "written in a personal way because I feel a sense of corporate guilt and responsibility that scientific knowledge should have been so mis-

Kathleen Lonsdale delivering a paper at the Dublin Institute for Advanced Studies, 1943. Courtesy of the *Journal of Chemical Education*.

used." She explored the correlation between international peace and world population needs with the personal perspective of being the youngest of ten children. She was strongly opposed to all nuclear weapons.

When Thomas retired they moved from London to the relatively quiet town of Bexhill-on-Sea. He assisted his wife in her prison and pacifist commitments. The first of their ten grandchildren was born on the day before the announcement was made that Kathleen Lonsdale would receive the title Dame Kathleen, a Dame Commander of the Order of the British Empire.

Kathleen Lonsdale had a number of important firsts. She and microbiologist Marjory Stephenson were the first two women elected as fellows of the Royal Society in 1945. Kathleen Lonsdale was the first woman professor at University College, London. In 1966 she was the first woman president of the International Union of Crystallography, and two years later the first woman president of the British Association for the Advancement of Science.

On April 1, 1971, Kathleen Lonsdale died of cancer. During her illness she was seen at least once sneaking across the street to get some references for her book on body stones which was in progress. When Dorothy Hodgkin visited her in the hospital, Kathleen asked her how important the bone marrow was. "Very important," was the reply.

In 1981, the chemistry building at University College was renamed the Kathleen Lonsdale Building. This is a fitting tribute to the first woman crystallographer to attain a world-wide reputation not only as a scientist, but as a humanist as well. Her willingness to be of service to others was exemplified in her careful work on the *International X-ray Tables* as well as in her prison reform and peace efforts. Her scientific achievements were many. She showed the planarity of the benzene ring and demonstrated the reality of sigma and pi electrons and their representation by molecular orbitals; she did important work on diamonds, diffuse scattering, and temperature factors; and when she was almost sixty, she began an exhaustive study on body stones. Kathleen Lonsdale was indeed an incredible worker, a methodical organizer, and a scientist of highest order.

ROSALIND FRANKLIN

Now let us turn to the career of the brilliant physical chemist Rosalind Franklin, who made major contributions to the studies of coal, DNA, and plant viruses.[38–44]

In December 1962, James Watson, Francis Crick, and Maurice Wilkins received the Nobel Prize for their discovery of the double helix structure of the DNA molecule. These men based their work in considerable part on experimental data obtained surreptitiously from the files of Rosalind Franklin. At the time of the award, Rosalind had been dead for four years and her pivotal contribution was not acknowledged.

Rosalind Franklin had an uncanny ability to work with poorly crystallized materials. Such samples are difficult to prepare, their X-ray diffraction patterns are tricky to record, and their analysis is usually next to impossible. She began her professional career as a physical chemist studying coal and graphite, then turned her attention to DNA, and finally made contributions to the structure of plant viruses.[60]

Rosalind Franklin was born in London on July 25, 1920, the daughter of well-to-do Jewish parents. She attended the fashionable St. Paul's Girls' School, where she received an excellent background in physics and chemistry. In the 1930s her father was active at the Center at Woburn House, which aided Jews who had escaped from Nazi Germany. Rosalind did volunteer work there while she was at school.

Shortly after Rosalind Franklin entered Cambridge University in 1938, much of the science faculty's time was absorbed in war research, which left the undergraduates, more so than usually, on their own. This suited her. After obtaining her undergraduate degree, she remained at Cambridge for her graduate work. Through French lessons, Rosalind Franklin met metallurgist Adrienne Weill,[137] a wartime refugee at Cambridge. At the age of twenty-two, Rosalind Franklin gave up her fellowship in order to

Rosalind Franklin in France in the early 1950s. Photo by Vittorio Luzzati. Courtesy of the *Journal of Chemical Education.*

take her first position as a physical chemist at the British Coal Utilization Research Association. Three years later, in 1945, she received her Ph.D. from Cambridge; the title of her dissertation was "The Physical Chemistry of Solid Organic Colloids with Special Relation to Coal and Related Materials."

In 1946 Rosalind Franklin wrote to Adrienne Weill, who by then had returned to Paris: "If ever you hear of anybody anxious for the services of a physical chemist who knows very little about physical chemistry, but quite a lot about the holes in coal, please let me know."[119] Weill suggested Marcel Mathieu at the Sorbonne, who in the 1920s had worked with William H. Bragg at the Royal Institution in London. At Mathieu's invitation Franklin came to Paris in February 1947 as a *chercheur* in the Laboratoire Central des Services Chimiques de l'Etat. She worked closely with crystallographer Jacques Mering and began using X-ray diffraction techniques to study the graphitizing and nongraphitizing carbons.[38] Rosalind Franklin particularly enjoyed the informal relationships that existed among the laboratory workers who ate in local bistros and spent much of their free time together.

Meanwhile, the crystallographic frontier lay in the biological substances. With mixed feelings, Franklin left France in 1951 and accepted Professor John Randall's offer of a three-year Turner-Newall Research Fellowship at

King's College, London. The newly formed biophysics department was beginning to work on DNA, which was generally thought to be associated with heredity; thus Franklin chose DNA for the subject of her work. DNA, like poorly crystallized graphite, gave poor diffraction patterns and demanded the X-ray techniques Franklin had used in France.

Life at King's College was not easy for Franklin. She found it difficult to work with her colleague, Maurice Wilkins. This problem was compounded by the "men's club" atmosphere of King's College which excluded her from almost anything but formal laboratory contacts with her fellow scientists. The contrast was great compared with the informal atmosphere she had enjoyed so much in France. She remained at King's College from January 1951 to the spring of 1953.

Franklin mobilized her invaluable experience with poorly crystallized materials and in eight months set up an Ehrenberg fine-focus X-ray tube with a Phillips microcamera for taking high-resolution photographs of single fibers of DNA. To further facilitate data collection she designed a microcamera specifically to photograph crystals inclined to the X-ray beam at several angles. Franklin then took an unrivaled series of X-ray diffraction photographs of DNA. Under carefully controlled humidity conditions, she discovered the hydrated B-form of DNA, with its diffuse pattern clearly indicating a helix. She defined the conditions under which the two distinct reversible phases, A and B, existed.[41,42] Previous DNA photographs were a mixture of the two phases. This careful experimental work and lucid analysis cleared up the confusion of earlier workers who had assumed that this A + B mixture of phases was a single phase. Franklin also showed that the sugar phosphate backbone was on the outside of the DNA molecule and that hydrogen bonding played an important role. She then concentrated her efforts on performing a Patterson analysis of the A-form of DNA, which showed a more detailed crystalline pattern. The mathematics involved is complex and subtle, so it was distinctly possible that the A-form might assume a very different shape (like rods or sheets inclined to the fiber axis) [see A. Klug, *Nature* 219 (1968):808] from the helix shown in the hydrated B-form.[40] Although her notebooks from this time indicate that she knew the B-form was helical and contained ten nucleotides per turn, she was not sure of the number of intertwined chains. The correct analysis of the A-form would yield more precise information. Because of the reversible phase change between A and B, the molecular model would have to be easily adaptable to crystallize in either form.

Meanwhile, at Cambridge, Crick and Watson were building their now-famous models of DNA. Their helical DNA model-building needed an experimental foundation. Wilkins secretly showed Watson the unique picture of the B-form that had been taken by Franklin. Also at this time, Professor Max Perutz gave Crick and Watson a privileged copy of Franklin's Medical Research Council report containing calculations from the spacings of the critical reflections on the B-form photograph. Watson and Crick then fitted the DNA molecule into the space defined by the helical repeat distance and helical diameter. Franklin's picture formed the major experimental link that culminated in Watson and Crick's solution of the structure of DNA. Their contribution was the splendid assembling of the double helix in which they matched the base pairs, adenine to thymine and guanine to cytosine.

By agreement, three papers on DNA appeared in the famous April 25, 1953 issue of *Nature:* the first by Watson and Crick (p.737), the second by Wilkins and his co-workers, A. R. Stokes and H. R. Wilson (p.739), and the third by Franklin and her graduate student, Raymond Gosling.[39] Watson and Crick's paper listed the helical pitch, the diameter of the helix, and the number of turns between repeats along the helix, but did not say where the data originated. Franklin rejoiced at Watson and Crick's success of the DNA structure and immediately, without suspicion, confirmed its agreement with her own experimental results. She never dreamed that their double helix model was based on her experimental evidence. Three months later she and Gosling published a confirmation of the helix in the A-form.[42] Of course science is built on many workers' research, but it is heinous that Franklin was never given credit for her crucial experimental information either during her lifetime or at the time the Nobel Prize was awarded in 1962.

Professional conditions between Franklin and Wilkins at King's deteriorated so badly that in the spring of 1953 she took her fellowship, but not the DNA problem, to Birkbeck College, London. J. D. Bernal had invited her to study the structure of plant viruses, especially the tobacco mosaic virus.[43] Again, these materials are of borderline crystallinity. At Birkbeck she had a number of successful collaborations, especially with Aaron Klug,[44] who won the 1982 Nobel Prize in Chemistry.

In the autumn of 1956, Rosalind Franklin learned that she had cancer. She refused to let up on her work and, even near the end, began a personally dangerous project on the structure of polio virus. On April 16, 1958, at the age of thirty-seven, she died [see obituary by J. D. Bernal, *Nature* 182 (1958):154].

Much of Franklin's life was sad, not only in its cruel shortness, but in the difficult position in which she found herself at King's College. Her contributions were many and brilliant. There were seventeen papers, many of them recognized as classics today, on carbon and coal. Her two years of research on DNA produced five papers associated with work that later won the Nobel Prize. During her last five years, she wrote seventeen papers on viruses. Her last three papers were published posthumously.

Watson, Crick, and Wilkins's three Nobel Prize lectures contain a total of 98 references, and not one of Franklin's papers is specifically mentioned. Wilkins makes the only textual reference to her in this casual remark: "Rosalind Franklin (who died some years later at the peak of her career) made

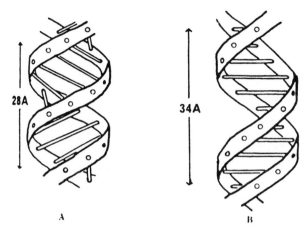

(A) DNA A-form: There are eleven nucleotides in the axial repeat distance of 28 angstroms. (B) DNA B-form: The ribbons symbolize the phosphate-sugar chains, and the rods the pairs of hydrogen-bonded bases holding the chains together. There are ten nucleotides in the axial repeat distance of 34 angstroms.

some very valuable contributions to the X-ray anyalysis." Only Wilkins included her in his acknowledgments [see *Nobel Lectures in Molecular Biology 1933–1975*, New York: Elsevier, 1975].

In 1968, Watson published *The Double Helix*, in which he told his version of the DNA story. This book was the first time the incident of the viewing of Franklin's B-form picture was publicly revealed. In an interview with Anne Sayre on June 15, 1970, Wilkins discussed the pirate viewing of this picture: "Perhaps I should have asked Rosalind's permission and I didn't. Things were very difficult. Some people have said that I was entirely wrong to do this without her permission, without consulting her at least, and perhaps I was."[119]

Later, protein crystallographer David Harker, head of the biophysics department of the Roswell Park Memorial Institute at Buffalo, New York, commented,

And the real tragedy in this affair is the very shady behavior by a number of people, as well as a number of unfortunate accidents, which resulted in the transfer of information in an irregular way. . . . I would never have consciously become involved in anything like this behavior, especially the transfer of information through a privileged manuscript. And I think these people are—to the extent that they did these things—outside scientific morals, as I know them.[122]

Rosalind Franklin should have been one of the 1962 Nobel Prize recipients had she lived. Her death simplified selection for the committee, since the Nobel Prize cannot be awarded posthumously and can be split at most among three persons. Exactly what would have happened if Rosalind Franklin had lived is impossible to guess.

DOROTHY WRINCH

Dorothy Wrinch was born of English parents in 1894 at Rosario, Argentina, where her father was an engineer for a British firm.[3,52,62,121,122]She was educated in England and won a scholarship to study mathematics at Girton College, Cambridge University. Her circle of friends in cluded Bertrand Russell, with whom she studied philosophy. After she received her Cambridge B.A. and M.A. degrees, she was appointed in 1918 to a lectureship in pure mathematics at University College, London. In 1922 she married John Nicholson, an Oxford physicist known for his work on atomic structure; their only child, Pamela, was born in 1928.

At Cambridge, and later at London and Oxford, Dorothy Wrinch was active in philosophical and mathematical circles. In addition to teaching mathematics, she addressed professional gatherings and wrote prolifically. By 1930, at the age of thirty-five, she had published sixteen philosophical papers on the scientific method, and twenty papers in pure and applied mathematics. Her additional credentials included both M.Sc. and D.Sc. degrees from the University of London, and an M.A. from Oxford University. In 1929 she was the first woman to receive a D.Sc. from Oxford.

That year, after having taught mathematics at Oxford for seven years, she unsuccessfully applied for a Rhodes Traveling Fellowship. Her sponsors frankly informed her that she failed because she was a woman. She filed two more applications: one for a Rockefeller fellowship to study mathematics at Göttingen, Germany, and another for a fellowship in sociology to travel in America. Although she was welcomed by the leading mathematician at Göttingen, she failed to get the Rockefeller, largely because foundation officials learned of the sociology proposal and considered it to be evidence that she was not solely devoted to mathematics. This was in fact true. Since the sociology application was also unsuccessful, she stayed in Oxford.

Nevertheless, in 1930, under the pseudonym Jean Ayling, she published a sociological study, *The Retreat from Parenthood*,[10] in which she sketched a broad plan for reorganizing medical services, home design, child care, and labor laws so that child rearing would be compatible with the professional lives of both women and men. At this time she made a definitive decision to concentrate her energies on biological architecture. At Cambridge, she was a member of Joseph Needham's Theoretical Biology Club, together with J. D. Bernal, J. H. Woodger, C. H. Waddington, and Dorothy Needham. They were interested in what is now know as molecular biology. Dorothy Wrinch described her new direction:

Until 1933, my work was research in mathematics and mathematical physics. I had, however, long had a consuming interest in structural problems in physiology and chemistry, and I had always hoped to find specialists in this field with whom I could develop certain ideas. It proved impossible to arrange such collaboration, since the mathematical point of view was difficult to link up with the point of view of the professional chemist. At this time, then, it became clear to me that I had but two choices, either to abandon the attempt to develop these ideas or to undergo apprenticeship in chemistry sufficiently extensive to enable me to formulate the ideas in a form suitable for development by specialist workers. I chose the latter course and spent a year's leave of absence from Oxford on the continent of Europe, beginning an apprenticeship in many different laboratories. Already in 1934 the work had progressed far enough to attract the Rockefeller Foundation, who in 1935 provided support for my project for 5 years. . . . [121]

By 1935, Dorothy Wrinch had produced an original and remarkable theory of the gene. At the Theoretical Biology Club, she suggested that the specificity of genes resides in the specificity of the amino acid sequences.[144,145] Thus she made the connection between the linear sequence of the genes and that of the amino acids in the polypeptide chain. She then went on to construct a molecular model of the chromosome. The cyclic tetranucleotides were assumed to be at right angles to the chromosome axis. This structure was shown not to be correct by experimental birefringence data. However, the sequence hypothesis was important.

Wrinch then turned her attention to the globular protein molecules and within a year or so produced a structural model.[146] Beginning with the single hypothesis that peptide chains would be polymerized into sheets by links between CO and NH groups, Dorothy Wrinch deduced that the sheets would fold into a series of closed octahedra, which she called cyclols. The series was described by the general formula $72 \times n$ where n was the integral number of amino acid residues. When Bergmann and Niemann [*J. Bio. Chem.*: 118 (1937):310–14] deduced that egg albumin had $288 = 72 \times 2^2$ residues in the molecule, Dorothy Wrinch felt her hypothesis was proven. Here was an example with $n = 2$. This clever deduction startled the emerging world of molecular biology. Controversy raged, with Nobel prize winners lining up on either side; Irving Langmuir was Wrinch's leading advocate and Linus Pauling her leading critic. The embroilment reached disastrous proportions with Wrinch being virtually blacklisted by most of the scientific community. Dorothy Wrinch felt personally attacked. In 1938, after a meeting with Linus Pauling, she wrote: "The fact that they are against cyclols in any fundamental and *a priori* way in itself rather gives the show away. Because it is undeniable that the theory had NOT yet been shown to be false and therefore only fools or men of evil wishes towards me will be against it *a priori* . . . ."[121] More than a decade was to pass before any correct protein structures were discovered. These proteins did

Dorothy Wrinch and Irving Langmuir examining a cyclol model in 1936. Photo from the Smith College Archives. Courtesy of the *Journal of Chemical Education.*

not contain cyclols; nonetheless, Dorothy Wrinch's ideas had stimulated much thought and work.

Dorothy Wrinch's personal life was not happy. In 1937 her marriage was dissolved, and two years later she moved with twelve-year-old Pamela to the United States. The Rockefeller Foundation supported Dorothy's work until 1940; she spent the academic year 1940–41 at the chemistry department of Johns Hopkins University. Pauling and Niemann [*J. Amer. Chem. Soc.* 61 (1939):1860–67] predicted from bond energy values and heats of combustion that a protein with the cyclol structure would be less stable than one with a polypeptide chain by about 28 kcal/mole of amino acid residues. Although the details of their calculations were incorrect, as Dorothy Wrinch pointed out at the time,[149] they were widely accepted. Dorothy Wrinch's fellowship was up, she could not find a job, and World War II had begun. She was worried about Pamela, who was quite sensitive over the controversy in which her mother was involved. Pamela wrote a letter to Pauling:

> Your attacks on my mother have been made rather too frequently. If you both think each other is wrong, it is best to prove it instead of writing disagreeable things about each other in papers. I think it would be best to have it out and

see which one of you is really right. There are many quarrels in the world alas!! Don't please let yours be one; it is these things that help to make the world a kingdom of misery!! [p. 180][122]

Dorothy Wrinch eventually found a position in 1941. With the help of O. C. Glaser of the Amherst College biology department, she was appointed a visiting professor at Amherst, Smith, and Mount Holyoke colleges. Not surprisingly, there was a fair amount of local opposition to her appointment. Some months later, Dorothy Wrinch and O. C. Glaser were married, and they settled permanently in Massachusetts.

After her year as a jointly appointed visiting professor, during which she gave seminars at all three colleges, Dorothy Wrinch was given a research position at Smith, nominally in the physics department. Her association with Smith lasted for thirty years, until her retirement in 1971. There she had a few graduate students, conducted seminars for students and faculty, lectured, and continued her research. In the summers, she taught and lectured at the Marine Biological Laboratory at Woods Hole, Massachussetts. Her work flowed at an incredible rate. In the 1940s she concentrated on developing techniques for interpreting X-ray data of complicated crystal structures, and applying these techniques to protein X-ray data which she was able to obtain from experimentalists. Numerous papers and an important monograph, *Fourier Transforms and Structure Factors*,[150] were published. She studied mineralogy, hoping to deduce significant features of protein structures by drawing analogies between the morphologies of their crystals and certain minerals. Notebook after notebook was filled with her criticisms and ideas on papers covering many branches of science. Her total list of publications eventually reached 192.

In the period following World War II, Dorothy Wrinch's life had almost taken a new turn. John von Neumann had invited her to become a consultant in his pioneering work on computers at Princeton because she had persuaded him that one of the major uses of computers would be to aid in the interpretation of the X-ray patterns of protein crystals. She was enthusiastic and full of ideas, but her hoped-for move to Princeton never came about. Surely if von Neumann really wanted her, he could have used his vast personal influence and power to obtain necessary funds.[122]

Dorothy Wrinch believed that proteins had the cyclol structure. She was convinced that the available experimental evidence had been incorrectly interpreted. Then one evening in 1954, a chemistry professor at Smith came across an article written two years earlier by Swiss chemist Arthur Stoll [*Prog. in Chem. of Org. Nat. Products* 9 (1952):114–174] which claimed that the cyclol structure had been found in the ergot alkaloids. Ergot is a parasitic fungus that thrives on cereals, especially on the ears of rye [see A. Stoll, A. Hoffman, and T. Petrizilka, *Helv. Chim. Acta*, 34 (1951):1544–78]. Ergot of rye is the starting material for many pharmaceutical preparations. The structure of the peptide portion of the ergot alkaloids gave Wrinch's theories experimental verification. The linkage between amine groups of neighboring peptide chains confirmed the cyclol theory. The ergot alkaloids are, in a sense, simple versions of proteins, and Dorothy Wrinch was absolutely delighted with this example.[151] Thirteen years after Pauling and Niemann had "proved" that the cyclol bond was too unstable to exist, it had been found! Alas, the scientific community had moved on to other fronts and it paid little attention to this vindication of the cyclol theory.

Dorothy Wrinch once told Marjorie Senechal, "First they said my structure couldn't exist. Then when it was found in nature they said it couldn't be synthesized in a laboratory. Then when it was synthesized, they said it wasn't important anyway." However, she persisted and was able to get funding from the Ziskind Foundation and, later, the National Science Foundation. She spent the rest of her life working out the details of her theory. Her daughter, Pamela Wrinch Schenkman, died in a fire in November 1975; Dorothy Wrinch died three months later at the age of eighty-two.

Dorothy Wrinch was brilliant, witty, ambitious, hard working, and at times was considered jealous, abrasive, and aggressive. Her considerable talents bridged many fields. She accomplished much, but was restless and dissatisfied. What might have happened if she had been less embittered, if her crossing of the boundaries between disciplines had been better accepted, if her cyclol theory had been evidenced more in nature, if she had not felt so strongly that she was fighting the male-female interface, or if her funding and job situation had been better, it is impossible to guess. Perhaps the zeal for attacking impossible and important problems, together with an ineptitude for the politics of science, would force conditions of disappointment on anyone. Michael Polanyi wrote to Dorothy Wrinch in 1948, "You and I have much in common in the manner we managed to make our scientific careers less dull than usual."[122] Indeed, hers was.

Isabella L. Karle

Isabella Helen Lugoski[63] was born in Detroit, Michigan, on December 2, 1921. Her parents had emigrated from Poland and spoke only Polish at home; she never heard or spoke English until she was in the first grade. Her mother was a seamstress and her father a house painter. Before Isabella entered grammar school, her mother had taught her in Polish to do the basic mathematical operations and to read and write. In school, young Isabella progressed with almost unbelievable speed; there were only fifteen and one-half years between her entering the first grade and her Ph.D. degree! She won a four-year undergraduate scholarship to the University of Michigan, where she became a teaching associate. Although she graduated from Michigan with highest honors, she could not get a graduate teaching assistantship in chemistry, because women had never held such a position. But because of her outstanding academic record, the American

Association of University Women awarded her a fellowship and she was able to begin her graduate studies.

As a result of a stimulating physical chemistry course with Lawrence O. Brockway at Michigan, Isabella Lugoski and fellow student Jerome Karle chose to study electron diffraction by gaseous molecules. Their apparatus, although excellent by 1940 standards, had homemade electronic circuits and difficult-to-maintain vacuum systems. After the electron beam passed through the gas, images of the diffracted, concentric rings were photographed. The diameters and different relative intensities of the rings depend on the geometric and topologic properties of the gas and the voltage of the electron beam. A mathematical analysis of the measurements of the positions and intensities of the rings made it possible to calculate the interatomic distances and angles of the molecules of the gas. Their computations were tedious; strips of paper with trigonometric functions strategically placed along the edge were lined up to aid the summations. A hand-cranked calculator was used for the arithmetic.

Jerome and Isabella were married in 1942 and finished their graduate degrees in 1943 and 1944 within a few months of each other. Isabella was twenty-two when she got her Ph.D. After graduation, they worked in Chicago on the Manhattan Project and then at the University of Michigan until the end of the war. When they began to search for permanent places in academia, nepotism rules prevented them from finding suitable positions for both in one place. At this point, the Naval Research Laboratory in Washington, D.C., offered them an opportunity to continue their research on gas electron diffraction. With the help of an excellent machine shop, they constructed two duplicate state-of-the-art diffraction instruments to be sure that at least one instrument was leak-proof. The system featured tandem electromagnetic lenses which enhanced the quality of the electron beam. Carbon dioxide was the first compound to have its vibrational parameters determined by electron diffraction.[65]

Isabella Karle's research in electron diffraction led naturally to an intense interest in structure analysis by X-ray diffraction. In the electron diffraction work, a non-negativity criterion was introduced to obtain an experimental function for the background intensity. This criterion was further developed in Isabella Karle's practical applications of a crystallographic technique called "direct methods," which revolutionized structure analysis. Before this technique, the first general systematic structural approach had been developed by A. L. Patterson in the 1930s. Patterson's method was known as the "heavy atom technique." An atom of high atomic number, relative to the rest of the atoms in the crystal, was first located and then the rest of the atoms were found by successive approximations. Many structures have been and are being solved in this way. However, a large number of compounds have nearly equally-weighted atoms such as carbon, oxygen, and nitrogen; for these substances the Patterson method often fails. "Direct

Isabella Karle at the electron diffraction instrument completed in 1948. Karle helped in the design and construction of the instrument. Courtesy of the *Journal of Chemical Education.*

methods" are mathematical techniques for phasing the diffraction pattern, to which Jerome Karle has made important contributions [*see Acta Cryst.* 3 (1950):181]. Isabella Karle applied the theory to a large number of structures and developed a systematic attack for the general problem. This method is called the "symbolic addition procedure."[67,69] To give an idea of the importance of her work, a National Science Foundation study found that her 1966 paper[69] outlining the theory of the "symbolic addition procedure" occurred sixth in the listing of one hundred chemical articles most cited in 1972.

Isabella Karle has applied these mathematical techniques to an interesting variety of substances. One example is the frog venom study.[70] Frog skin secretions have been used by South American natives as arrow-tip poisons. The toxins in these secretions block specific nerve impulses, making these toxins ideal for medical studies in nerve transmission. Isabella Karle studied the crystals distilled from the frog venom. She found that different species of frogs produced toxins with very different structures, even in frogs living on opposite sides of the same stream.

Isbella Karle also studied the effect of radiation on DNA in human cells. Radiation causes genetic damage by breaking the bonds between the atoms; abnormalities result when the new bonds connect in a variety of different ways. One of the substances she studied was irradiated thymine,[71] a substance related to the production of antibodies. She has also identified structures of photon-generated dimers of nucleic acid bases, which could be reversed readily to undamaged monomers, and of polymers which could not be readily reversed.

Another example of Isabella Karle's work is the study of the cyclic polypeptide molecule valinomycin,[72] which transports potassium ions across membranes in biological systems and also is an antibiotic for a type of tuberculosis. From a purely crystallographic point of view this molecule is the largest—by a factor of two—published structure analyzed to date without the heavy atom technique. Valinomycin has 156 independent nonhydrogen atoms and 180 hydrogen atoms.

Isabella Karle is the leading authority on solid state peptide structures.[73] She demonstrated that although the folding of the peptide backbone is quite unpredictable, the conformation is independent of the polarity of the solvent. She has also been investigating the structures of antamanide,[74] a cyclic decapeptide isolated from poisonous mushrooms. This substance is an antitoxin which counteracts the deadly toxin phalloidin also found in the same mushroom. Antamanide selectively transports sodium and calcium ions across membranes.

Most recently she solved the structure of enkephalin, a natural analgesic occurring in the brain. The crystal consists of four enkephalin molecules, each with a different conformation, and a large number of solvent molecules surrounding the peptides for a total of more than 210 independent carbon, nitrogen, and oxygen atoms.

Isabella Karle's laboratory includes postdoctoral students as well as a large number of visiting scholars from all over the world. She is a welcome lecturer particularly because of her detailed examples of the "symbolic addition procedure." In 1976 she was president of the American Crystallographic Association and that same year received the Garvan Medal of the American Chemical Society. Isabella Karle won the 1985 Chemical Pioneer Award given by the American Institute of Chemists. Her publication list has passed two hundred. In 1986 Jerome Karle won the Nobel Prize in Chemistry with Herbert Hauptman. The Karles have three daughters, each with successful scientific careers.

Isabella Karle developed the practical aspects of a mathematical theory of crystallography which revolutionized the types and complexity of problems which may be attacked by crystal structure analysis. Her research led to an exponential increase in the speed and facility with which the questions of identification and structural configuration can be answered.

DOROTHY CROWFOOT HODGKIN

Only once in a lifetime, if ever, does a scientist solve a problem of such difficulty that afterwards the discipline is forever changed. With her X-ray diffraction analyses of penicillin, vitamin B-12, and insulin, Nobel laureate Dorothy Crowfoot Hodgkin did this not once but three times.[57]

Dorothy Mary Crowfoot was born on May 12, 1910, in Cairo, Egypt, where her father, John Winter Crowfoot, was an inspector for the Ministry of Education. He was a classical scholar and archaeologist who later held important positions in the Sudan and at the British School of Archaeology in Jerusalem. In 1928, just before she left to study chemistry at Somerville College, Oxford University, Dorothy helped her parents excavate Byzantine churches at Jerash in Transjordan. During her first year at the university, she spent her spare time completing drawings of the mosaic pavements of the Byzantine churches.

In 1932 Dorothy Crowfoot left Oxford for a two-year stay at Cambridge with J. D. Bernal, who had just begun work on the sterols.[14] She describes the Cambridge lunches:

> Every day, one of the group would go and buy fresh bread from Fitzbillies, fruit and cheese from the market, while another made coffee on the gas ring in the corner of the bench. One day there was talk about anaerobic bacteria at the bottom of the lake in Russia, or Leonardo da Vinci's engines of war or about poetry or printing. We never knew to what enchanted land we would be taken next. More serious scientific discussions took place in the 'Space Groups' an informal series of colloquia in crystallography.[53]

With Bernal she studied early preparations of vitamin B-1, vitamin D, and several of the sex hormones. In the summer of 1934 they took the first single crystal photographs of a protein, the digestive enzyme pepsin.[12]

Back in Oxford Dorothy Crowfoot began a crystal analysis of the halogen derivatives of cholesterol. She returned to Cambridge whenever she could, because, in her relative solitude at Oxford, she missed the excitement of Bernal's laboratory.

Dennis Riley, Dorothy Crowfoot's first graduate student, describes his unusual situation in 1937: only two years before there had been a scandal when a women's college had allowed mixed tea parties on Sundays. He persisted in his desire to work with Dorothy Crowfoot. "This, at the time, was quite revolutionary and several eyebrows were lifted. Here was I, a member of a prestigious college [Christ Church], choosing to do my fourth year's research in a new borderline subject with a young female who held no university appointment but only a fellowship in a women's college" [p. 17].[27] Riley became excited about Dorothy Crowfoot's work in sterols after she gave a talk at his invitation to the undergraduate chemistry club. Women could not be admitted into the senior chemistry club and Dorothy Crowfoot had never before been asked to address it.

Dorothy married historian Thomas Hodgkin in 1937. After the war he became active in African studies in both Oxford and Ghana, which neces-

Dorothy Hodgkin at the Twelfth International Congress on Crystallography, at Ottawa, August 1981. Courtesy of the *Journal of Chemical Education*.

sitated their maintaining two households, one in England and one in Africa. Their three children were Luke (1938), Elizabeth (1941), and Toby (1946).

Late in the summer of 1942, Dorothy Hodgkin obtained one of the first crystalline degradation products of penicillin, penicillamine hydrochloride.[25] Molecular weight measurements were made on this and several other degradation products. By the following year, salts of a few penicillin derivatives were crystallized. Structural information would aid in the synthesis and production of the penicillin needed so desperately during the war. Barbara Rogers-Low (now a professor in the biochemistry department at Columbia University) was a graduate student with Dorothy Hodgkin. The project required the integrated efforts of both crystallographers and biochemists, and such cooperation was unusual at the time. Three-dimensional electron-density maps of the sodium and potassium salts of benzylpenicillin were calculated on an old IBM punch card machine in an abandoned government building. The essential chemical structure was determined by early summer 1945.[25] The biochemists had established many of the features; however, the importance of this work was that in attacking the analysis by using many different crystallographic techniques, the whole scope of what was possible in crystallography was enlarged.

Her next major work was the study of vitamin B-12, which is important in the understanding and control of pernicious anemia.[48, 49] In 1926 it was found that the disease could be controlled if the patient would eat daily one-half pound of raw liver. The active ingredient in the liver turned out to be vitamin B-12. The first X-ray photographs, taken in 1948, showed that the determination of the structure of vitamin B-12 would be an enormous, probably impossible, undertaking. Dorothy Hodgkin coordinated the data collected from several derivatives by different co-workers over a period of six years.[48] In 1949 June Broomhead (now Lindsey) collected three-dimensional intensity data of a crystal of the vitamin, which was coated to inhibit drying, and then in 1951 Clara Brink (now Shoemaker) used crystals that were surrounded by their mother liquid. M. Jennifer Harrison (now Kamper) studied the crystal structure of a chlorine substituted vitamin B-12 in 1955. This is a sample of a few of the women workers on the project. The mathematics were extremely tedious; it was not unusual to spend many months on one map of part of the structure. One day Kenneth Trueblood of UCLA visited Oxford from California. He was doing pioneer work on SWAC (National Bureau of Standards Western Automatic Computer), one of the first high-speed computers. Trueblood offered to help with the calculations, and the work became known as "structure analysis by post and cable." Vitamin B-12 was the largest organic molecule whose structure was determined before 1956. The analysis disclosed several surprising, unanticipated features of the molecule.

Jenny Pickworth was a graduate student on the Oxford end. Many years before, Dorothy Hodgkin had been Jenny Pickworth's tutor at Somerville College, Oxford. After a year's work with spectroscopist H. W. Thompson, Jenny Pickworth started graduate work with Dorothy Hodgkin when good crystals of hexacarboxylic acid derived from vitamin B-12 had just become available. For her D.Phil., Jenny Pickworth solved the structure of this ring system of the B-12, showing the shape in three dimensions and the chemical formula.[48] In 1955 she took a postdoctoral fellowship to work with Robert Corey at the California Institute of Technology and married Don Glusker that year. When she requested a letter of recommendation from Dorothy Hodgkin to work for a commercial firm in Philadelphia, Dorothy Hodgkin instead wrote to A. L. Patterson. In 1956 he invited Jenny Glusker to the Institute for Cancer Research, where she remains to this day. Her research interests are in enzyme mechanisms[47] and the interactions of polycyclic carcinogens, mutagens, and antitumor agents. She has published about one hundred papers, has written six books, received the Garvan Medal in 1979, and was president of American Crystallographic Association that year.

Max Perutz won the Nobel Prize in 1962 along with John Kendrew for Perutz's work on hemoglobin and Kendrew's work on myoglobin. Perutz wrote:

I felt embarrassed when I was awarded the Nobel Prize before Dorothy, whose great discoveries had been made with such fantastic skill and chemical insight and had preceded my own. The following summer I said as much to the Swedish crystallographer Gunner Hagg when I ran into him in a tram in Rome. He encouraged me to propose her, even though she had been proposed before. In fact, once there had been a news leak that she was about to receive the Nobel Prize, but it proved false; Dorothy never mentioned that disappointment to me until long after. Anyway, it was easy to make out a good case for her; Bragg and Kendrew signed it with me, and to my immense pleasure it produced the desired result soon after. [pp. 10–11][27]

In 1964 Dorothy Hodgkin was the sole recipient of the Nobel Prize in chemistry for her work in determining the structure of biochemical compounds of primary importance.[49]

Dorothy Hodgkin's third major achievement was the determination of the structure of the protein insulin, the result of more than thirty years of research that followed the taking of the first X-ray picture in 1935. Proteins are polymers built up from specific sequences of twenty different amino acids. Insulin contains fifty-one amino acids. Fredrick Sanger, a biochemist, determined, for the first time, the exact ordering of the chain. Dorothy Hodgkin unraveled the unique three-dimensional arrangement. The functioning of proteins depends on the optimal positioning of the amino acid side-groups, which lead to hydrogen bonds, disulfide bridges, appropriate solubilities, and, in general, energetically favorable structures. Calculations comparing wet insulin with air-dried insulin first showed that the protein molecule was a rigid unit.[26] The molecules rotated about a threefold axis as the crystal dried. A whole series of insulin crystals—zinc-free insulin, zinc insulin, cadmium insulin, and lead insulin—were measured and compared. By 1969 Dorothy Hodgkin was able to show that the long peptide chains are folded into compact molecules which form a hexamer of six molecules around two zinc atoms.[4] The hexamers crystallize even within the pancreatic cells that synthesize insulin.

At the time of the announcement of her Nobel Prize in October 1964, Dorothy Hodgkin was in Accra, Ghana, with her husband, Thomas. Luke, twenty-five, was teaching mathematics at the University of Algiers; Elizabeth, twenty-three, was on the faculty of a girls' school in Zambia; and Toby, eighteen, was in voluntary service in India. The previous year Thomas Hodgkin's cousin Alan Hodgkin had shared the Nobel Prize in physiology and medicine. Thus, the three Hodgkin children have a Nobel laureate on each side of their family tree.

During this work I have become aware of all sorts of strange inconsistencies. For example if one looks up the name "Dorothy Hodgkin" in *The New Columbia Encyclopedia* (1975) she is not there, but "Alan Hodgkin" is. Alan Hodgkin received a one-third share of the 1963 Nobel Prize and

Dorothy Hodgkin was the sole recipient of the 1964 Nobel Prize. Crick, Watson, and Wilkins are also in this encyclopedia.

Dorothy Hodgkin is an informal person. At Oxford she is called Dorothy, a very unusual custom at a European university. Many women have worked with her. In addition to Jenny Pickworth (Glusker), Maureen Mackay, Diana Pilling, and Helen Stoeckli-Evans were associated with the vitamin B-12 work. Marjorie M. Harding, Ann F. Kennedy, Eleanor Dodson, Pauline Cowan, and Margaret Joan Adams contributed to the insulin project.[4,22,23] Beryl Oughton worked on gramicidin.

One cannot but help make a connection between the mosaic patterns on the floors of the Byzantine churches that young Dorothy Crowfoot drew and her later contributions in unraveling the marvelous secrets revealed in the beautiful three-dimensional patterns of penicillin, vitamin B-12, and insulin. As J. Dunitz eloquently puts it:

> Dorothy had an unerring instinct for sensing the most significant structural problems in this field, she had the audacity to attack these problems when they seemed well-nigh insoluble, she had the perseverance to struggle onward where others would have given up, and she had the skill and imagination to solve these problems once the pieces of the puzzle began to take shape. [p. 59][27]

CONCLUSION

In this chapter we have looked at the question of both the nature of crystallography and the number of women in crystallography. Women do have a high visibility in crystallography. This impression is, in part at least, because a few outstanding women like Dorothy Hodgkin and Kathleen Lonsdale are constantly before us. In addition, a number of other women have made important contributions. Rosalind Franklin's studies on coal, DNA, and plant viruses explored new areas in the semicrystalline materials. Isabella Karle, by her mathematical applications of "direct methods," has solved structures of biological importance that were unattainable before her work. Dorothy Wrinch inspired much discussion and a new avenue of approach to the proteins. At the grass-roots level there is a continual outpouring of work from women. This has been carefully substantiated in the references for this chapter, which have been designed so that each reference contains at least one woman as an author. The journals chosen present the most prestigious of the scholarly publications.

But when actual numbers are tabulated, women only account for 14 percent of all crystallographers. By comparison to other physical sciences— physics has only 2 percent—crystallography appears to do well. Nevertheless, overpopulation by women is not the case. The achievements of a few women were so brilliant as to exaggerate the small numbers.

In many ways crystallography is the scientific legacy of the Braggs. They had women in their laboratories and the lineage we have traced shows how important this was in ultimately determining the large contribution women have made to this field. Because of the Braggs, doors were opened. Positive action taken by a few individuals can make the difference!

N O T E S

The two most useful references cited in this study were:

1. Ewald, P. P. *Fifty Years of X-ray Diffraction.* NVA Oosthoesk's Uitgevers-maatschappij, Utrech, The Netherlands (1962).

2. *World Directory of Crystallographers,* Sixth Edition. A. L. Bednowitz, general editor. D. Reidel Publishing Company, Dordrecht, Holland (1981).

The following references all contain women as authors and are indicated in the text:

3. Abir-Am, Pnina. "Synergy or Clash: Disciplinary and Marital Trajectories in the Career of Mathematical Biologist Dorothy Wrinch, 1894–1976," In *Uneasy Careers and Intimate Lives: Women in Science, 1789–1979,* P. Abir-Am and D. Outram, eds. Rutgers University Press (1987).

4. Adams, Margaret Joan, T. L. Blundell, Eleanor Dodson, M. Vijayan, E. N. Baker, Marjorie Harding, Dorothy Hodgkin, B. Rimmer, and S. Sheat. "Structure of Rhombohedral 2-zinc Insulin Crystals." *Nature London* 224 (1969):491–95.

5. Allen, Natalie C. B., and T. H. Laby. "Sensitivity of Photography Plates to X-rays." *Nature* 103 (1919):177.

6. Allen, Natalie C. B. "The Crystal Structure of Benzil." *Phil. Mag.* 3 (1927):1037.

7. Armstrong, Elizabeth. "Relation between Secondary Dauphinee Twinning and Irradiation-coloring in Quartz." *Am. Min.* 31 (1946):456–61.

8. Astbury, W. T., and Kathleen Yardley. "Tabulated Data for the Examination of the 230 Space Groups by Homogeneous X-rays." *Phil. Trans.* A224 (1924):221.

9. Astbury, W. T., and Florence Bell. "Some Recent Developments in the X-ray Study of Proteins and Related Structures." *Cold Spring Harbor Symp.* 6 (1938):109.

10. Ayling, Jean (pseudonym of Dorothy Wrinch). *The Retreat from Parenthood.* Kegan Paul, Trench, Trubner and Co. Ltd. (1930).

11. Bernal, J. D., W. T. Astbury, and Thora C. Marwick. "X-ray Analysis of the Structure of the Wall of Valonia Ventricosa." *Proc. Roy. Soc. London* B109 (1932):433–450.

12. Bernal, J. D., and Dorothy Crowfoot. "X-ray Photographs of Crystalline Pepsis." *Nature London* 133 (1934):794.

13. Bernal, J. D., and Helen Megaw. "The Function of Hydrogen in Intermolecular Forces." *Proc. Roy. Soc. London* A151 (1935):384–420.

14. Bernal, J. D., Dorothy Crowfoot, and I. Fankuchen. "X-ray Crystallography and the Chemistry of the Sterols." *Trans. Roy. Soc. London* A239 (1940):135.

15. Bijvoet, J. M., N. H. Kolkmeijer, and Caroline H. MacGillavry. *Roentgen-analyse von Krystallen.* Springer, Berlin (1940).

16. Brink, Clara, and Caroline H. MacGillavry. "Structure of K2CuC13 and Isomorphous Substances." *Acta Cryst.* 2 (1949):158.

17. Brink, Clara, and D. P. Shoemaker. "A Variation on the Sigma-Phase Structure: The Crystal Structure of the P Phase, Mo-Ni-Cr." *Acta Cryst.* 8 (1955):734.

18. Carlisle, C. H., Helen Scouloudi, and Marianne Spier. "Recent Observations on the Structure of Ribonuclease." *Proc. Roy. Soc. London* B141 (1953):85–89.

19. Caroe, Gwendoline Mary Bragg. *William Henry Bragg.* Cambridge University Press, Cambridge, England (1978).

20. Cauchois, Yvette. "X-ray Absorption Spectra of Some Bivalent Nickel Salts and Their Aqueous Solutions." *Comp. rend.* 224 (1947):1156–58.

21. Codding, Penelope W., and K. Ann Kerr. "Triphenylphosphine Sulfide." *Acta Cryst.* B34 (1978):3785–87.

22. Cowan, Pauline M., and Dorothy Crowfoot. "Observations on Peptide Chain Models in Relation to Crystallographic Data in Gramicidin B and Insulin." *Proc. Roy. Soc. London* B141 (1953):89–92.

23. Cowan, Pauline M., S. McBavin, and A. C. T. North. "The Polypeptide Chain Configuration of Collagen." *Nature* 176 (1955):1062–64.

24. Cox, E. G., G. A. Jeffrey, and Mary R. Truter. "Crystal Structure of Nitronium Perchlorate." *Nature* 162 (1948):259.

25. Crowfoot, Dorothy, C. W. Bunn, Barbara Rogers-Low, and A. Turner-Jones. "The X-ray Crystallographic Investigation of the Structure of Penicillin." In *Chemistry of Penicillin*, H. T. Clark et al., editors. Princeton University Press, Princeton, New Jersey (1949), pp. 310–367.

26. Crowfoot, Dorothy, and D. P. Riley. "X-ray Measurements on Wet Insulin Crystals." *Nature* 144 (1939):1011–12.

27. Dodson, G., Jenny Glusker, and D. Sayre, editors. *Structure Studies on Molecules of Biological Interest, a Volume in Honour of Professor Dorothy Hodgkin.* Clarendon Press, Oxford (1981).

28. Donnay, J. D. H., and Gabrielle E. Hamburger Donnay. "The One-dimensional Crystal: I General." *Acta Cryst.* 2 (1949):366–69.

29. Donnay, Gabrielle H. "The One-dimensional Crystal: II A Graphical Method for Computing Structure Factors." *Acta Cryst.* 2 (1949):370–71.

30. Donnay, Gabrielle H., and M. J. Buerger. "Determination of the Crystal Structure of Tourmaline." *Acta Cryst.* 3 (1950):379–88.

31. Donnay, Gabrielle H., and J. D. H. Donnay. "The Crystallography of Bastnaesite, Parisite, Roentgenite and Synchisite." *Am. Min.* 38 (1953):932–63.

32. Donnay, Gabrielle H. "Roentgenite, a New Mineral from Greenland." *Am. Min.* 38 (1953):868–70

33. Dornberger-Schiff, Kate. "Patterson and Fourier Projections of Single Molecules of Hemoglobin." *Acta Cryst.* 3 (1950):143–46.

34. Dornberger-Schiff, Kate. "Structure of Hemoglobin of Patients with Sickle-cell Anemia." *Ach Geschwulstforsch* 6 (1954):192–96.

35. Dornberger-Schiff (Boll), Kate. "Order-disorder Structures (OD-structures)." *Acta Cryst.* 9 (1956):593–601.

36. Dornberger-Schiff (Boll), Kate. "The OD-structure (order-disorder structure) of Purpurogallin." *Acta Cryst.* 10 (1957):271–77.

37. Elam, C. F. (Mrs. G. H. Tipper). "Etching of Copper by Oxygen." *Trans. Faraday Soc.* 32 (1936):1604–14.

38. Franklin, Rosalind E. "Crystallite Growth in Graphitizing and Nongraphitizing Carbons." *Proc. Roy. Soc.* A209 (1951):196–218.

39. Franklin, Rosalind E., and R. G. Gosling. "Molecular Configuration in Sodium Thymonucleate." *Nature* 171 (1953):740–41.

40. Franklin, Rosalind E., and R. G. Gosling. "The Structure of Sodium Thymonucleate Fibres. I. The Influence of Water Content." *Acta Cryst.* 6 (1953):673–77.

41. Franklin, Rosalind E., and R. G. Gosling. "The Structure of Sodium Thymonucleate Fibres. II. The Cylindrically Symmetrical Patterson Function." *Acta Cryst.* 6 (1953):678–85.

42. Franklin, Rosalind E., and R. G. Gosling. "Evidence for a 2-chain Helix in Crystalline Structure of Sodium Deoxyribonucleate." *Nature* 172 (1953):156–57.

43. Franklin, Rosalind E. "Structure of Tobacco Mosaic Virus." *Nature* 175 (1955):379.

44. Franklin, Rosalind E., A. Klug, and K. C. Holmes. "On the Structure of Some Ribonucleoprotein Particles." *Trans. Faraday Soc.* 25 (1958):197.

45. Gilchrist, Helen S., and Berta Karlik. "Separation of Normal Long Chain Hydrocarbons by Fractional Distillation in High Vacuum Systems." *J. Chem. Soc.* (1932):1992–95

46. Gilchrist, Helen S. "The Preparation and Constitution of Synthetic Facts Containing a Carbohydrate Chain." *Rept. Brit. Assoc. Adv. Sc.* (1922):357.

47. Glusker, Jenny. "Citrate Conformation and Chelation: Enzymatic Implications." *Acc. Chem. Res.* 13 (1980):345–52.

48. Hodgkin, Dorothy, M. Jennifer Kamper, June M. Lindsey, Maureen MacKay, Jenny Pickworth, J. H. Robertson, Clara B. Shoemaker, J. G. White, R. J. Prosen, and K. N. Trueblood. "The Structure of Vitamin B-12. I. An Outline of the Crystallographic Investigation of Vitamin B-12." *Proc. Roy. Soc. London* A242 (1957):228–83.

49. Hodgkin, Dorothy. "X-ray Analysis of Complicated Molecules." *Science* 150 (1965):979 (Nobel Prize lecture).

50. Hodgkin, Dorothy, and D. P. Riley. "Ancient History of Protein X-ray Analysis." In *Structural Chemistry and Molecular Biology*, A Rich et al., editors. W. H. Freeman and Company, San Francisco (1968), pp. 15–28.

51. Hodgkin, Dorothy M. C. "Kathleen Lonsdale." *Biographical Memoirs of FRS* 21 (1975):447–84.

52. Hodgkin, Dorothy M. C. "Obituary of Dorothy Wrinch." *Nature* 260 (1976):564.

53. Hodgkin, Dorothy M. C. "John Desmond Bernal." *Biographical Memoirs of FRS* (1981):17–84

54. James, R. W., and Elsie M. Firth. "X-ray Study of Heat Motions of the Atoms in a Rock Salt Crystal." *Proc. Roy. Soc. London* A117 (1927):62–87.

55. Julian, Maureen M. "X-ray Crystallography and the Work of Dame Kathleen Lonsdale." *The Physics Teacher* 19 (1981):159–65.

56. Julian, Maureen M. "Kathleen Lonsdale and the Planarity of the Benzene Ring.: *J. Chem. Educ.* 58 (1981):365–66.

57. Julian, Maureen M. "Profiles in Chemistry: Dorothy Crowfoot Hodgkin, Nobel Laureate." *J. Chem. Educ.* 59 (1982):124–25.

58. Julian, Maureen M. "Profiles in Chemistry: Kathleen Lonsdale 1903–1971." *J. Chem. Educ.* 59 (1982):965–66.

59. Julian, Maureen M. "Eamon de Valera, Esin Schrodinger and the Dublin Institute for Advanced Studies." *J. Chem. Educ.* 60 (1983):199–200.

60. Julian, Maureen N. "Profiles in Chemistry: Rosalind Franklin, From Coal to DNA to Plant Viruses," *J. Chem. Educ.* 60 (1983):660–62.

61. Julian, Maureen M. "Eighteen Years of Chemical Pioneers," *The Chemist* 5 (1983):22.

62. Julian, Maureen M. "Profiles in Chemistry: Dorothy Wrinch and a Search for the Structure of Proteins." *J. Chem. Educ.* 61 (1984):890–92.

63. Julian, Maureen M. "Profiles in Chemistry: Isabella Karle and a Mathematical Breakthrough in Crystallography." *J. Chem. Educ.* 63 (1986):66–67.

64. Julian, Maureen M. "Crystallography in the Laboratory of William Henry Bragg." *Chemistry in Britain* 72, 8 (1986):66, 67.

65. Karle, Isabella L., and J. Karle. *J. Chem. Phys.* 17 (1949):1052.

66. Karle, Isabella L., H. Hauptman, J. Karle, and A. B. Wing. "Crystal and Molecular Structure of p,p'-dimethyloxybenzo-phenone by the Direct Probability Method." *Acta Cryst.* 11 (1958):257–63.

67. Karle, Isabella L., and J. Karle. *Acta Cryst.* 16 (1963):969.

68. Karle, Isabella L., J. Karle, T. B. Owen, and J. L. Hoard. "The Structure of a Saturated Dimer of Hexaflourobutadiene." *Acta Cryst.* 18 (1965):345.

69. Karle, J., and Isabella L. Karle. "The Symbolic Addition Procedure for Phase Determination for Centrosymmetric and Noncentrosymmetric Crystals." *Acta Cryst.* 21 (1966):849.

70. Karle, Isabella, and J. Karle. "The Structural Formula and Crystal Structure of O-p-Bromobenzoate Derivative of Batrachotoxinin A, a Frog Venom and Steroidal Alkaloid." *Acta Cryst.* B25 (1969):428.

71. Karle, Isabella. "Crystal Structure of a Thymine-Thymine Adduct from Irradiated Thymine." *Acta Cryst.* B25 (1969):2119.

72. Karle, Isabella. "The Conformation of Valinomycin in a Triclinic Crystal Form." *J. Amer. Chem. Soc.* 97 (1975):4379.

73. Karle, Isabella. "X-ray Analysis: Conformation of Peptides in the Crystalline State." In *The Peptides*, E. Gross and J. Meinhofer, editors. Academic Press, New York (1981), Vol. 4, pp. 1–5.

74. Karle, Isabella. "Antamanide: Alkali Metal Ion Complexation, Channel Formation and Effects of Polarity of Solvent." Proc 6th Amer. Peptide Symp., E. Gross and J. Meinhofer, eds. Pierce Chemical Co. (1979), pp. 681–90.

75. Kennard, Olga, and C. Rimington. "Identification and Classification of Some Porphyrins on the Basis of Their X-ray Diffraction Patterns." *Biochem. J.* 55 (1953):105–109.

76. Kennard, Olga, K. Ann Kerr, D. G. Watson, J. K. Fawcett, and L. Riva di Sanserverino. "Crystal Structure of (+ −)-1-acetyl-16-methyl-aspidospermidine 4-methiodide." *Chem. Com.* 24 (1967):1286–87.

77. Kerr, K. Ann, and J. M. Robertson. "Crystal and Molecular Structure of 1,1'-benaphthyl." *J. Chem. Soc.* B9 (1969):1146–49.

78. Knaggs, I. Ellie, Berta Karlick, and C. F. Elam. *Tables of Cubic Crystal Structure of Elements and Compounds.* Adam Hilger Ltd., London (1932).

79. Knaggs, I. Ellie, and Kathleen Lonsdale. "The Structure of Benzil." *Nature* 143 (1939):1023.

80. Kranjc, Katarina. "A Theoretical Possibility of Correcting the Collimation Error in Small-angle X-ray Scattering." *Acta Cryst.* 7 (1954):709–10.

81. Lange, J. J. de., J. M. Robertson, and Ida Woodward. "X-ray Crystal Analysis of Trans-Azobenzene." *Proc. Roy. Soc. London* A171 (1939):398.

82. Lonsdale, Kathleen. "Preliminary Results of Hexamethylbenzene." *Nature* 122 (1928):810.

83. Lonsdale, Kathleen. "The Structure of the Benzene Ring in C6(CH3)6." *Proc. Roy. Soc. London* A123 (1929):494–515.

84. Lonsdale, Kathleen. "X-ray Evidence of the Structure of the Benzene Ring." *Trans. Faraday Soc.* 25 (1929):352–66.

85. Lonsdale, Kathleen. "Magnetic Anisotropy and Electronic Structure of Aromatic Molecules." *Proc. Roy. Soc. London* A159 (1937):149–61.

86. Lonsdale, Kathleen. "Diamagnetic Anisotropy of Organic Molecules." *Proc. Roy. Soc. London* A171 (1939):541–68.

87. Lonsdale, Kathleen. "Extra Reflections from the Two Types of Diamond." *Proc. Roy. Soc. London* A179 (1942):315–20.

88. Lonsdale, Kathleen. "Crystal Photography with Divergent X-rays." *Nature London* 151 (1943):52.

89. Lonsdale, Kathleen. "Diamonds, Natural and Artificial." *Nature London* 153 (1944):669–72.

90. Lonsdale, Kathleen. "Divergent Beam X-ray Photography of Crystals." *Phil. Trans. Roy. Soc. London* A240 (1947):219–50.

91. Lonsdale, Kathleen. *Crystals and X-rays.* Bell, London (1948).

92. Lonsdale, Kathleen, and N. F. M. Henry. *International Tables for X-ray Crystallography.* Vol. I. Kynoch Press, Witton Birmingham (1952).

93. Lonsdale, Kathleen. *Is Peace Possible?* Penguin Special (1957).

94. Lonsdale, Kathleen. "Intermolecular Distances and Diamagnetic Anisotropy in Crystals as Measures of the Polarity of Benzene and Borazole Substitutes." *Nature London* 184 (1959):1060.

95. Lonsdale, Kathleen, and J. Kasper. *International Tables for X-ray Crystallography.* Vol. II. Kynoch Press, Witton Birmingham (1959).

96. Lonsdale, Kathleen, H. Judith Milledge, and E. Nave. "X-ray Studies of Synthetic Diamonds." *Min. Mag. London* 32 (1959):185–201.

97. Lonsdale, Kathleen, Caroline H. MacGillavry, and G. D. Riech. *International X-ray Tables for X-ray Crystallography.* Vol. III. Kynoch Press, Witton Birmingham (1962).

98. Lonsdale, Kathleen, E. Nave, and J. F. Stephens. "X-ray Studies of a Single Crystal Chemical Reaction: Photo-oxide of Anthracene to (Anthraquinone, anthrone)." *Phil. Trans. Roy. Soc. London* A261 (1966):1–31.

99. Lonsdale, Kathleen. "Human Stones." *Scientific American* 219 (1968):104–111.

100. Lonsdale, Kathleen, D. June Sutor, and Susan Wooley. "Composition of Urinary Calculi by X-ray Diffraction." *Br. J. Urol.* 33 (1968):402.

101. Lonsdale, Kathleen. "Formation of Lonsdaleite from Single Crystal Graphite." *Am. Min.* 56 (1971):333–36.

102. MacGillavry, Caroline H., and H. J. Vos. "Anomale Untergrundschwarzung in Weissenberg Diamrammen." *Z. Krist.* A105 (1943):257–67.

103. MacGillavry, Caroline H., and B. Strijk. *Nature* 157 (1946):135.

104. MacGillavry, Caroline H., and B. Strijk. *Physica* 11 (1946):369.

105. MacGillavry, Caroline H., and B. Strijk. *Physica* 12 (1946):129.

106. MacGillavry, Caroline H. "On the Derivation of Harker-Kasper Inequalities." *Acta Cryst.* 3 (1950):214–17.

107. MacGillavry, Caroline H. *Fantasy and Symmetry: The Periodic Drawings of M. C. Escher.* Harry N. Abrams, Inc., New York (1976).

108. Manghi, E., C. A. de Caroni, M. R. de Benyacar, and M. J. de Abeledo. "Optical Properties of p-Dichlorobenzene." *Acta Cryst.* 23 (1967):205–208.

109. Martinez Carrera, Sagrario. *Acta Cryst.* 9 (1956):145–50.

110. Martinez Carrera, Sagrario. *Acta Cryst.* 20 (1966):783.

111. Marwick, Thora C. "The Space Group of Strychnine." *Nature* 126 (1930):438.

112. Marwick, Thora C., and W. T. Astbury. "Crystal Structure of Cellulose." *Nature* 127 (1931):12–13.

113. Megaw, Helen D. *Crystal Structures: A Working Approach.* W. B. Saunders Co., Philadelphia (1973).

114. Pepinsky, R., and Caroline H. MacGillavry. "Phase Limiting Relations Following from a Known Maximum Value of the Electron Density." *Acta Cryst.* 4 (1951):284.

115. Pickett, Lucy, Mary L. Sherrill, and Belle Otto. "Isomers of 2-pentene." *J. Am. Chem. Soc.* 51 (1929):3023–33.

116. Przybylska, Maria. "The Structure of (+)-Des-(oxymethylene)-Lycoctoninc." *Acta Cryst.* 14 (1961):424.

117. Przybylska, Maria. "The Crystal Structure of (−)-N-Methyl-Gelsemicine Hydriodide." *Acta Cryst.* 15 (1962):301–309.

118. Przybylska, Maria. "The Crystal and Molecular Structure of (+)-Hetisine Hydrobromide." *Acta Cryst.* 16 (1963):871.

119. Sayre, Anne. *Rosalind Franklin and DNA.* W. W. Norton and Co., New York (1975).

120. Scouloudi, Helen. "The Crystal Structure of Myoglobin IV Seal Myoglobin." *Proc. Roy. Soc. London* A258 (1960):181–201.

121. Senechal, Marjorie. "A Prophet Without Honor: Dorothy Wrinch, Scientist, 1894–1976." *Smith Alumnae Quarterly,* April 1977.

122. Senechal, Marjorie. *Structures of Matter and Patterns in Science.* Schenkman Publishing Company, Cambridge, Massachusetts (1980).

123. Shoemaker, Clara B., and D. P. Shoemaker. "Tetraedrisch dicht gepackte Strukturen von Legierunger der Ubergangsmetalle." *Monatshefte für Chemie* 102 (1971):1643–66.

124. Shoemaker, Clara B. "Relationships between Structures Derived from Complex Halogenides R3MX3." *Z. Krist.* 137 (1973):225–39.

125. Shoemaker, Clara B., and D. P. Shoemaker. "Structure of the I Phase, $V_{41}Ni_{36}Si_{23}$, a Pseudo Superstructure." *Acta Cryst.* B37 (1981):1–8.

126. Slater, Rose C. L. M. "Triiodide ion in Tetraphenylarsonium Triiodide." *Acta Cryst.* 12 (1959):187–96.

127. Stora, Cecile. "Patterson-Fourier Analysis of the Iodides of Palmityl and Stearycholine." *Compt. rend.* 228 (1949):324–26.

128. Sutor, D. June. "Structure of Pyrimidines and Purines VII: Crystal Structure of Caffeine." *Acta Cryst.* 11 (1958):453–58.

129. Takagi, Mieko. "Electron Diffraction Study of Liquid-Solid Transition of Thin Metal Films." *J. Phys. Soc. Japan* 9 (1954):359.

130. Takagi, Mieko, and A. R. Lang. "X-ray Bragg Reflection, 'Spike' Reflection and Ultra-Violet Absorption Topography of Diamond." *Proc. Roy. Soc.* A281 (1964):310.

131. Templeton, D. H., Lieselotte K. Templeton, J. C. Phippips, and K. D. Hodgson. *Acta Cryst.* A36 (1980):436–42.

132. Templeton, Lieselotte K., D. H. Templeton, and R. P. Phizackerley. *J. Am. Chem. Soc.* 102 (1980):1185.

133. Templeton, D. H., and Lieselotte K. Templeton. *Acta Cryst.* A36 (1980): 237- .

134. Templeton, Lieselotte K., D. H. Templeton, R. P. Phizackerley, and K. D. Hodgson. *Acta Cryst.* A38 (1982).

135. Truter, Mary, D. W. J. Cruickshank, and G. A. Jeffrey. "The Crystal Structure of Nitronium Perchlorate." *Acta Cryst.* 13 (1960):855–62.

136. Truter, Mary. *Proc. Roy. Soc.* 252 (1960):205–17; *Proc. Roy. Soc.* 254 (1960):218–28; Proc. Roy. Soc. 266 (1962):527–46.

137. Weill, Adrienne R. "X-ray Study of Temper Brittleness in a Steel of Low Nickel and Chromium Content." *Compt. rend.* 230 (1950):652–54.

138. Wood, Elizabeth. "The Question of Phase Transition in Silicon." *J. Phys. Chem.* 60 (1956):508–509.

139. Wood, Elizabeth. *Crystal Orientation Manual.* Columbia University Press, New York (1963).

140. Wood, Elizabeth. *Crystals and Light.* D. Van Nostrand Company, New York (1964).

141. Wood, Elizabeth. *Science for the Airplane Passenger.* Ballantine Books, New York (1968).

142. Woodward, Ida. "X-ray Studies of the Porphines." *J. Chem. Soc.* (1940):601–603.

143. Wooster, W. A., and Nora Wooster. "A Graphical Method of Interpreting Weissenberg Photographs." *Z. Krist.* (A)84 (1933):327–31.

144. Wrinch, Dorothy M. "Chromosome Behaviour in Terms of Protein Pattern." *Nature* 134 (1934):978–79.

145. Wrinch, Dorothy M. "On the Molecular Structure of Chromosomes." *Protoplasma* 25 (1936):550–69.

146. Wrinch, Dorothy M. "On the Pattern of Proteins." *Proc. Roy. Soc.* 160A (1937):59.

147. Wrinch, Dorothy M. "The Structure of the Insulin Molecule." *Science* 88 (1938):148–49.

148. Wrinch, Dorothy M. "The Geometry of Discrete Vector Maps." *Phil. Mag.* 27 (1939):98.

149. Wrinch, Dorothy M. "The Cyclol Hypothesis." *Nature* 145 (1940):669.

150. Wrinch, Dorothy M. *Fourier Transforms and Structure Factors.* Am. Soc. for X-ray and Electron Diffraction Monograph No. 2, Cambridge, Mass. (1946).

151. Wrinch, Dorothy M. "Structure of Bacitracin." *Nature* 179 (1957):536–37.

152. Yardley, Kathleen. "The Crystalline Structure of Succinic Acid, Succinic Anhydride and Succinimide." *Proc. Roy. Soc. London* A105 (1924):451–67.

153. Yardley, Kathleen (Mrs. Lonsdale). "An X-ray Study of Some Simple Derivatives of Ethane." *Proc. Roy. Soc. London* A118 (1928):449–97.

I had the privilege of working for the late Kathleen Lonsdale. The following people have directly contributed to this chapter: Dorothy Hodgkin, Isabella Karle, Judith Milledge, June Sutor, Doris Evans, Caroline H. MacGillavry, P. P. Ewald, Anne Sayre, David Sayre, Marjorie Senechal, Helen Scouloudi, Clara Shoemaker, Jenny Glusker, Mary Truter, Olga Kennard, Maria Przbylska, Mieko Takagi, K. Ann Kerr, A. Vos, Kenneth Trueblood, Jerome Karle, Lieselotte and David Templeton, James V. and Enid Silverton, Gabrielle and Jose Donnay, Katrina Kranjc, Sagrario Martinez Carrera, Elizabeth Moore, Vivian Cody, Elizabeth Wood, Gwendolin Bragg Caroe, Helen Ghiradella, Mary Virginia Orna, Llyn Sharp, D. Cruickshank, George Jeffrey, John H. Robertson, Henry Lipson, Arnold Beevers, Ted Benfey, Martin Buerger, James Speakman, J. M. Robertson, G. V. Gibbs, Kathleen and Francis Julian, and Carl L. Julian.

PATRICIA FARNES

Afterword

The women who contributed to this volume come from diverse academic and social backgrounds. Some, like many of the women scientists cited in these chapters, spent their undergraduate years in women's colleges, while others attended state universities or small coeducational colleges. The Ivy League has sprinkled representation in terms of graduate degrees and faculty appointments. Nearly all of the authors have a longstanding background of interest in writing about, teaching about, and talking about issues relevant to women in science, but few of us, until pursuing this assignment, had critically addressed historical perspectives about women in our respective disciplines. Here we have provided a different dimension from statistical analyses and generalizations about our forebears. That it was for the most part possible to recruit women to represent these various scientific fields is in itself a statement of sorts about where we are now. It is a landmark of being able to take the time to look back at our prehistory before we look forward.

What about the future? All of the scientific disciplines represented here share a common historical dynamic—the slow entrance of women into the prerequisite educational mainstreams. Now that women have access to educational and professional opportunities, can't we expect some more equitable representation in the sciences? This might prove to be the case, but there are no guarantees about that.

Historically, women were prevented from entering the study of sciences by nineteenth-century social and medical mythology which presupposed constitutional intellectual deficits and detriment to childbearing functions as a result of education. Currently, there is a curious *déjà vu* in new controversies which are surfacing about the capabilities of women in the scientific-technologic arena. Are there, for instance, inherent right- and left-brain differences between males and females? The well-known dropout rate of female students in higher mathematics courses is attributed by some to deficiencies in the female brain relative to abstract reasoning and spatial relationships. Others take the position that such differences in mathematics interest and performance are based upon nurtural rather than natural influences. Controversies of this sort lead to virtually insoluble dilemmas because of uncontrollable variables.

It is important for all of us to keep an ear open for subtle replays of the nineteenth-century attitudes which could affect women's educational and professional opportunities. One hundred years ago, for example, all events relating to women's reproductive functions were viewed as abnormal. There is a trend within our own time to reestablish such a perspective. Examples include the now notorious premenstrual syndrome, the increasing application of technology in pregnancy and childbirth, and the notion that menopause is a hormonal deficiency disease. We are hearing resounding echoes of the old message that women's normal physiologic functions are fundamentally abnormal.

We are even hearing again that professional training and achievement may prove detrimental to childbearing functions. Questions are being raised about the wisdom of postponing childbearing until after careers are well established. There are suggestions that infertility may increase during the fourth decade. News items have appeared linking endometriosis (an infertility-related disorder) to pregnancy postponement, and even to professional achievement.

In our view, the best way to deal with such issues is to be aware of them, and of the nonfeminist bias which pervades many of these arguments and positions, even in our own time. The perspective of this book which relates to these points is clear: here are women who, perhaps being unaware of the stereotypes, did not fit them. There will be many more such women.

Contributors

MICHELE L. ALDRICH, Archivist and Manager of Computer Services for the American Association for the Advancement of Science, has coauthored or coedited five books and numerous articles on women in science and on the history of science. She has chaired the Women's Committee of the History of Science Society and the History Division of the Geological Society of America.

PATRICIA FARNES (1931–1985) was Director of Hematology at Rhode Island Hospital and Associate Professor of Medical Science at Brown University. She authored numerous papers and abstracts on topics of medical research as well as on women's place in the health care system. She taught courses at Brown University and at the University of Rhode Island. Her interest in women's contributions to science stemmed from a lifelong concern about the medical profession's treatment of women.

JUDY GREEN, Marymount University, Virginia, and Rutgers University, Camden, has published articles in mathematical logic and in the history of mathematics. She and Jeanne LaDuke are completing a study of American women who received Ph.D.'s in mathematics prior to 1940.

CYNTHIA C. IRWIN-WILLIAMS, Executive Director, Social Sciences Center, Desert Research Institute, University of Nevada System, Reno, Nevada, has served as president of the Society for American Archaeology, and as an officer of the American Quaternary Association and of the American Anthropological Association. She has conducted extensive archaeological research in the Western United States and Mexico. Her publications include *The Oshara Tradition: Origins of Anasazi Culture; Hell Gap: The Sequence of Paleo-Indian Occupation on the High Plains;* and *The Structure of Chaco Society in the Southwest.*

LORELLA M. JONES, Professor of Physics at the University of Illinois at Urbana-Champaign, is a Fellow of the American Physical Society with substantial experience in both teaching and research. She is author or coauthor of about fifty papers on high-energy phenomenology and of the textbook *An Introduction to Mathematical Methods of Physics.*

MAUREEN M. JULIAN, Adjunct Professor and Senior Research Scientist, Molecular Orbital Section, Department of Geological Science, Virginia Polytechnic Institute and State University, is currently doing theoretical calculations on silicon nitride, writing a textbook on chemistry to be published by Scott, Foresman, and Company, and completing a biography of Dame Kathleen Lonsdale.

GABRIELE KASS-SIMON, Associate Professor of Zoology at the University of Rhode Island, received her B.A. from the University of Michigan, an M.A.

from Columbia University in English literature, and a D.Phil. in zoology from the University of Zürich, Switzerland. Her major research is in invertebrate neurobiology.

JEANNE LaDUKE, DePaul University, has published articles in abstract harmonic analysis and in the history of American mathematics. She and Judy Green are completing a study of American women who received Ph.D.'s in mathematics prior to 1940.

PAMELA E. MACK, Assistant Professor of History of Technology at Clemson University, has published articles on the history of the space program. She is currently completing a history of NASA's Landsat project and working on a history of science at Mount Holyoke College with Miriam R. Levin.

JANE A. MILLER, Assistant Professor of Chemistry and Education at the University of Missouri, St. Louis, has published articles in the *Journal of Chemical Education*, the *2YC Distillate*, and in the American Chemical Society's van't Hoff-LeBel Centennial volume. She has also prepared an educational tape on women in chemistry for the American Chemical Society. She has served as chair of the History of Chemistry Division of the American Chemical Society and the Midwest Junto for the History of Science.

MARTHA MOORE TRESCOTT, Research Associate, College of Engineering, University of Illinois at Urbana-Champaign, is currently completing *New Images, New Paths: Women Engineers in American History, in Their Own Words*. She is also the author of *The Rise of the American Electrochemicals Industry, 1880–1910: Studies in the American Technological Environment* and editor of *Dynamos and Virgins Revisited: Women and Technological Change in History*.

NAME INDEX

Abbreviations used in the Name Index

A	= archaeology	P	= physics
G	= geology	B	= biology
AS	= astronomy	MD	= medicine
M	= mathematics	C	= chemistry
E	= engineering	CR	= crystallography

SUBJECT INDEX